Roman A. Valiulin

Organic Chemistry: 100 Must-Know Mechanisms

Also of Interest

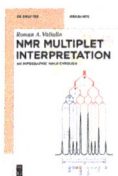

NMR Multiplet Interpretation.
An Infographic Walk-Through
Valiulin, 2019
ISBN 978-3-11-060835-9, e-ISBN 978-3-11-060840-3

Introductory Organic Chemistry.
Organic Reactivity and Reactions
Lewis, 2023
ISBN 978-3-11-067478-1, e-ISBN 978-3-11-067481-1

Organic Chemistry.
Fundamentals and Concepts
McIntosh, 2022
ISBN 978-3-11-077820-5, e-ISBN 978-3-11-077831-1

Organometallic Chemistry.
Fundamentals and Applications
Haiduc, Silaghi-Dumitrescu, 2022
ISBN 978-3-11-069526-7, e-ISBN 978-3-11-069527-4

Host–Guest Chemistry.
Supramolecular Inclusion in Solution
Wagner, 2020
ISBN 978-3-11-056436-5, e-ISBN 978-3-11-056438-9

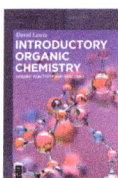

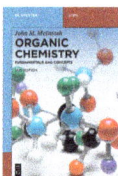

Roman A. Valiulin

Organic Chemistry: 100 Must-Know Mechanisms

2nd Edition

DE GRUYTER

Author
Dr. Roman A. Valiulin
Informed Educational Insights LLC
CA, USA

ISBN 978-3-11-078682-8
e-ISBN (PDF) 978-3-11-078683-5
e-ISBN (EPUB) 978-3-11-078701-6

Library of Congress Control Number: 2023932398

Bibliographic information published by the Deutsche Nationalbibliothek
The Deutsche Nationalbibliothek lists this publication in the Deutsche Nationalbibliografie;
detailed bibliographic data are available on the Internet at http://dnb.dnb.de.

© 2023 Walter de Gruyter GmbH, Berlin/Boston
Cover image: Roman A. Valiulin (graphics), Ani_Ka/iStock/Getty Images Plus (background)
Typesetting: Integra Software Services Pvt. Ltd.
Printing and binding: CPI books GmbH, Leck

www.degruyter.com

Nothing in life is to be feared, it is only to be understood. Now is the time to understand more, so that we may fear less.

— Marie Curie

Second Edition

This edition of *Organic Chemistry: 100 Must-Know Mechanisms* is an enriched and improved version of the first book with over 40 new illustrations. It builds upon the existing 100 fundamental mechanisms mentioned in the previous book and adds other mechanisms related to the original 100 with engaging, supplementary examples. The book is also fine-tuned with features that can help broaden its usefulness to earlier-stage students of chemistry and related sciences.

Formal **Lewis** (dot) structures are added to all the previously mentioned intermediates in each mechanism, keeping track of the movement of electrons and making the schemes more comprehensive. This improvement is a valuable enhancement for those still learning organic chemistry and expands usefulness to undergraduate students and college students in STEM fields whose area of major concentration is not chemistry. The original 100 mechanistic schemes are also visually improved: the chemical structures are more vivid and consistent throughout the book. Additionally, newly formed bonds are highlighted in red color and accentuated using bold lines for each final product or key intermediate.

In the first version, many related mechanisms were only mentioned by name. The second edition expands upon those examples and has numerous new mechanistic schemes. For example, the reader will find a new illustration of the **Bouveault–Blanc** *reduction* mechanism of esters and ketones, a separate **Wolff** *rearrangement,* which was previously mentioned only as a part of the **Arndt–Eistert** *synthesis* mechanism. Other new illustrations include the **Dakin** *reaction* mechanism, the **Myers–Saito** *cyclization* mechanism, the **Pomeranz–Fritsch** *reaction* mechanism, an example of *the* **Benzidine** *rearrangement* mechanism, the **Delépine** *reaction* mechanism, the **Peterson** *olefination* mechanism, the **Kabachnik–Fields** *reaction* mechanism, the **Petasis** *reaction* mechanism, the **Stetter** *reaction* mechanism for aromatic and aliphatic aldehydes, the **Fischer** *esterification* mechanism – and its comparison to the **Mitsunobu** *ester synthesis* and the stereochemical outcome of both reactions involving a chiral alcohol – the **Ullmann** *biaryl ether coupling* mechanism catalyzed by Cu(I) complex with a neutral bidentate ligand, the **Reimer–Tiemann** *reaction* mechanism, and the **Clemmensen** *reduction* mechanism. The *nontraceless* and *traceless* **Staudinger** *ligation* mechanisms are also highlighted, making them especially relevant after the announcement of the 2022 *Nobel Prize* in Chemistry for the development of click chemistry and bioorthogonal chemistry [30g, 30h].

Moreover, many of the original mechanistic schemes depicted in the first edition of this book were general, covering a vast scope of chemical structures and often using a general R-group representation, instead of a particular example of an actual organic compound. This edition is enhanced with a variety of real-case examples, such as the bromination and nitration of an aromatic ring (an example of the aromatic electrophilic substitution), the **Beckmann** *rearrangement* mechanism of cyclohexanone oxime, the **Chichibabin** *amination* mechanism of quinoline, a sequence of the **Cope** *rearrangements*

https://doi.org/10.1515/9783110786835-202

involving (3*R*,4*R*)-3,4-dimethylhexa-1,5-diene, (2*E*,4*R*,5*R*,6*E*)-4,5-dimethylocta-2,6-diene, (2*Z*,4*R*,5*R*,6*E*)-4,5-dimethylocta-2,6-diene, and (2*Z*,4*R*,5*S*,6*E*)-4,5-dimethylocta-2,6-diene, several variations of the ***Diels–Alder*** *cycloaddition* reactions using various *dienes* and *dienophiles,* the ***Favorskii*** *rearrangement* mechanism of 2-chlorocyclohexan-1-one, the ***Grob*** *fragmentation* mechanism of (1*R*,3*S*)-3-chloro-1-methylcyclohexan-1-ol, the ***Bischler–Napieralski*** *cyclization* mechanism of *N*-phenethylacetamide, the ***Polonovski*** *reaction* mechanism (*N*-demethylation) of a morphinan derivative, and the ***Suzuki*** *cross-coupling* mechanism catalyzed by either **Pd(dppf)Cl$_2$** or tetrakis(triphenylphosphine)palladium (0): **Pd(PPh$_3$)$_4$**. Also noteworthy, an educational example of the *ozonolysis* reaction mechanism of (–)-α-fenchene and anomalous (interrupted) *ozonolysis* reaction mechanisms are presented as well, in addition to the synthesis of cubane-1,4-dicarboxylic acid (with the key ***Favorskii*** *rearrangement* transformation step), the rearrangement mechanism of bicyclo[2.2.2]octane system (an example of the ***Hell–Volhard–Zelinsky*** *reaction*), and two plausible mechanisms of *adamantane* rearrangement undergoing a sequence of numerous ***Wagner–Meerwein*** *rearrangement* steps.

This edition continues in the tradition of the first: presenting information efficiently by using clear, balanced, and intuitive visuals and infographic diagrams. Like a stone sculpture, this version is a refined and more finely chiseled version of the first. The goal is to build upon what worked well, update the content where needed, and to add key pieces of information or notation, with the ultimate objective of making the book more useful to more students of chemistry and the sciences. Of course, we cannot promise perfection, because it, like an asymptote, is unreachable, but we hope that you will find this version to be a valuable addition or update to your scientific library.

Preface and Overview

Pedagogical Principles. At first, every body of knowledge that is new to us seems to have boundless complexity and creates the initial impression of incomprehensibility and even fear. Organic chemistry provides an excellent example of this phenomenon. The discipline is replete with complex and initially abstract concepts, as a result, the information may seem overwhelming, particularly for the young chemist. But as with most new subjects, consistent study and practice reveals patterns, commonalities, rules, and an apparent logic. Eventually, an "architecture" becomes more apparent as we grow to become more experienced chemists. To develop this intuition, it requires close study, repetition, and breadth of exposure. A significant element of that learning is intrinsic and simply requires time and immersion. However, to help with the development of this intuition, an organic chemist would also be wise to focus on mechanisms for organic reactions as a foundation or anchoring point. This, in combination with deep study, can help organize knowledge into skill and expertise. An understanding of reaction mechanisms provides a solid foundation for the field and a scaffold for further study and life-long learning. Mechanisms are highly useful because they can logically explain how a chemical bond in a molecule was formed or broken and help to rationalize the formation of the final synthetic target or an undesired side-product. Moreover, as we parse an increasing number of mechanisms, we begin to see the similarities and an invisible conceptual "thread" then forms in our mind's eye that was not previously apparent. It helps to organize thinking and brings sense to the otherwise foreign concepts such as reactive intermediates, transition states, charges, radicals, and mechanistic arrows.

The Approach. To help galvanize – and perhaps catalyze – the organic chemist's inductive ability and to provide a "go-to" reference for closer study, this book strives to present an abridged summary of some of the most important mechanisms. In today's terms, these are 100 MUST-KNOW mechanisms. The author draws upon scientific knowledge developed through undergraduate and graduate years, including postdoctoral research and study focused on organic synthesis. With a keen awareness of the incremental learning process, the book curates and presents mechanisms by category, starting with the fundamental and basic mechanisms (e.g., *nucleophilic substitution* or *elimination*), mechanisms associated with the most well-known named reactions (e.g., the **Diels–Alder** *reaction* or the **Mitsunobu** *reaction*). Additionally, the collection is complemented with historically important mechanisms (e.g., the *diazotization* or the *haloform reaction*). Finally, it includes some mechanisms dear to the author's heart, which he deems elegant or simply "cool" (e.g., the **Paternò–Büchi** *cycloaddition* or the *alkyne zipper reaction*).

Organization. The mechanisms are organized alphabetically by chapter for ease of reference, and numbered from 1 to 100. The dedicated student will consistently proceed through every single mechanism, giving each one time to study, practice with, memorize, and ponder. At the same time, the book can be used as a quick visual

https://doi.org/10.1515/9783110786835-203

reference or as a starting point for further research and reading. The 100 mechanisms are selected for being classic and famous, core or fundamental, and useful in practice. Of course, a good degree of personal intuition is involved in the selection and it is definitely not a dogmatic ordering or a comprehensive anthology. The book is intended to be a visual guide as distinguished from a traditional textbook. The presentation of each mechanism constitutes a complete InfoGraphic (or "MechanoGraphic") and provides distilled information focusing on key concepts, rules, acronyms, and terminology. It heavily focuses on the basic core – the starting amount of information, the extract – that a good organic chemist can commit to memory and understanding. Starting initially as a daily micro-blog post with a "hash tag" (#100MustKnowMechanisms) that gained a lot of support from students and chemists around the world, the book is really intended to bring together an array of mechanisms, organize them, provide additional historical context, and enable a conceptual space where the reader can focus on learning them as well as serve as a desk-reference or a "flip-book."

The book is color-coded: each key reaction is enclosed in a dark blue frame; each key mechanism (the centerpiece of the book) is presented in a red frame; other reactions and mechanisms related to the core 100 mechanisms covered in this book are usually summarized in gray or black frames. The book also collects a few useful rules, facts, and concepts that are presented in green frames. The reader may find several star diagrams, representing synthetic diversity, for example, throughout the book as well. Relevant comments and clarifications can be found in footnotes.

Sources. The underlying information stays very close to information usually covered in classic or key organic chemistry textbooks [1]. More specialized literature may be necessary in some cases (for organometallic or photochemical transformations, for example) [2]. The reader is also encouraged to familiarize themselves with some other supporting bibliography [3]. Where appropriate, it also references texts that the author trusts and cites for further in-depth study if the reader so chooses. Since this book strives to be an abridged visual illustration, students are encouraged to use other, more comprehensive books on the subject, especially those related to the *named reactions* in organic chemistry [4]. Additionally, open online sources, when thoughtfully selected, can also be very useful [5]. Such sources may be mentioned here when the information was deemed accurate, thorough, and supported by the references. This is further supplemented by the author's aggregate knowledge and education gained through college, graduate school, and postdoctoral academic research. The author also found the encyclopedia of organic reagents [6] to be an extremely useful "go-to" starting point in his personal experience and professional career, especially when embracing a new chemistry topic or using a new reagent. Moreover, each *MechanoGraphic* is supported by a reference to the likely first original publication where the related reaction or mechanism was first mentioned (see the time-scale after each mechanism). Finally, several key and fundamental reviews, publications on recently elucidated mechanisms, and other research articles are referenced, as needed. The author uses his best judgment in each case. However, even though the provided information was carefully checked and presented in agreement with standard and accepted chemistry rules, this does not guarantee that it is free of all errors. A further caveat, the variety of text and scholarly references does not imply a comprehensive and chronological review of the literature and history – it is not a global historic review of mechanisms from 1800 to 2023. Mechanisms and our understanding of them can also change as this book is being prepared and the corresponding literature revised. Thus, the reader should supplement the use of the book with primary source reading and deeper study through a comprehensive textbook prepared by a cohort of experienced professors and experts. Here, the most common and known pathways, those that do not violate basic standard chemistry rules and that are frequently referenced in the classic and contemporary literature, are summarized visually.

A Few Things to Keep in Mind. It is also important that the reader remain flexible and mindful that mechanisms are represented based on our current understanding, taking into consideration basic chemistry rules, valency, electron pushing rules, charge preservation, Lewis dot structures, and so on. They may not be the most "cutting-edge" or up-to-date (e.g., cross-coupling reactions that may not be well-understood). They may also be substrate-dependent and each reaction may undergo a slightly different pathway. Thus, the reader should not treat the book as a dogmatic guide, and should keep an open mind for new data, creativity, and view the book as a part of a continuous debate in the subject.

Background Knowledge. To fully benefit from the book, the reader should have basic knowledge of organic chemistry. Figures are presented with an assumption that the reader understands common terms and symbols. Thus, basic concepts are not introduced or explained. Undergraduate students, graduate students, scientists, teachers, and professors in the discipline should be able to utilize the book. The book can also serve as a good condensed "refresher" for the experienced organic chemist who wants to "zero-in" on the most basic and fundamental core mechanisms as judged by the author.

The Inspiration and Further Reading. The author heavily draws upon his personal experience as a student of chemistry and later an academic researcher. Never having taken a formal course on mechanisms in organic chemistry, he approached the material initially through memorization as opposed to derivation. The first impression was fear and a sense of being overwhelmed. However, after many years of experience, more obvious patterns, trends, rules, and dependencies appear to have crystallized, providing an inductive ability to navigate and identify the mechanisms behind reactions. This personal experience has definitely shaped the teaching philosophy of the book and is further enhanced by the efficient way in which information can be conveyed through visuals and space. Moreover, as most individuals have a predisposition for visual learning, this book is more intuitively aligned with the way that we seem to learn the fastest. It strives to be a focused collection of the most useful, basic, and fundamental mechanisms. Started initially as a microblog post, the discussion, engagement, and interest it sparked indicated a clear need for a more carefully prepared, organized, and curated presentation in a format that could be placed in a physical library and easily internalized. The author hopes the book serves as a good starting point for the developing chemist who may need the most guidance and encouragement. No doubt it may stimulate constructive discussion, but nevertheless this will ultimately encourage and challenge everyone to learn, to search for a different answer, to think critically, and grow as a chemist and stay sharp as a scientist. Finally, knowledge is a fractal-like concept, the closer we look the more detail we see and learn. Here, we strive to reach a reasonable asymptote of precision and comprehensiveness given the purpose of the book. Further core reading [1], reference of primary and secondary sources [2–4], online sources [5, 6], as well as actual experimentation and practice will help paint the complete picture and prepare the organic chemist to be a well-rounded and informed scientist.

Contents

List of Acronyms and Abbreviations

≡	Identical to (a depiction of a chemical structure)
1°	Primary (e.g., carbocation) or first generation (e.g., catalyst)
2°	Secondary (e.g., carbocation) or second generation (e.g., catalyst)
3°	Tertiary (e.g., carbocation) or third generation (e.g., catalyst)
Ac	Acetyl
acac	Acetylacetonate
Ad$_E$2	**Bi**molecular electrophilic addition
Ad$_E$3	**Tri**molecular electrophilic addition
ADMET	Acyclic diene metathesis (polymerization)
AIBN	Azo*bis*isobutyronitrile; 2,2′-azo*bis*(2-methylpropionitrile)
Alk = R	Alkyl group
anti	From opposite sides (in *anti*-addition or *anti*-elimination)
APA	3-Aminopropylamine; 1,3-diaminopropane
aq	Aqueous (work-up)
Ar	Aryl; aromatic ring
B (B⁻)	General Brønsted–Lowry base (proton acceptor)
B$_2$pin$_2$	*Bis*(pinacolato)diboron; 4,4,4′,4′,5,5,5′,5′-octamethyl-2,2′-bi-1,3,2-dioxaborolane
9-BBN	9-Borabicyclo[3.3.1]nonane
BH (BH⁺)	General Brønsted–Lowry acid (proton donor)
Bn	Benzyl
Boc	*Tert*-butoxycarbonyl; *t*-butoxycarbonyl
Bs	Brosyl; 4-bromobenzenesulfonyl
Bu	Butyl (if not specified = *n*-Bu)
CHD	1,4-Cyclohexadiene
CM = XMET	(Olefin) cross-metathesis
con	Conrotatory
3-CR (MCR)	3-Component reaction (multi-component reaction)
4-CR (MCR)	4-Component reaction (multi-component reaction)
CuAAC	Copper(I)-catalyzed azide-alkyne cycloaddition
CuTC	Copper(I) thiophene-2-carboxylate
Cy	Cyclohexyl
Cy$_2$BH	Dicyclohexylborane
DABCO	1,4-Diazabicyclo[2.2.2]octane
DBU	1,8-Diazabicyclo[5.4.0]undec-7-ene
DCC	*N,N′*-Dicyclohexylcarbodiimide; 1,3-dicyclohexylcarbodiimide
DCM	Dichloromethane; methylene chloride
DEAD	Diethyl azodicarboxylate
DIAD	Diisopropyl azodicarboxylate
DIBAL = DIBAL-H	Diisobutylaluminum hydride = (*i*-Bu)$_2$AlH
dis	Disrotatory
DMAP	4-Dimethylaminopyridine; 4-(dimethylamino)pyridine
DMP	Dess–Martin periodinane
DMSO	Dimethyl sulfoxide
E-	*Entgegen* (*trans*- or opposite)
e⁻	Electron
E (or E⁺)	Electrophile
E1	**Uni**molecular elimination

https://doi.org/10.1515/9783110786835-205

E1cB (E1cb)	**Uni**molecular elimination conjugate base
E2	**Bi**molecular elimination
EDC = EDCI	1-Ethyl-3-(3'-dimethylaminopropyl)carbodiimide hydrochloride; N-(3-dimethylaminopropyl)-N'-ethylcarbodiimide hydrochloride
EDCI = EDC	1-Ethyl-3-(3'-dimethylaminopropyl)carbodiimide hydrochloride; N-(3-dimethylaminopropyl)-N'-ethylcarbodiimide hydrochloride
EDG (= ERG)	Electron donating group (same as ERG)
E$_i$	Internal or intramolecular elimination
eq	Equivalent (e.g., 2 eq = 2 equivalents; 2 moles)
ERG (= EDG)	Electron releasing group (same as EDG)
Et$_2$BH	Diethylborane
EWG	Electron withdrawing group
EYM	Enyne metathesis
Grubbs 1°	The first generation Grubbs catalyst
Grubbs 2°	The second generation Grubbs catalyst
H$_3$B•THF	Borane–tetrahydrofuran complex; borane tetrahydrofuran complex
H$_3$B•Me$_2$S = BMS	Borane–dimethyl sulfide complex; borane dimethyl sulfide complex
HATU	N-[(Dimethylamino)-1H-1,2,3-triazolo[4,5-b]pyridin-1-ylmethylene]-N-methylmethanaminium hexafluorophosphate N-oxide; 1-[bis(dimethylamino) methylene]-1H-1,2,3-triazolo[4,5-b]pyridinium 3-oxide hexafluorophosphate
HBTU	O-Benzotriazol-1-yl-N,N,N',N'-tetramethyluronium hexafluorophosphate; 3-[bis (dimethylamino)methyliumyl]-3H-benzotriazol-1-oxide hexafluorophosphate
HET = ^{HET}Ar	Heterocycle; heteroaromatic ring; heteroaryl
HOAt = HOAT	1-Hydroxy-7-azabenzotriazole; 3-hydroxy-3H-1,2,3-triazolo[4,5-b]pyridine
HOBt = HOBT	1-Hydroxybenzotriazole
HOMO	Highest occupied molecular orbital
hv	Light (direct irradiation) or excited state
I$_i$(BR)	Intermediate (biradical)
I$_i$(RP)	Intermediate (radical pair)
IBX	2-Iodoxybenzoic acid; o-iodoxybenzoic acid
IC	Internal conversion
Ipc$_2$BH	Diisopinocampheylborane
IpcBH$_2$	Monoisopinocampheylborane
ISC	Intersystem crossing
KAPA	Potassium 3-aminopropylamide
L	Ligand or leaving group
(*l*)	Liquid (as in liquid ammonia: NH$_3$ (l))
LA	Lewis acid
LAPA	Lithium 3-aminopropylamide
LDA	Lithium diisopropylamide = (i-Pr)$_2$NLi
L$_m$Pd	Palladium(0) cross-coupling catalyst
L$_n$Pd	Low-coordinate palladium(0) cross-coupling catalyst
LUMO	Lowest occupied molecular orbital
M	Metal
[M]	Metal catalyst (not specified)
M^{+3} = M(III)	Oxidation state (oxidation number) of an element (e.g., Cu^{+2} = Cu(II); Pd^0 = Pd(0))
M^{3+}	Charge (e.g., Ti^{3+} in TiCl$_3$ versus Ti^{+3} = Ti(III))
m-CPBA (MCPBA)	*Meta*-chloroperbenzoic acid; m-chloroperbenzoic acid; 3-chloroperbenzoic acid

MCR	Multicomponent reaction
Mes	Mesityl (from mesitylene = 1,3,5-trimethylbenzene)
Ms	Mesyl; methanesulfony = SO_2Me
n	Nonbonding (molecular) orbital
NACM	Nitrile-alkyne cross-metathesis
NBS	*N*-Bromosuccinimide; 1-bromo-2,5-pyrrolidinedione
N-HBTU	1-[Bis(dimethylamino)methylene]-1*H*-benzotriazolium hexafluorophosphate 3-oxide
NiAAC	Nickel-catalyzed azide–alkyne cycloaddition
NMM	*N*-Methylmorpholine; 4-methylmorpholine
NMO	*N*-Methylmorpholine *N*-oxide; 4-methylmorpholine *N*-oxide
Ns	Nosyl; 4-nitrobenzenesulfonyl or 2-nitrobenzenesulfonyl
Nu (or Nu⁻)	Nucleophile
NuH	General Brønsted–Lowry acid (proton donor, like BH)
[O]	General oxidant (e.g., $2KHSO_5 \cdot KHSO_4 \cdot K_2SO_4$)
O-HBTU	*N*-[(1*H*-Benzotriazol-1-yloxy)(dimethylamino)methylene]-*N*-methylmethanaminium hexafluorophosphate
p [sp, sp², sp³]	p Orbital
P	Product (in photochemical reactions)
PCC	Pyridinium chlorochromate
PDC	Pyridinium dichromate
Ph	Phenyl
Ph₃P = TPP	Triphenylphosphine
PhthNH	Phthalimide (Phth = phthaloyl)
pK_a	Acidity constant = $-\log_{10}(K_a)$
Pr	Propyl (if not specified = *n*-Pr)
Py	Pyridine
R	Reactant; starting material (in photochemical reactions)
R (–R₁, –R₂, –R′, –R″, . . .)	(Radical) group; alkyl group; substituent; (molecular) fragment
R*	Excited reactant (in photochemical reactions)
RCAM	Ring-closing alkyne metathesis
RCEYM	Ring-closing enyne metathesis
RCM	Ring-closing metathesis
R$_L$	Large group (substituent)
ROM	Ring-opening metathesis
ROMP	Ring-opening metathesis polymerization
R$_S$	Small group (substituent)
RuAAC	Ruthenium-catalyzed azide–alkyne cycloaddition
s [sp, sp², sp³]	s Orbital
S₀	Ground state
S₁	First (energy level) singlet excited state
S₂	Second (energy level) singlet excited state
S$_E$Ar = S$_E$(Ar) = S$_E$2Ar	(**Bi**molecular) aromatic electrophilic substitution = arenium ion mechanism
³sens	Sensitized irradiation (to the triplet excited state)
SET	Single electron transfer
Sia₂BH	Disiamylborane; *bis*(1,2-dimethylpropyl)borane
S$_N$1	**Uni**molecular nucleophilic substitution
S$_N$2	**Bi**molecular nucleophilic substitution
S$_N$Ar = S$_N$2Ar	(**Bi**molecular) aromatic nucleophilic substitution

$S_{RN}1$	**Uni**molecular radical nucleophilic substitution
syn	From the same side (in *syn*-addition or *syn*-elimination)
T_1	First (energy level) triplet excited state
T_2	Second (energy level) triplet excited state
TBAF	Tetrabutylammonium (tetra-*n*-butylammonium) fluoride = *n*-Bu$_4$NF
Tf	Triflyl; trifluoromethanesulfonyl = SO$_2$CF$_3$
TFA	Trifluoroacetic acid
TFAA	Trifluoroacetic anhydride
THF	Tetrahydrofuran
Thx$_2$BH$_2$	Thexylborane (2-methylpentan-2-yl)borane
TLC	Thin-layer chromatography
TMEDA	*N,N,N′,N′*-Tetramethylethylenediamine; 1,2-*bis*(dimethylamino)ethane
TMS	Trimethylsilyl = SiMe$_3$
TPAP	Tetrapropylammonium (tetra-*n*-propylammonium) perruthenate = (*n*-Pr)$_4$NRuO$_4$
TPP = Ph$_3$P	Triphenylphosphine
Ts	Tosyl; *p*-toluenesulfonyl
X (in –X)	Halogen or a general leaving group (see L)
X (in =X)	Variable atom; variable group (usually O or N)
XMET = CM	(Olefin) cross-metathesis
Z-	*Zusammen* (*cis-* or same)
Z (in –Z)	Variable group (often EWG)
α	Alpha position (first position)
β	Beta position (second position)
γ	Gamma position (third position)
Δ	Temperature; heat or ground state (in photochemical reactions)
δ+	Partial positive charge (low electron density)
δ–	Partial negative charge (high electron density)
π	Involving a π-bond (e.g., π-complex)
1π e$^-$, 2π e$^-$, . . .	Number of electrons in a π-orbital
σ	Involving a σ-bond (e.g., σ-complex)
σ*	(Antibonding) sigma star (molecular) orbital
Φ_{ISC}	Quantum yield (for intersystem crossing)

1 Electrophilic Addition Mechanism

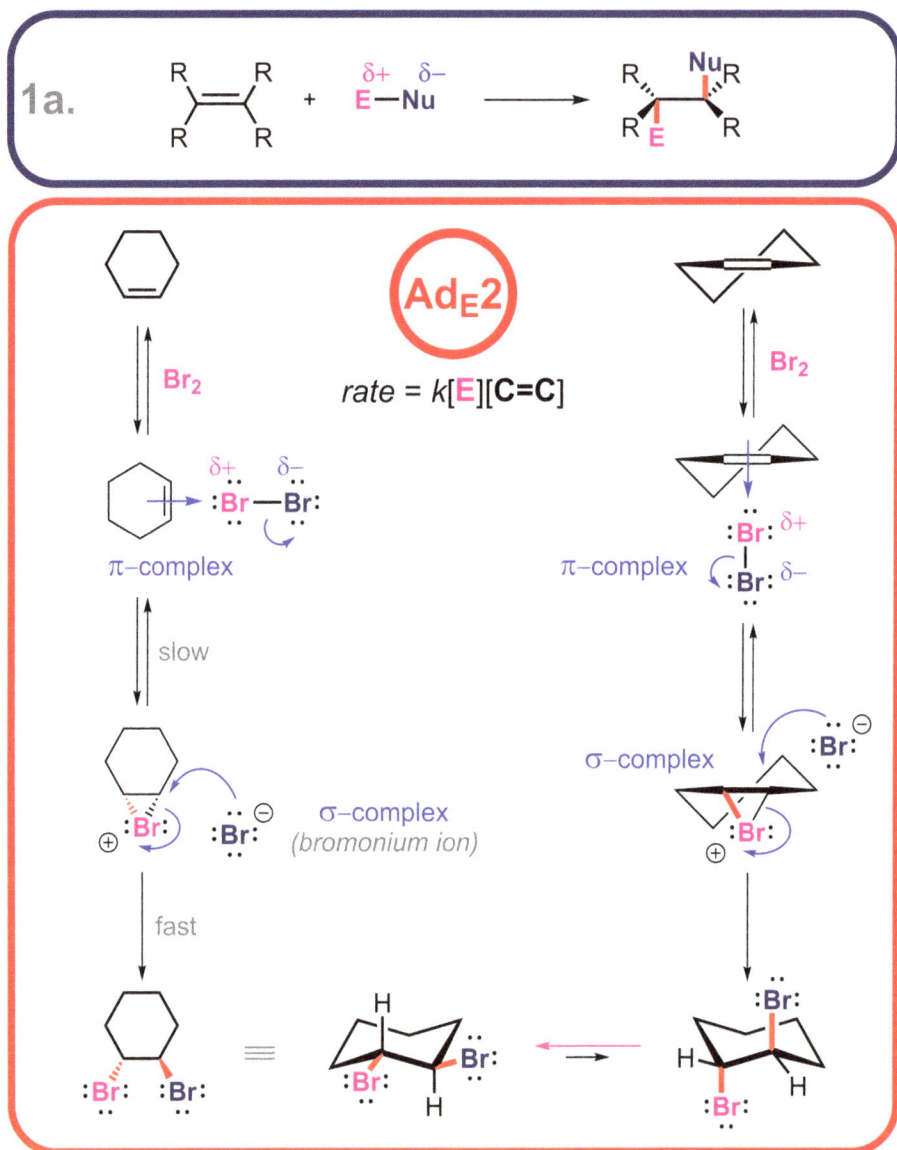

Fig. 1.1: Bimolecular electrophilic addition mechanism ($\textbf{Ad}_E2$).[1]

---

1 Symbol $\textbf{Ad}_E2$ stands for **Ad**dition **E**lectrophilic **Bi**molecular (**2**), that is, the rate of the reaction is *second order*, and the rate-determining step (i.e., the *slow* step) depends on the concentration of two reactants. In the bromination of cyclohexene, it is the *electrophile* ($\textbf{E}$ or $\textbf{Br}_2$) and *alkene* ($\textbf{C=C}$): $rate = k[\textbf{E}]^1[\textbf{C=C}]^1$.

https://doi.org/10.1515/9783110786835-001

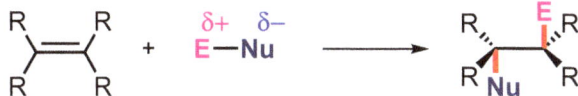

$$rate = k[\text{HCl}]^2[\text{C=C}]$$

Ad$_E$3 Mechanism I:

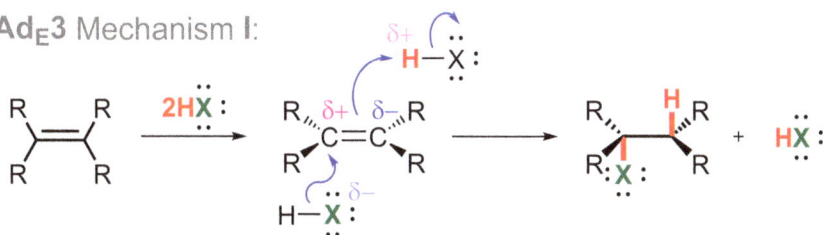

Ad$_E$3 Mechanism II:

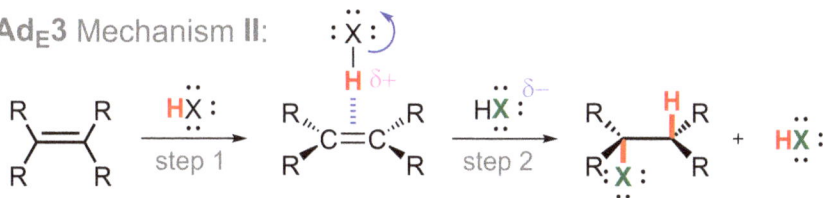

Fig. 1.2: Trimolecular electrophilic addition mechanism (Ad$_E$3).[2]

---

2 Symbol **Ad$_E$3** stands for **Ad**dition **E**lectrophilic **Tri**molecular (**3**), that is, the rate of the reaction is *third order*, and the rate-determining step (i.e., the *slow* step) depends on the concentration of three reactants. In this less common example, it is the two *electrophiles* (2HX or HCl + HCl) and *alkene* (C=C): *rate* = $k[\text{HCl}]^1[\text{HCl}]^1[\text{C=C}]^1 = k[\text{HCl}]^2[\text{C=C}]^1$. In Mechanism I, the collision of all three components is less probable and simultaneous. In more probable Mechanism II, a complex between the first HX and alkene is formed first (step 1), followed by step 2 (addition of the second HX).

2 Nucleophilic Substitution Mechanism

2a.

$$RL \xrightarrow[\text{slow}]{k_1} R^{\oplus} + L^{\ominus}$$

$$R^{\oplus} + Nu^{\ominus} \xrightarrow[\text{fast}]{k_2} RNu$$

S_N1

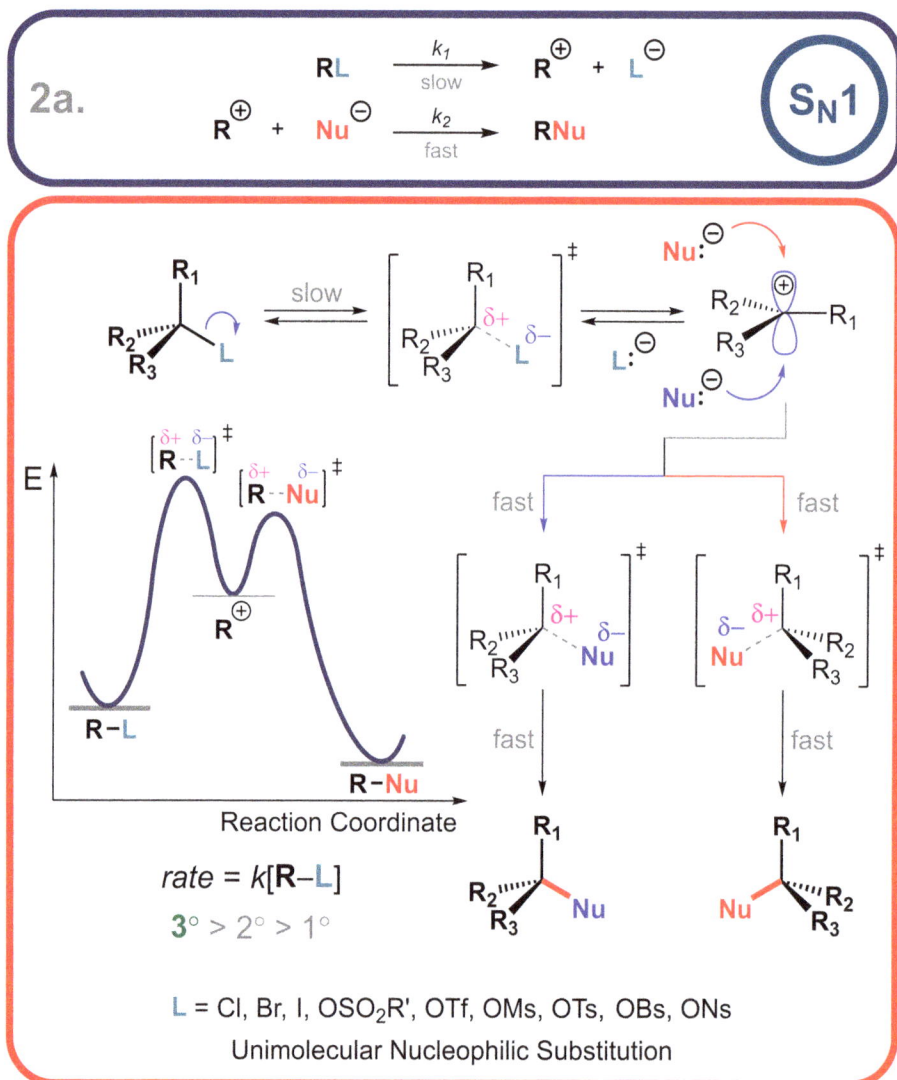

$$rate = k[\mathbf{R{-}L}]$$

$$3° > 2° > 1°$$

L = Cl, Br, I, OSO_2R', OTf, OMs, OTs, OBs, ONs

Unimolecular Nucleophilic Substitution

Fig. 2.1: Unimolecular nucleophilic substitution mechanism (S_N1).[3]

---

3 Symbol S_N1 stands for **S**ubstitution **N**ucleophilic **Uni**molecular (**1**), that is, the rate of the reaction is *first order*, and the rate-determining step (i.e., the *slow* step) depends on the concentration of one reactant. In this example, it is the *starting material* (substrate) containing a leaving group (**RL**): *rate* = $k[\mathbf{RL}]^1$.

https://doi.org/10.1515/9783110786835-002

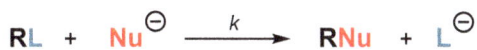

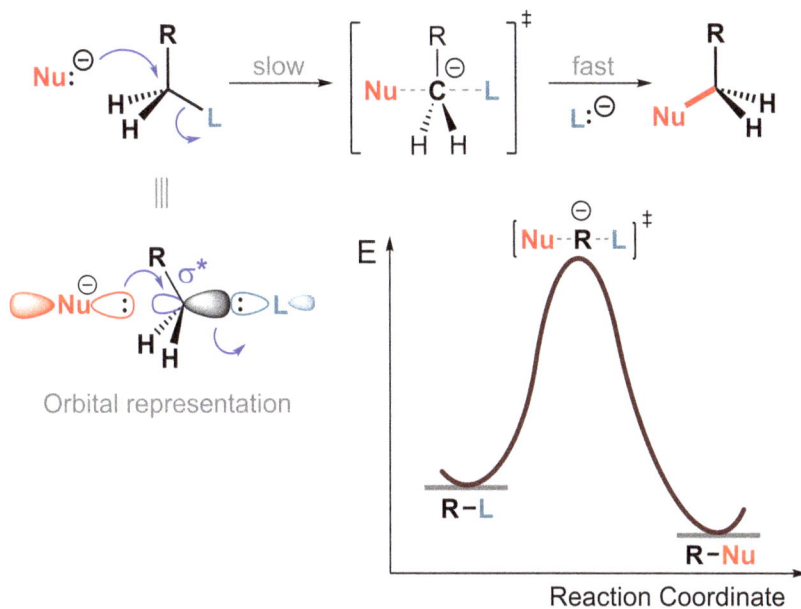

Orbital representation

L = Cl, Br, I, OSO$_2$R', OTf, OMs, OTs, OBs, ONs

Bimolecular Nucleophilic Substitution

$$rate = k[\text{Nu}][\text{R–L}]$$

1° > 2° > 3°

Fig. 2.2: Bimolecular nucleophilic substitution mechanism (**S$_N$2**).[4]

---

4 Symbol **S$_N$2** stands for **S**ubstitution **N**ucleophilic **Bi**molecular (2), that is, the rate of the reaction is **second order**, and the rate-determining step (i.e., the *slow* step) depends on the concentration of two reactants. In this example, it is the **nucleophile** (**Nu**) and the **starting material** (**RL**): *rate* = k[Nu]^1[RL]^1.

3 Aromatic Electrophilic Substitution Mechanism

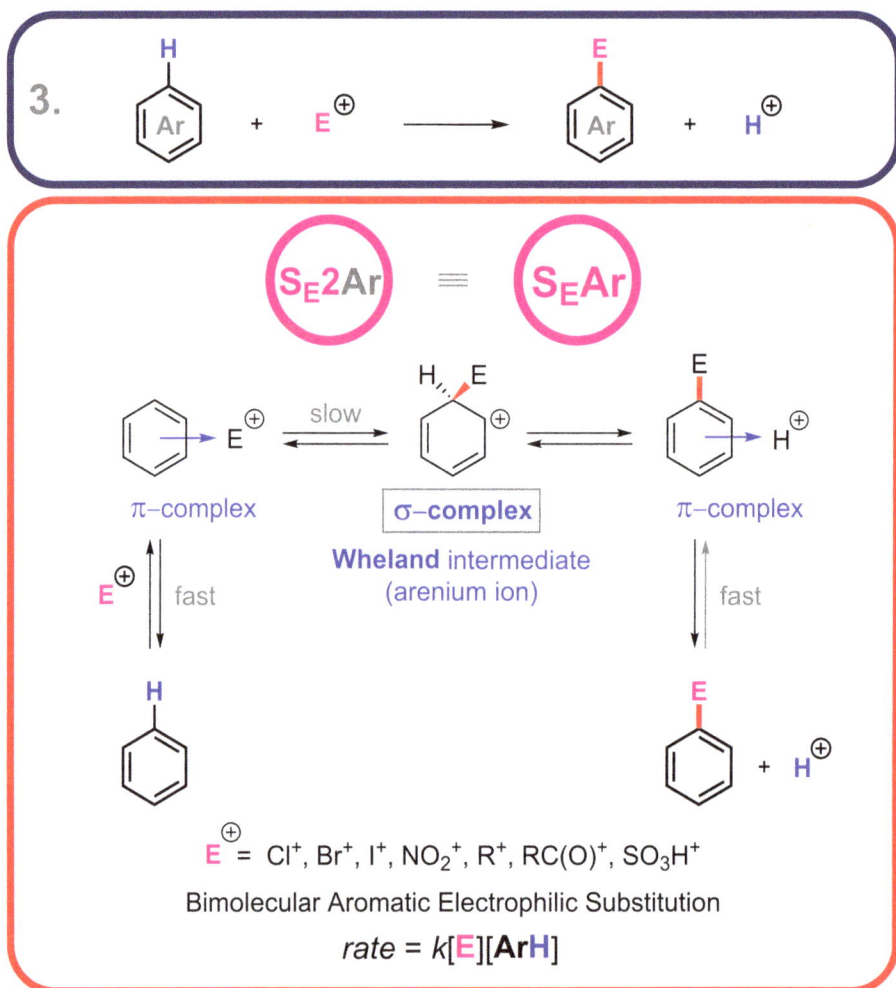

Fig. 3.1: The arenium ion mechanism (S_EAr).[5]

---

[5] Symbol S_EAr or $S_E(Ar)$ stands for Substitution Electrophilic Arenium (ion) (often confused with Aromatic), that is, the **arenium ion** mechanism. In this example, it is a **Bi**molecular (2) reaction, that is, the rate of the reaction is **second order**, and the rate-determining step (i.e., the *slow* step) depends on the concentration of two reactants. It is the **electrophile** (E) and **arene** (ArH): *rate* = $k[E]^1[ArH]^1$. To emphasize that it is a bimolecular mechanism, sometimes S_E2Ar or $S_E2(Ar)$ notation is used (the use of simple S_E2 symbol can be confusing, since it can also apply to an Aliphatic Electrophilic Substitution).

https://doi.org/10.1515/9783110786835-003

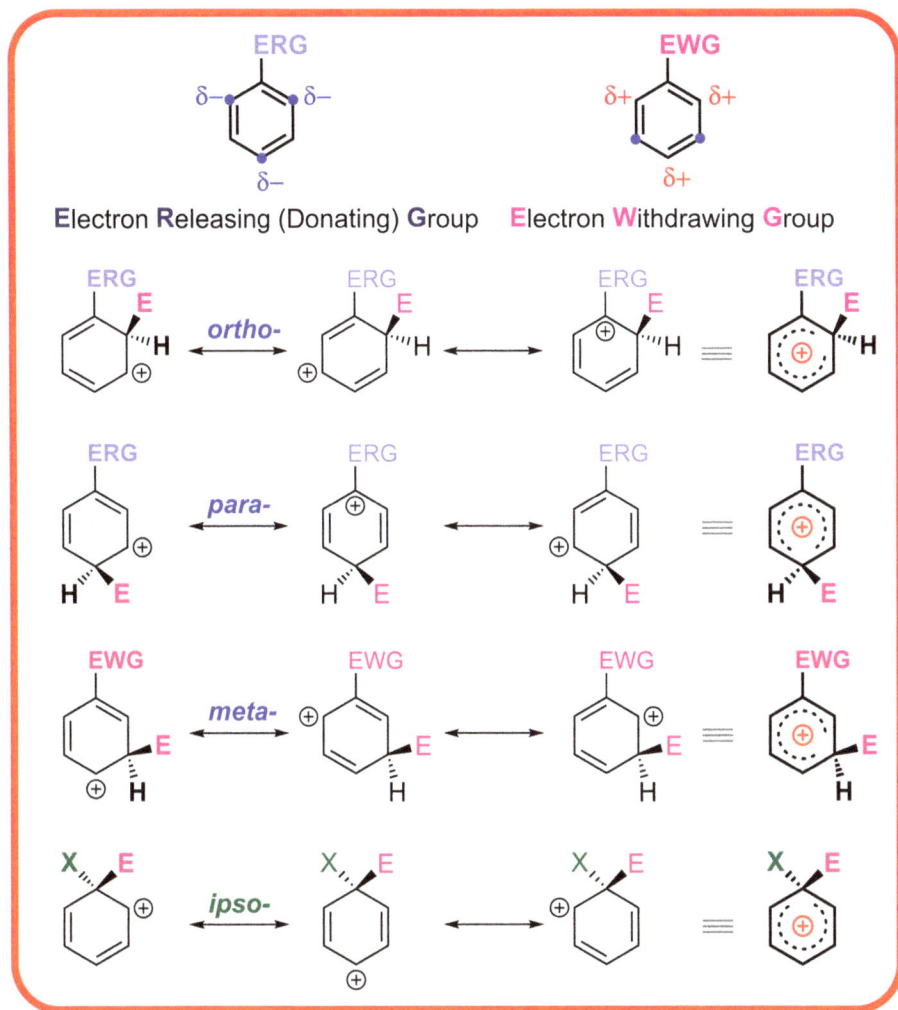

Fig. 3.2: The orientation of substitution with substrates containing EWG and ERG.[6]

---

6 In this book the terms "electron releasing group" (ERG) and "electron donating group" (EDG) are used interchangeably. Please note, *ipso-substitution* is provided only for the comparison with *ortho-*, *para-*, and *meta-substitution*.

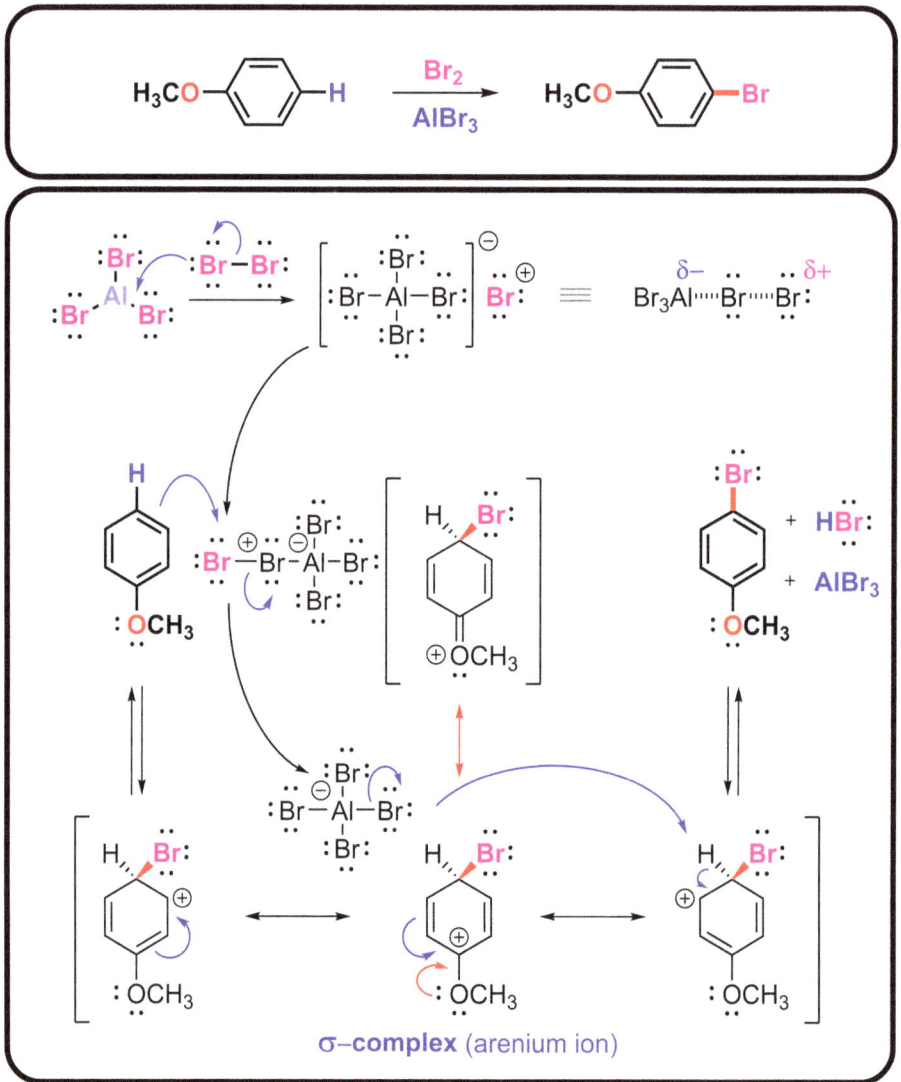

---

7 An example of an aromatic electrophilic substitution: bromination of anisole (a substrate with an ERG).

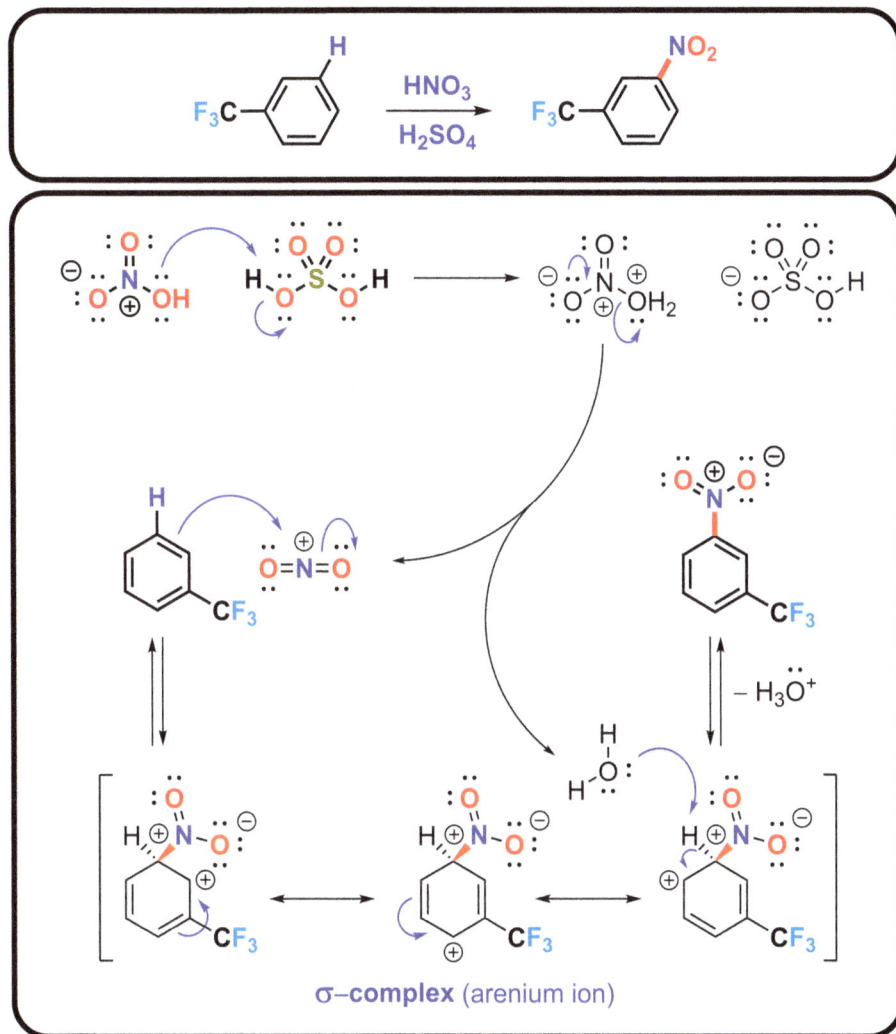

Fig. 3.4: Nitration of (trifluoromethyl)benzene.[8]

---

8 An example of an aromatic electrophilic substitution: nitration of (trifluoromethyl)benzene (a substrate with an EWG).

4 Aromatic Nucleophilic Substitution Mechanism

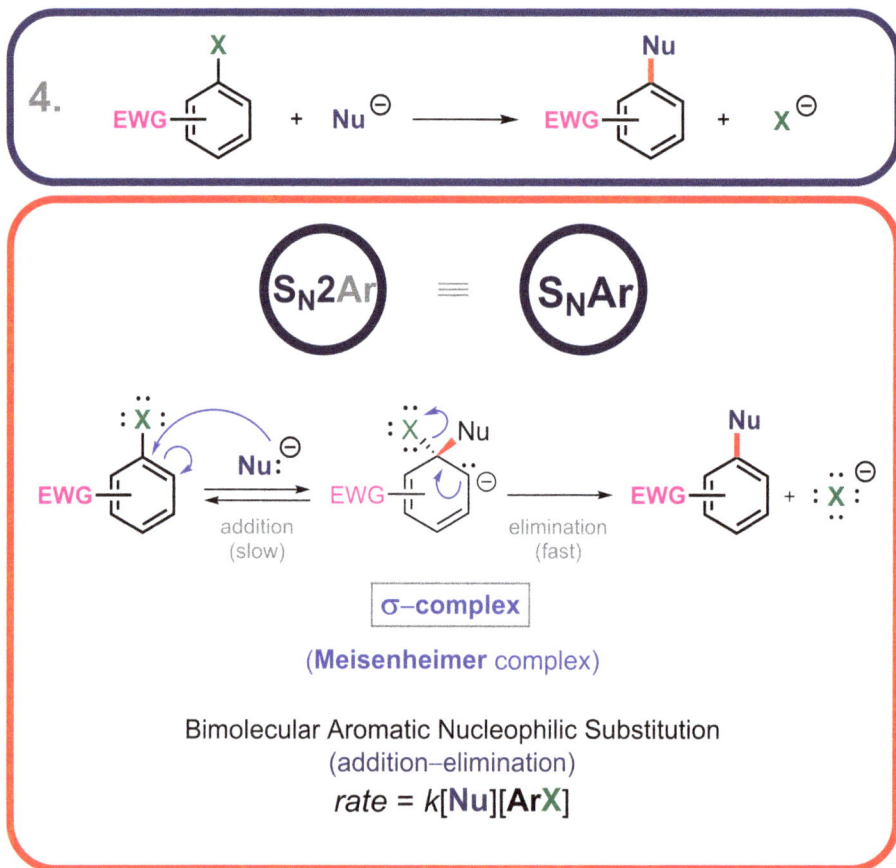

Fig. 4.1: Bimolecular aromatic nucleophilic substitution (addition–elimination) mechanism (S_NAr).[9]

---

9 Symbol S_N**Ar** stands for **S**ubstitution **N**ucleophilic **Ar**omatic; it is also called the *addition–elimination* mechanism. In this example, it is a **Bi**molecular (2) reaction, that is, the rate of the reaction is *second order* and the rate-determining step (i.e., the *slow* step) depends on the concentration of two reactants. It is the *nucleophile* (**Nu**) and *arene* (**ArX**): *rate* = $k[\mathbf{Nu}]^1[\mathbf{ArX}]^1$. To emphasize that it is a bimolecular mechanism, sometimes S_N**2Ar** notation is used.

https://doi.org/10.1515/9783110786835-004

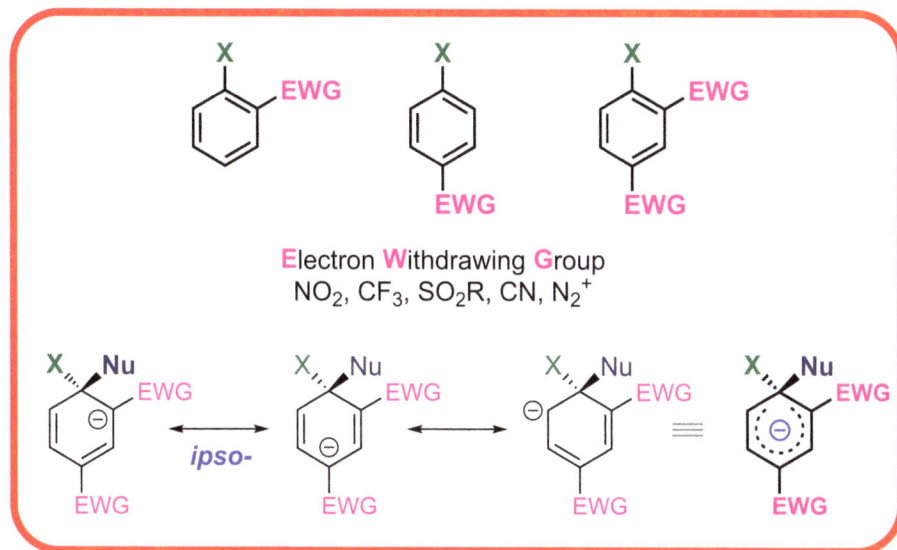

Fig. 4.2: Typical activated S_NAr substrates.[10]

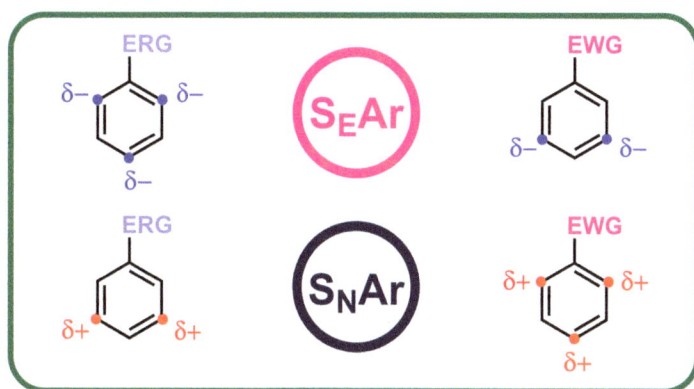

Fig. 4.3: The orientation of substitution in S_EAr and S_NAr.[11]

---

10 A typical S_NAr substrate usually contains an activating electron withdrawing group (EWG) and a leaving group (X).

11 In the S_EAr reaction, an EWG group orients (directs) the substitution in the *meta*-position and an ERG (EDG) directs the substitution in the *ortho*-position and/or *para*-position. However, in the S_NAr reaction, this trend is reversed: an EWG group orients (directs) the substitution in the *ortho*-position and/or *para*-position and ERG (EDG) directs the substitution in the *meta*-position.

5 Aromatic Radical Nucleophilic Substitution Mechanism

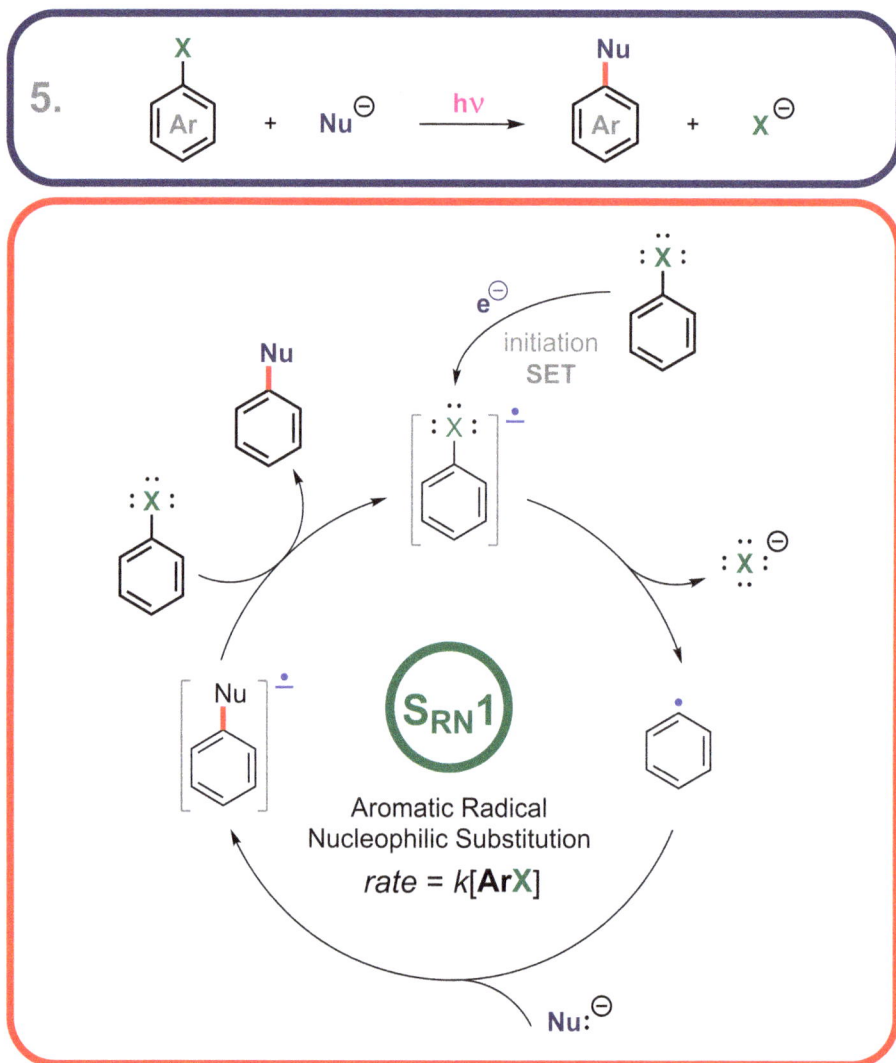

Fig. 5.1: Unimolecular aromatic radical nucleophilic substitution mechanism ($S_{RN}1$).[12]

---

12 Symbol $S_{RN}1$ stands for **S**ubstitution **R**adical **N**ucleophilic **U**nimolecular (**1**), that is, the rate of the reaction is ***first order***, and the rate-determining step (the *slow* step) depends on the concentration of one reactant. In this example, it is the ***starting material*** that contains a leaving group (**ArX**): *rate* = $k[\mathbf{ArX}]^1$.

https://doi.org/10.1515/9783110786835-005

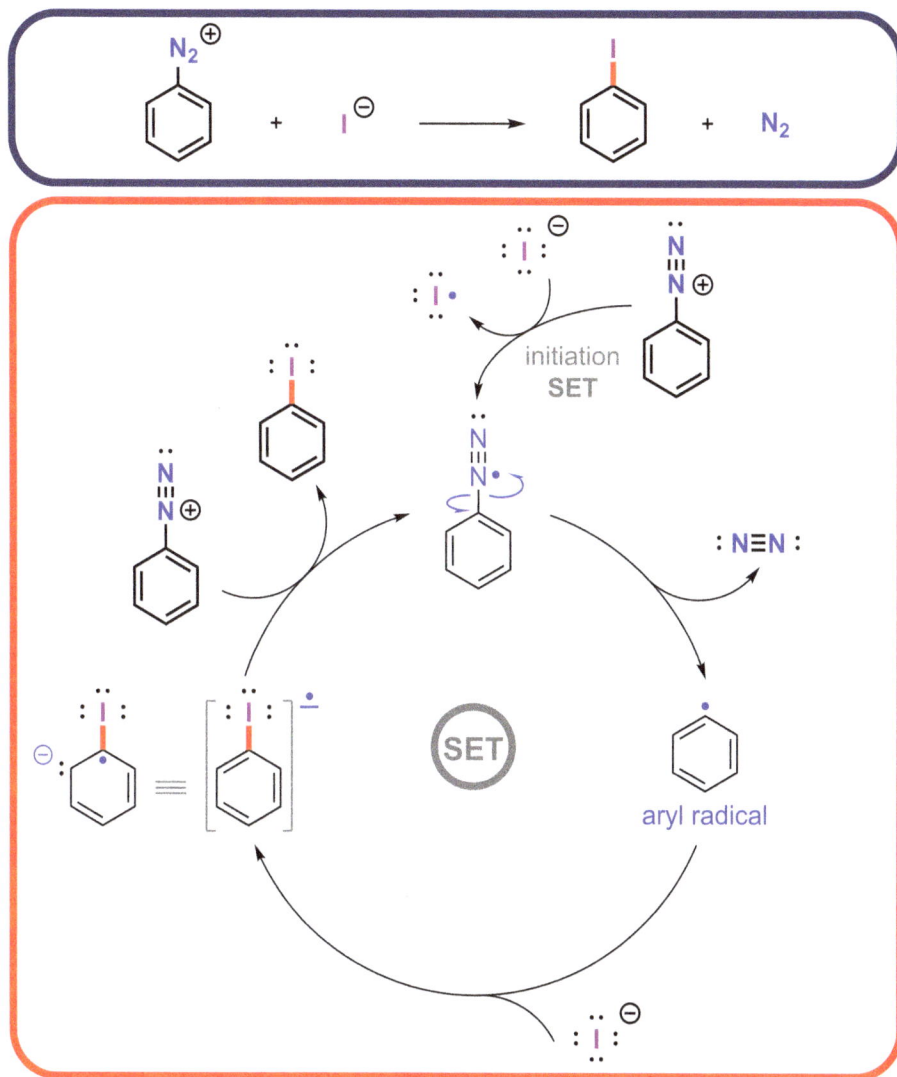

Fig. 5.2: Replacement of the diazonium group by iodide.[13]

---

13 The substitution of a diazonium group by iodide is an example of the **SET** (Single Electron Transfer) mechanism. Please note that the $S_{RN}1$ mechanism and the **SET** mechanism are closely related and are not differentiated in this book. Jerry March [1a] distinguishes the $S_{RN}1$ mechanism (the initial attack of the aromatic substrate occurs by an electron donor) from the SET mechanism (the initial attack occurs by a nucleophile). The Sandmeyer reaction mechanism (not shown) is related (see https://doi.org/10.1002/cber. 18840170219 and https://doi.org/10.1002/cber.188401702202, accessed December 5, 2019).

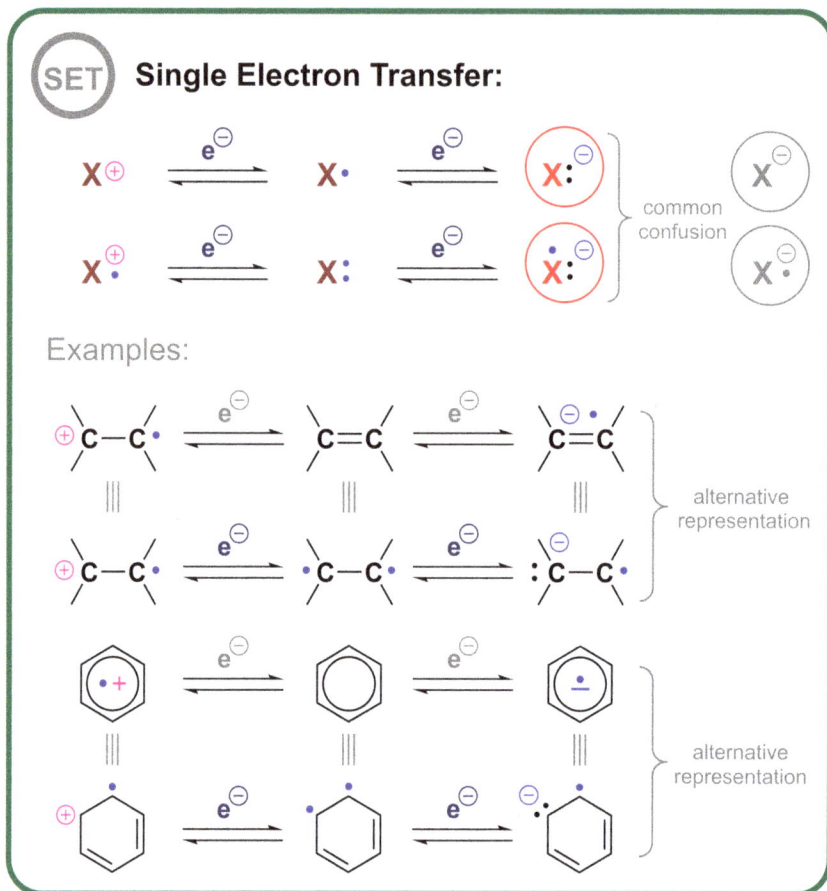

Fig. 5.3: Lewis electron dot structures of radical species involved in SET.[14]

14 This figure summarizes the Lewis (electron) dot structures of various SET processes: **cation** → **radical** → **anion** or **cation radical** → **diradical** or **lone pair** → **anion radical**, and provides several common examples. Please note that in the literature *cation radical* is often called *radical cation* and *anion radical* is called *radical anion*. In some instances, a *lone pair* associated with an *anion* or *anion radical* is not represented for clarity (sometimes this simplification causes confusion).

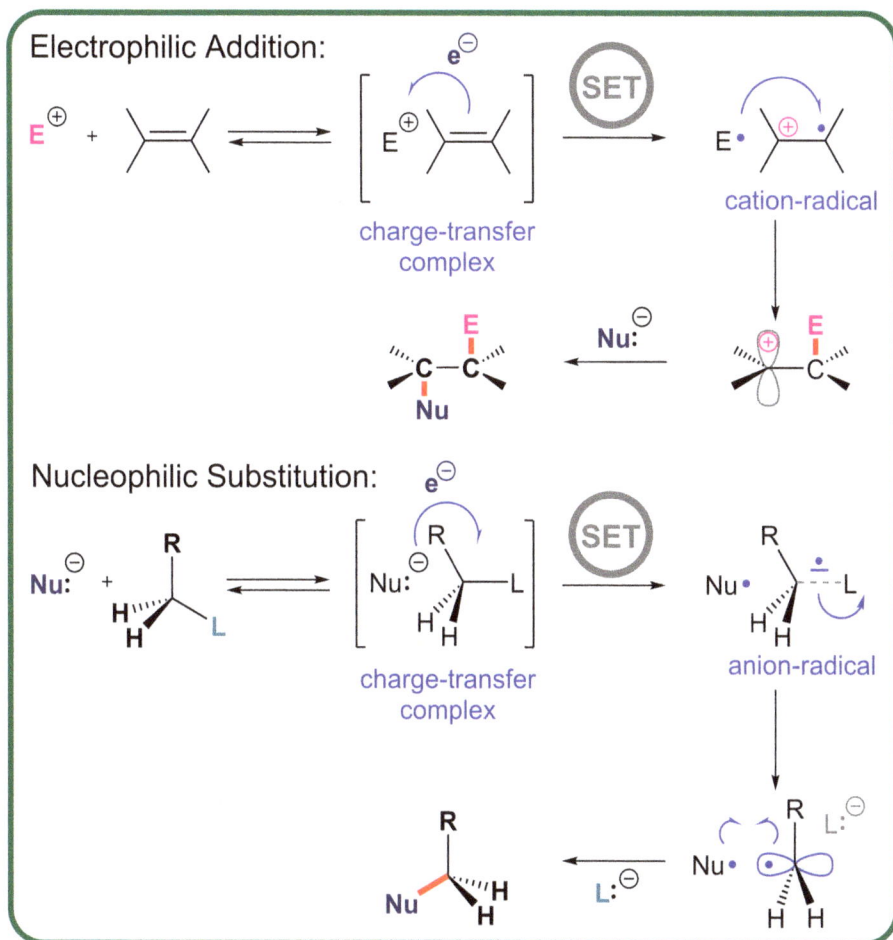

Fig. 5.4: The single electron transfer mechanism (**SET**) examples.[15]

---

[15] An example of *electrophilic addition* described by the SET mechanism: a single electron transfer from an alkene to an electrophile and the formation of a *cation radical* (radical cation). An example of *nucleophilic substitution* described by the SET mechanism: a single electron transfer from a nucleophile to a substrate and the formation of an *anion radical* (radical anion) [3].

6 Elimination Mechanism

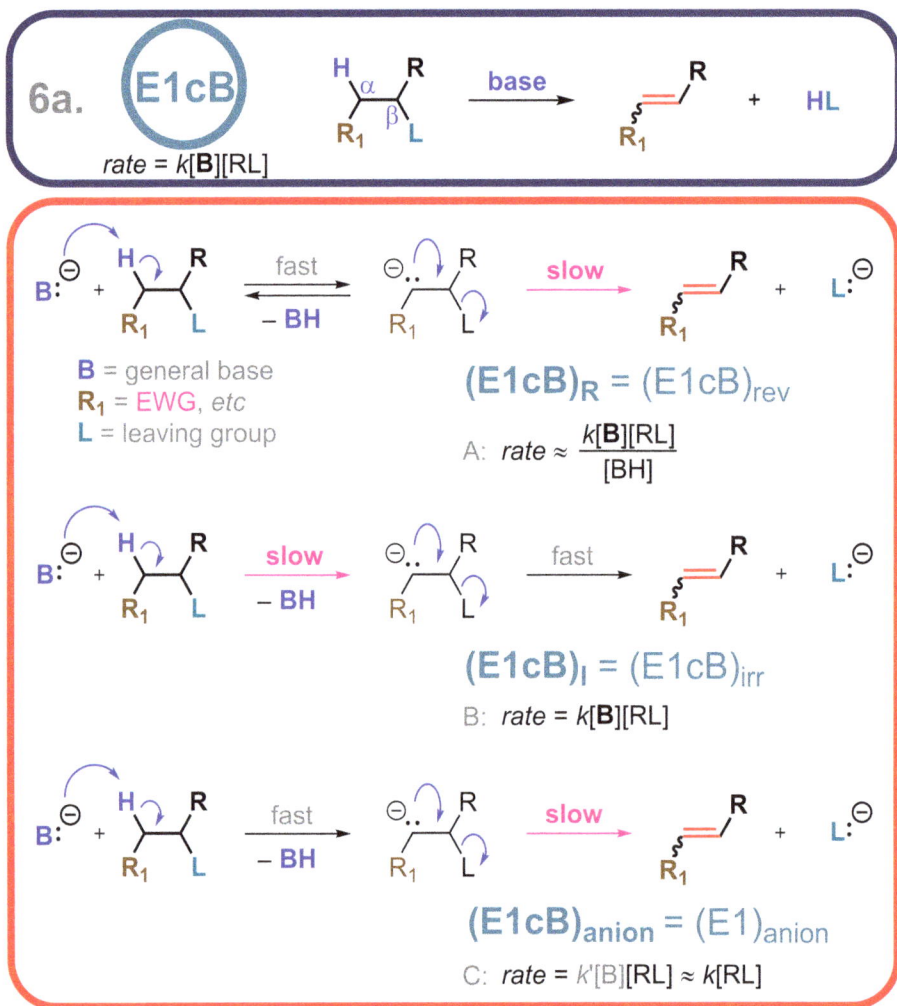

6a. **E1cB**

$rate = k[\mathbf{B}][RL]$

B = general base
R₁ = EWG, *etc*
L = leaving group

$(E1cB)_R = (E1cB)_{rev}$

A: $rate \approx \dfrac{k[\mathbf{B}][RL]}{[BH]}$

$(E1cB)_I = (E1cB)_{irr}$

B: $rate = k[\mathbf{B}][RL]$

$(E1cB)_{anion} = (E1)_{anion}$

C: $rate = k'[B][RL] \approx k[RL]$

Fig. 6.1: Unimolecular β-elimination mechanism (**E1cB**).[16]

---

16 Symbol **E1cB** (**E1cb**) stands for **E**limination **Uni**molecular (**1**) **c**onjugate **B**ase (**b**ase); it is also called the ***carbanion*** mechanism [McLennan DJ. The carbanion mechanism of olefin-forming elimination. *Q. Rev. Chem. Soc.* **1967**, 21 (4), 490–506]. The mechanism consists of two steps: the formation of a carbanion (step 1) and subsequent elimination (step 2). (Scenario A) Step 1 is fast and reversible (**R** or **rev**) and step 2 is rate-determining (slow): $(E1cB)_R = (E1cB)_{rev}$. Here, the rate of the reaction is ***second order*** and the rate-determining step depends on the concentration of two reactants, that is, the ***base*** (**B**) and ***substrate*** (**RL**): $rate \approx k[\mathbf{B}]^1[\mathbf{RL}]^1/[BH]$. (Scenario B) Step 1 is slow and irreversible (**I** or **irr**)

https://doi.org/10.1515/9783110786835-006

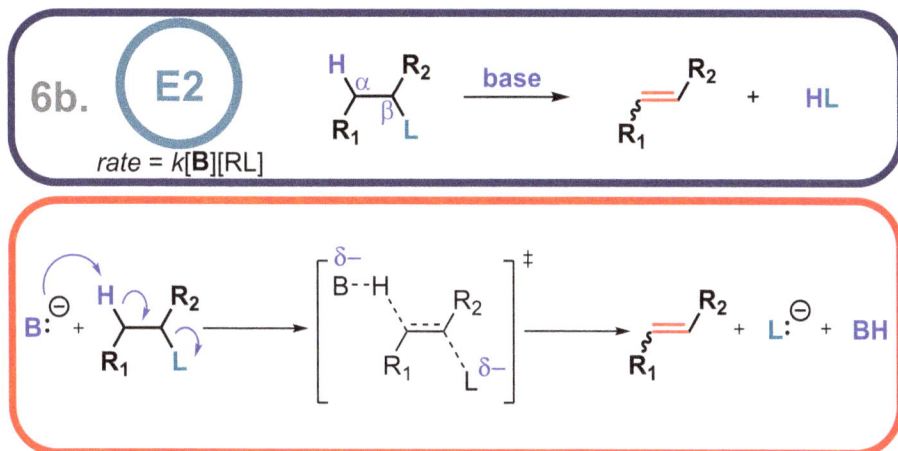

Fig. 6.2: Bimolecular β-elimination mechanism (**E2**).[17]

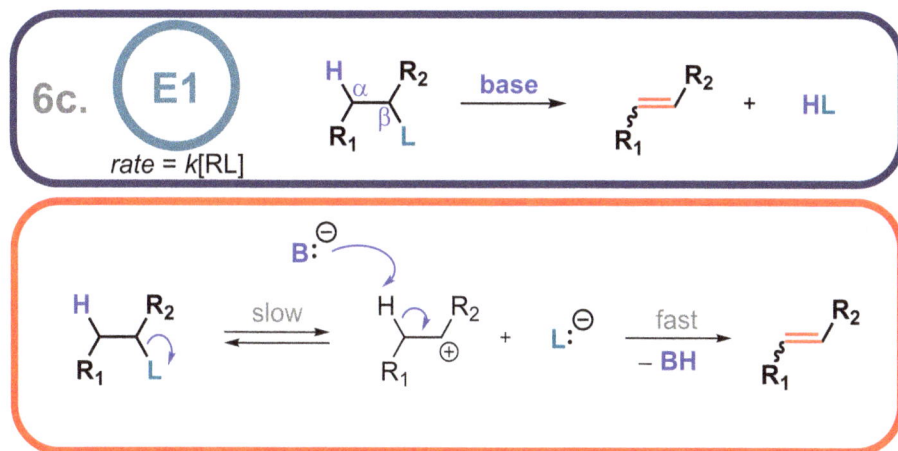

Fig. 6.3: Unimolecular β-elimination mechanism (**E1**).[18]

(rate-determining) and step 2 is fast: **(E1cB)$_I$** = **(E1cB)$_{irr}$**. Here, the rate of the reaction is **second order** and the rate-determining step depends on the concentration of two reactants, that is, the **base** (B) and **substrate** (RL): rate = $k[\mathbf{B}]^1[\mathbf{RL}]^1$. (Scenario C) Step 1 is fast and step 2 is rate-determining (slow): **(E1cB)$_{anion}$** = **(E1)$_{anion}$**. Here, the rate of the reaction is **first order** and the rate-determining step depends on the concentration of one reactant, that is, the **substrate** (RL): rate ≈ $k[\mathbf{RL}]^1$.

17 Symbol **E2** stands for Elimination **Bi**molecular (2), that is, the rate of the reaction is **second order** and the rate-determining step (i.e., the *slow* step) depends on the concentration of two reactants. In this example, it is the **base** (B) and the **substrate** (RL): rate = $k[\mathbf{B}]^1[\mathbf{RL}]^1$.

18 Symbol **E1** stands for Elimination **Uni**molecular (1), that is, the rate of the reaction is **first order** and the rate-determining step (i.e., the *slow* step) depends on the concentration of one reactant. In this example, it is the **substrate** (RL): rate = $k[\mathbf{RL}]^1$.

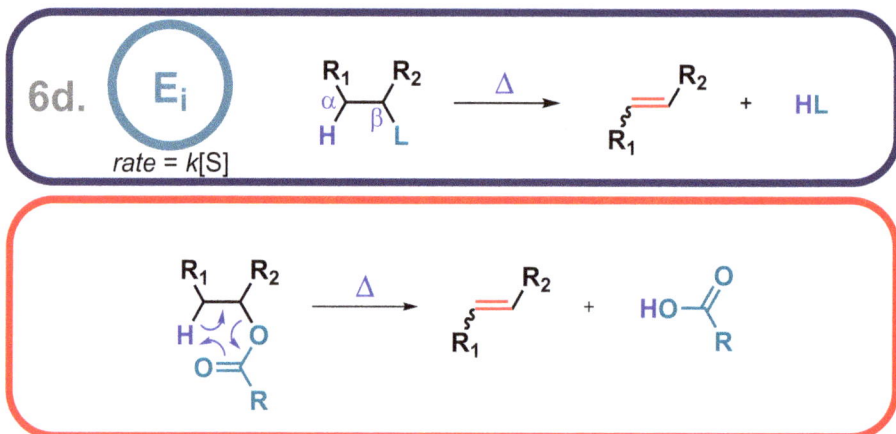

Fig. 6.4: Internal or intramolecular β-elimination mechanism (**E$_i$**).[19]

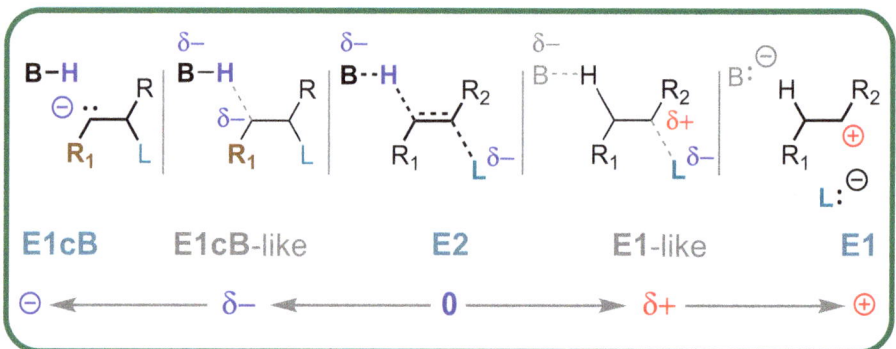

Fig. 6.5: **E1cB**, **E2**, and **E1** mechanisms.[20]

---

19 Symbol **E$_i$** stands for Elimination Internal or Intramolecular. The rate of the reaction is ***first order*** and the rate-determining step (i.e., the *slow* step) depends on the concentration of one reactant. In this example, it is the ***substrate*** (**S**): *rate* = $k[S]^1$.

20 The **E1cB** mechanism is also called the carbanion mechanism; its transition state is the most extreme case with a full negative charge. The **E2** mechanism is simultaneous and the transition state lies in the middle. A typical E2 reaction often competes with an S_N2 reaction and vice versa. The **E1** mechanism is exactly the opposite of E1cB and its transition state has a positive charge. A typical E1 reaction often competes with an S_N1 reaction and vice versa.

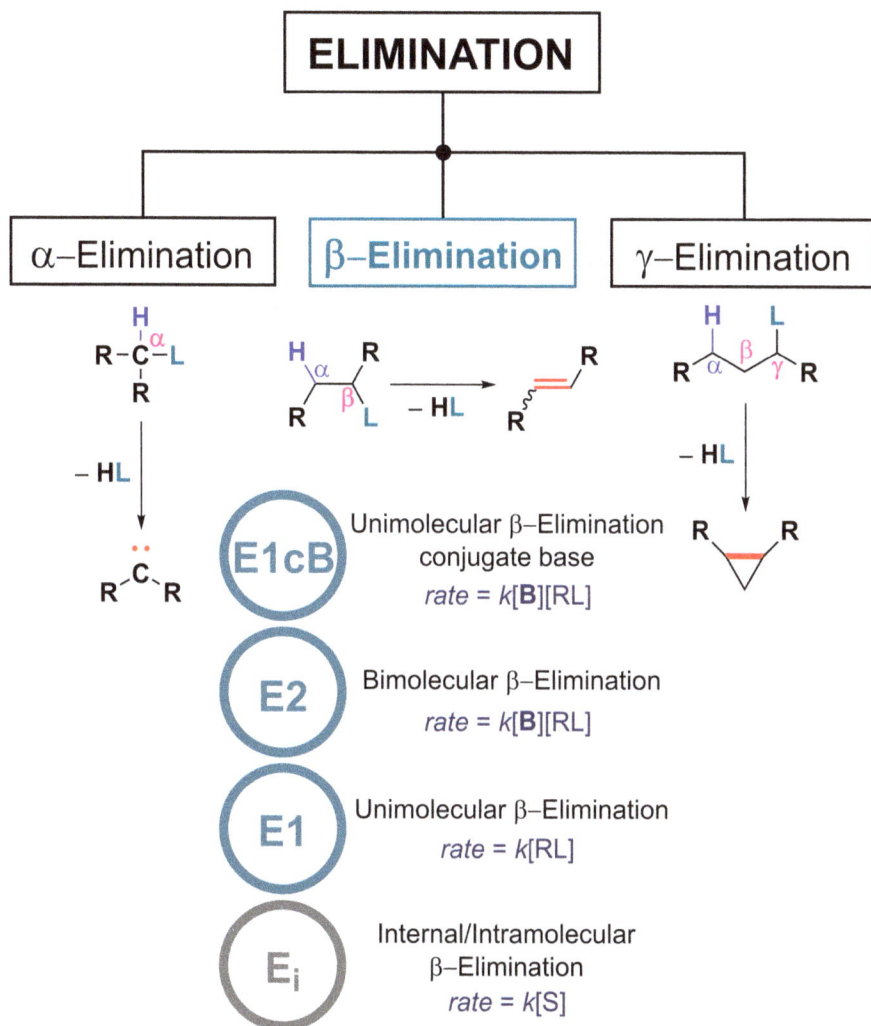

Fig. 6.6: The classification of characteristic elimination reactions.[21]

---

21 Only the key *β-elimination* examples are covered in this book.

7 Acyloin Condensation

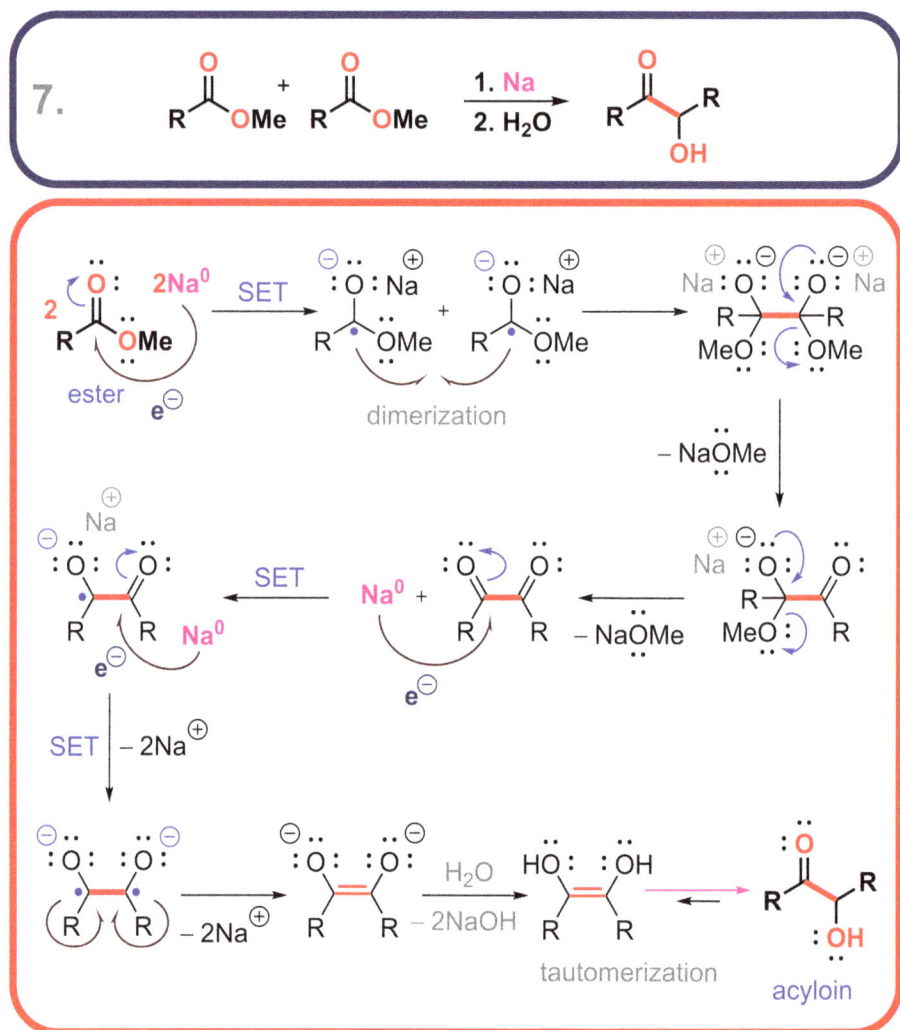

Fig. 7.1: The *acyloin condensation* mechanism.[22]

---

22 The reaction is also called the *acyloin **ester** condensation*. Please note that an *acyloin* is an α-hydroxy ketone.

https://doi.org/10.1515/9783110786835-007

Bouveault–Blanc Reduction

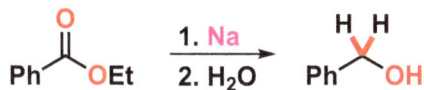

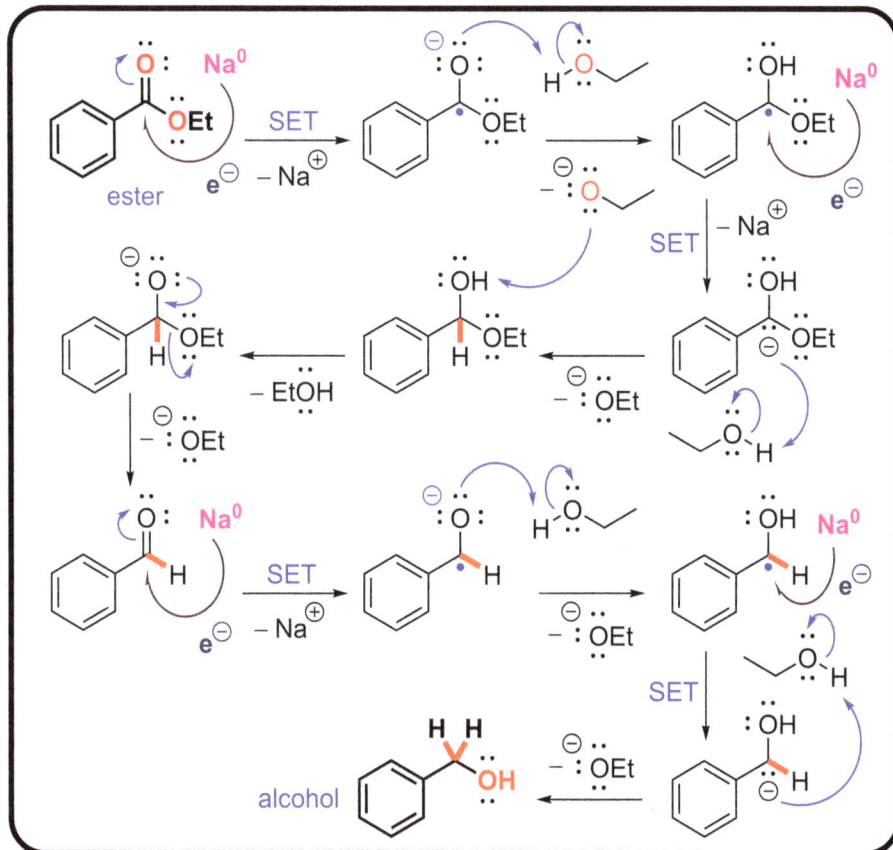

Fig. 7.2: The **Bouveault–Blanc** reduction mechanism (ester reduction).[23]

---

23 Several reactions are mechanistically related to the *acyloin condensation*: the **Bouveault–Blanc** *reduction* of esters [1a, 7a].

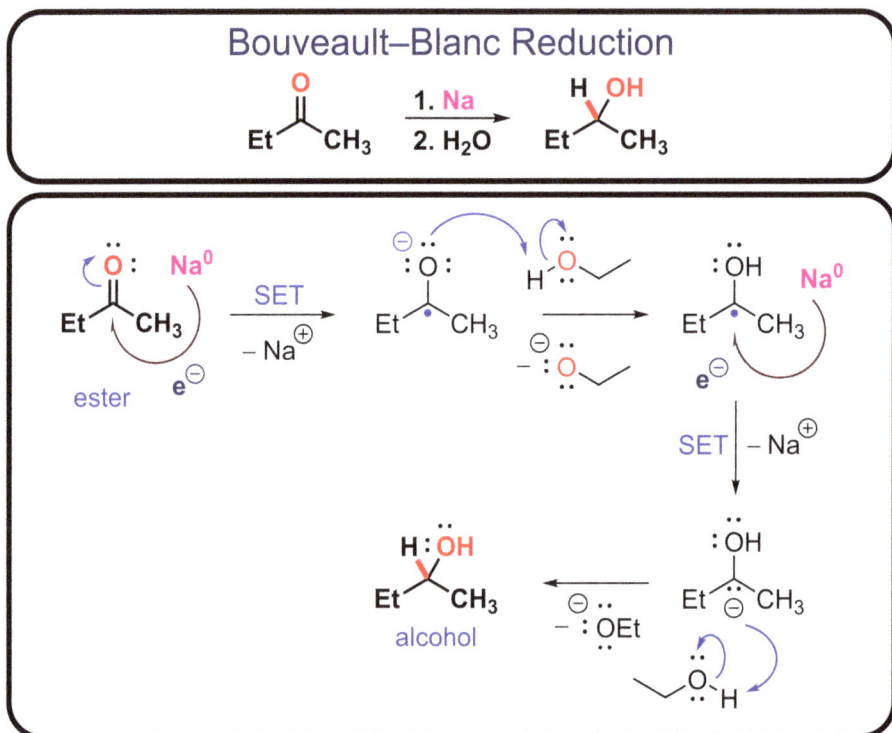

Fig. 7.3: The *Bouveault–Blanc reduction* mechanism (ketone reduction).[24]

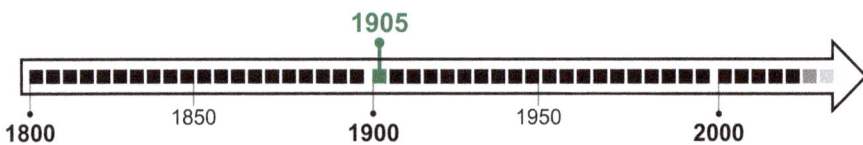

Fig. 7.4: The discovery of the *acyloin condensation*.[25]

---

24 Several reactions are mechanistically related to the *acyloin condensation*: the **Bouveault–Blanc** *reduction* of ketones [1a, 7a].

25 The reaction was likely first described around 1905 [7b].

Pinacol Coupling

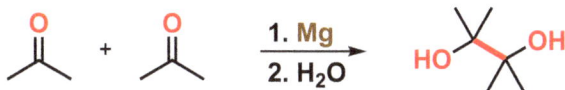

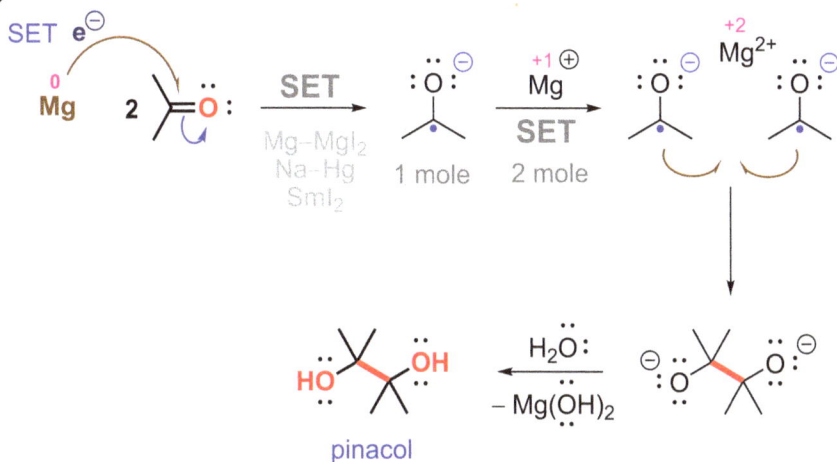

Fig. 7.5: The *pinacol coupling* mechanism.[26]

---

26 Several reactions are mechanistically related to the *acyloin condensation*: the *pinacol coupling* and the **McMurry** *coupling* (both covered in Chapter 57). The *benzoin condensation* (covered in Chapter 15) undergoes a different mechanism, but it also yields α-hydroxy ketones containing aromatic groups (*benzoins*).

8 Alkyne Zipper Reaction

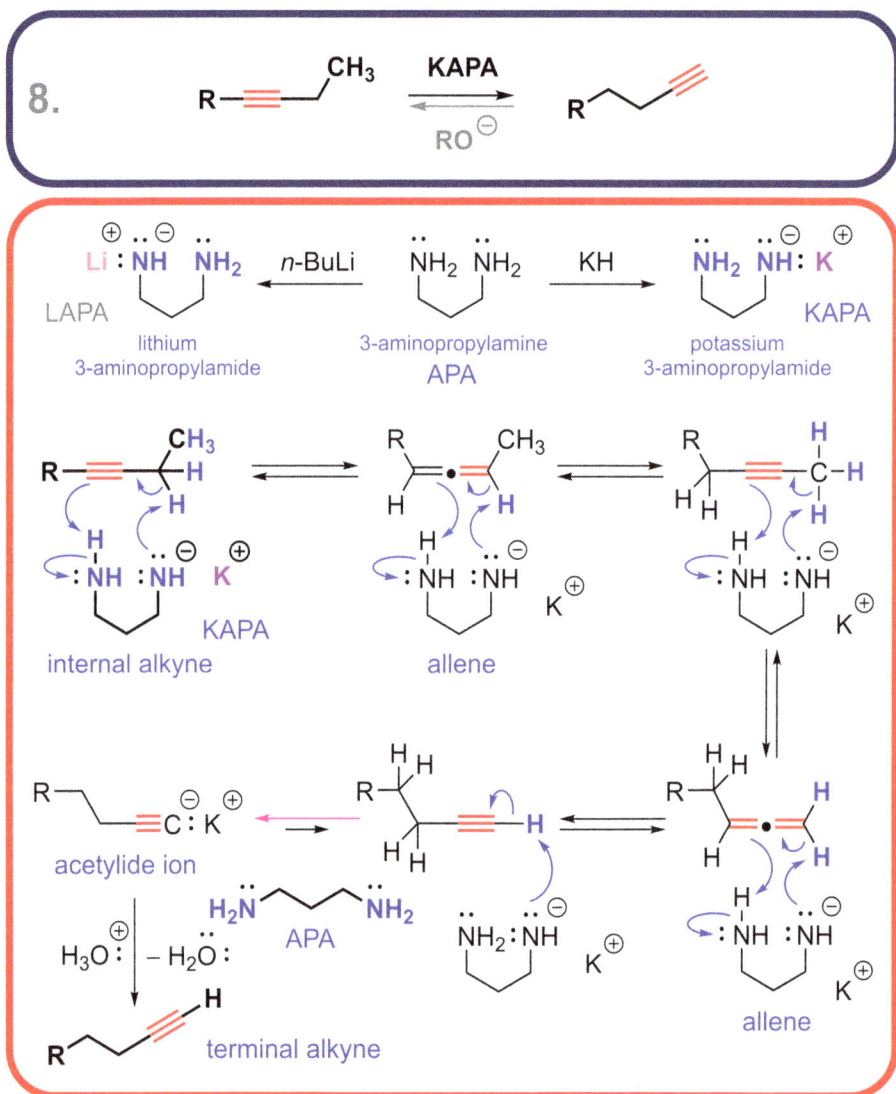

Fig. 8.1: The *alkyne zipper reaction* mechanism.[27]

---

27 The reaction is also called the *alkyne isomerization reaction* or the *alkyne–allene rearrangement*.

https://doi.org/10.1515/9783110786835-008

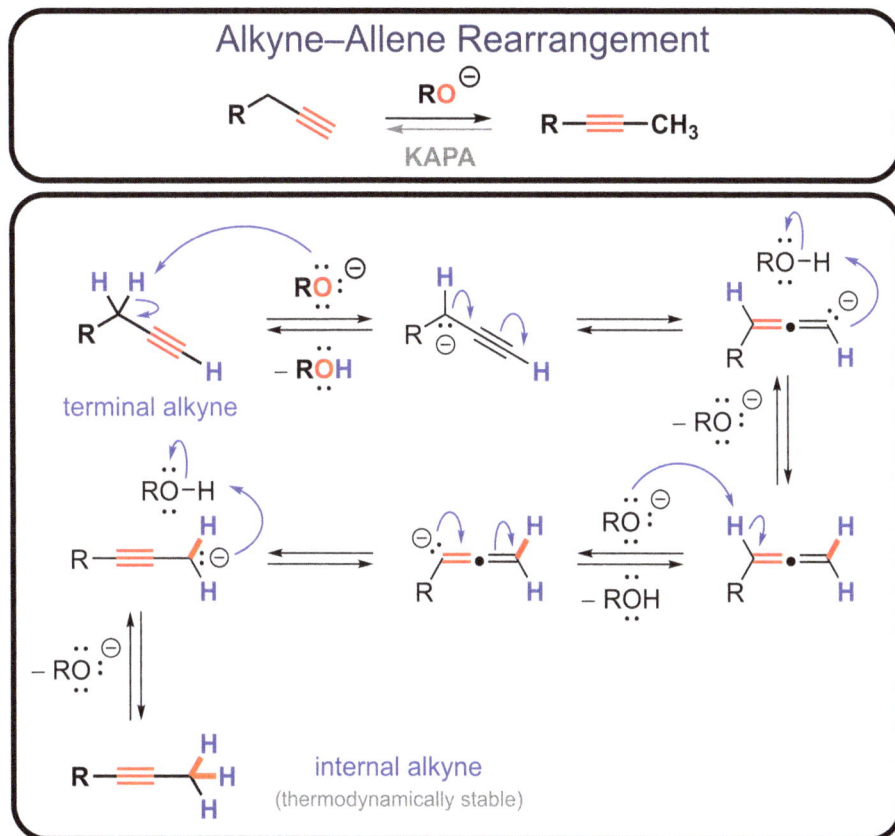

Fig. 8.2: The *alkyne–allene rearrangement* mechanism.[28]

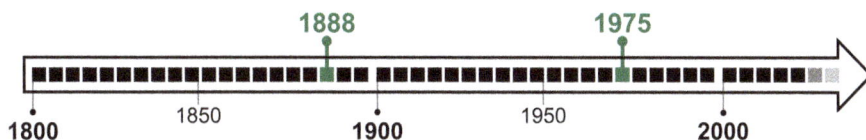

Fig. 8.3: The discovery of the *alkyne zipper reaction*.[29]

---

28 The *alkyne zipper reaction* with KAPA yields thermodynamically _less_ stable *terminal alkyne*, whereas the typical *alkyne–allene rearrangement* usually produces thermodynamically _more_ stable *internal alkyne*. Both reactions are reversible.

29 The reaction was likely first mentioned around 1888 by A. Favorsky (Favorskii) (in Russian A. E. Фаворский) [8a, 8b, 8c], the variation presented here was likely first described around 1975 [8d].

9 Arbuzov Reaction

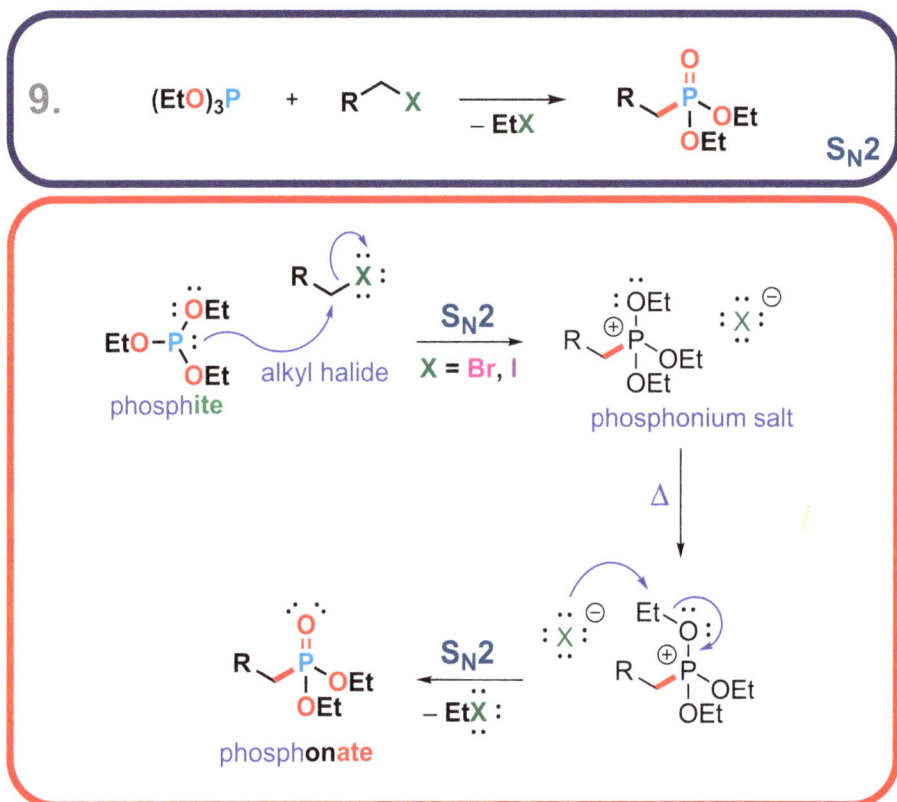

Fig. 9.1: The **Arbuzov** reaction mechanism.[30]

---

30 The **Arbuzov** reaction is an example of the bimolecular **nucleophilic substitution** (S_N2), covered in Chapter 2. It is also referred to as the **Michaelis–Arbuzov** reaction or the **Michaelis–Arbuzov** rearrangement.

https://doi.org/10.1515/9783110786835-009

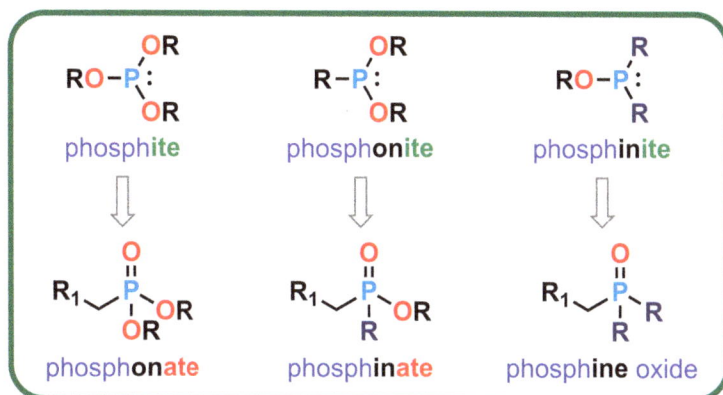

Fig. 9.2: The nomenclature of selected organophosphorus(III) and (V) compounds.[31]

Horner–Wadsworth–Emmons Olefination

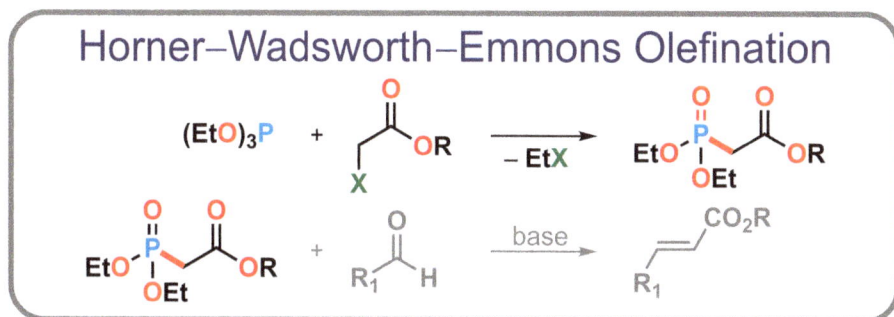

Fig. 9.3: The *HWE* olefination.[32]

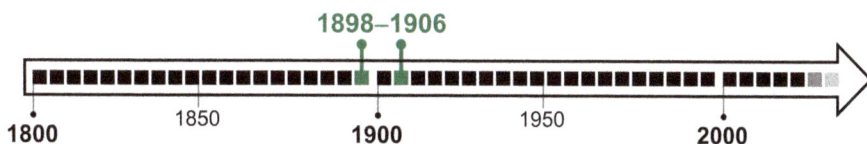

Fig. 9.4: The discovery of the *Arbuzov reaction*.[33]

---

31 A selected example of the complex organophosphorus nomenclature: the organophosphorus(III) compounds have a common suffix -*ite* [phosphites P(OR)$_3$, phosphonites P(OR)$_2$R] and the organophosphorus(V) compounds have a common suffix -*ate* [phosphonates PO(OR)$_2$R, phosphinates PO(OR)R$_2$] [9a].

32 The *phosphonates* produced in the *Arbuzov reaction* are essential in the *Horner–Wadsworth–Emmons (HWE) olefination* (covered in Chapter 50).

33 The reaction was likely first described around 1898 by Michaelis [9b] and around 1906 by Arbuzov [9c].

10 Arndt–Eistert Synthesis

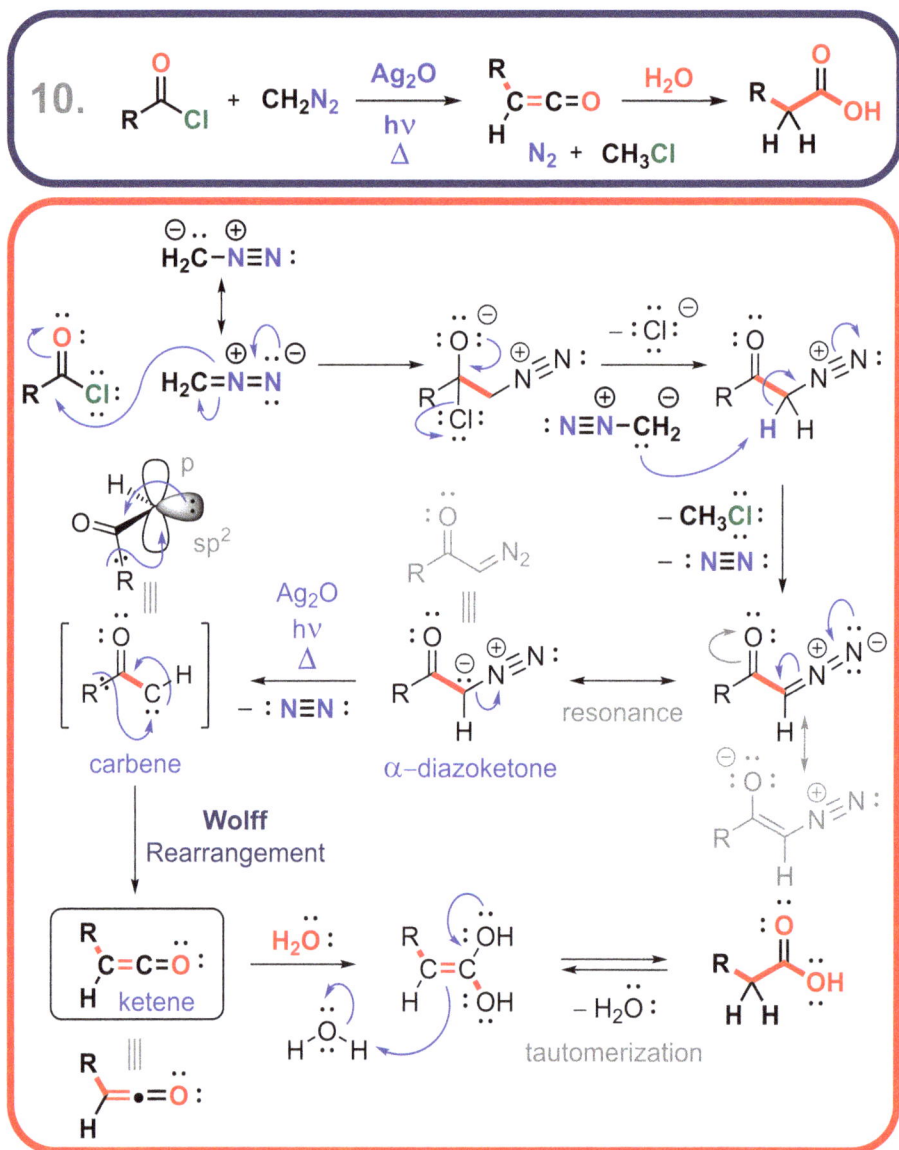

10. $R-C(=O)-Cl$ + CH_2N_2 $\xrightarrow[\Delta]{Ag_2O,\ h\nu}$ $R,H-C=C=O$ $\xrightarrow{H_2O}$ $R,H_2-C(=O)-OH$
  N_2 + CH_3Cl

$\overset{\ominus}{H_2C}-\overset{\oplus}{N\equiv N:}$

$H_2C=N=N$

$R-C(=O)-Cl$

$R-C(O^{\ominus})(Cl)-N=\overset{\oplus}{N}:$ $:N\equiv N-CH_2$

$-:Cl:$

$R-C(=O)-\overset{\oplus}{N}=N:$, $H\ H$

carbene

sp^2

$R-C(=O)-C-H$ $-:N\equiv N:$

Ag_2O $h\nu$ Δ $-:N\equiv N:$

$R-C(=O)-\overset{\ominus}{C}(H)-\overset{\oplus}{N}\equiv N:$ α–diazoketone

$R-C(=O)-N_2$

resonance

$-CH_3Cl:$ $-:N\equiv N:$

$R-C(=O)-C(H)=\overset{\oplus}{N}=\overset{\ominus}{N}:$

$R-C(O^{\ominus})=C(H)-\overset{\oplus}{N}\equiv N:$

Wolff Rearrangement

$R,H-C=C=O:$ ketene

$R,H-C=\bullet=O:$

$H_2O:$

$R,H-C=C(OH)(OH)$

$H-O-H$

$-H_2O:$ tautomerization

$R-C(=O)-OH$, $H\ H$

Fig. 10.1: The **Arndt–Eistert** synthesis mechanism.[34]

34 The **Arndt–Eistert** synthesis is also called the **Arndt–Eistert** reaction (homologation). The **Wolff** rearrangement (α-diazoketone) is part of the **Arndt–Eistert** synthesis mechanism [10a].

https://doi.org/10.1515/9783110786835-010

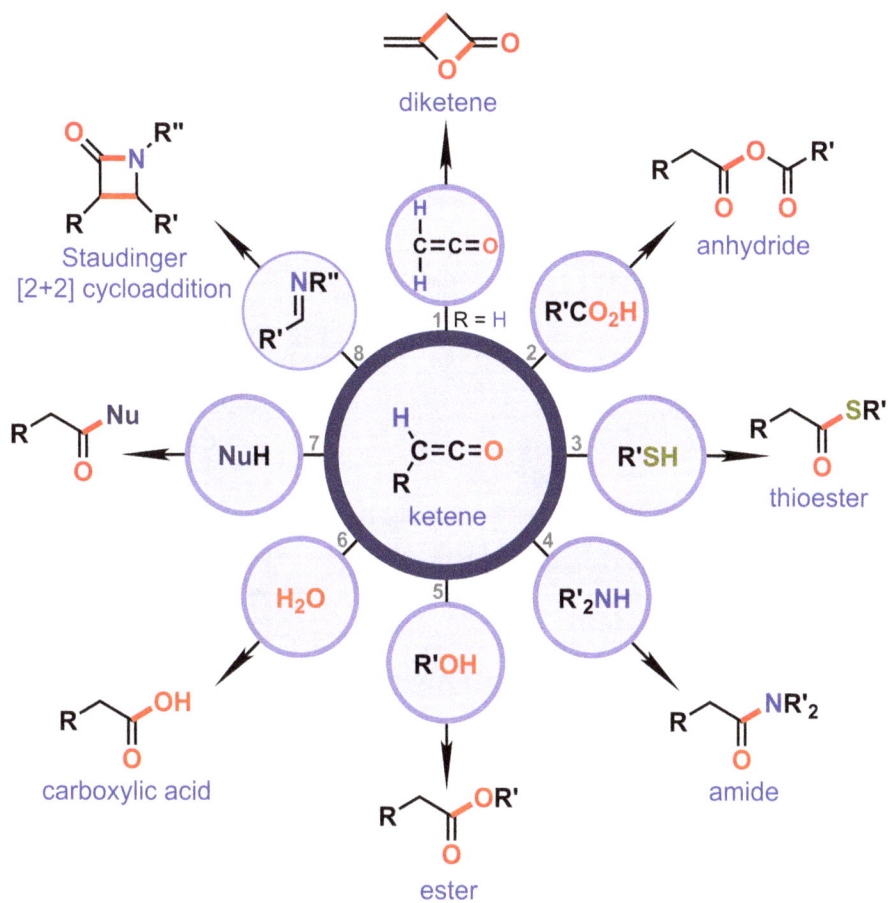

Fig. 10.2: The synthetic versatility of ketenes.[35]

35 The *ketenes* formed during the ***Arndt–Eistert*** *synthesis* can either be trapped by a variety of nucleophiles, or undergo [2 + 2] cycloaddition including dimerization.

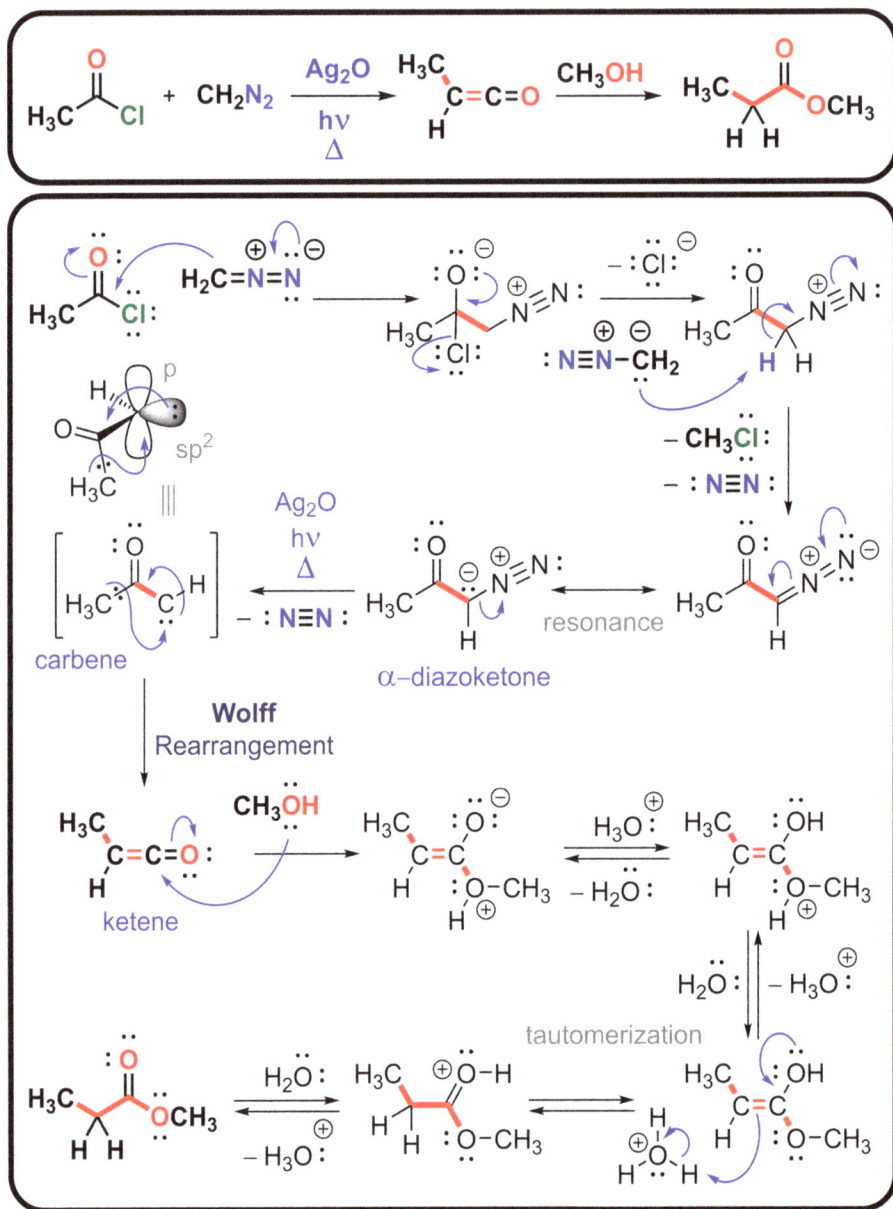

Fig. 10.3: The **Arndt–Eistert** reaction mechanism.[36]

---

[36] The *ketene* (prop-1-en-1-one) formed during the **Arndt–Eistert** synthesis is trapped by methanol.

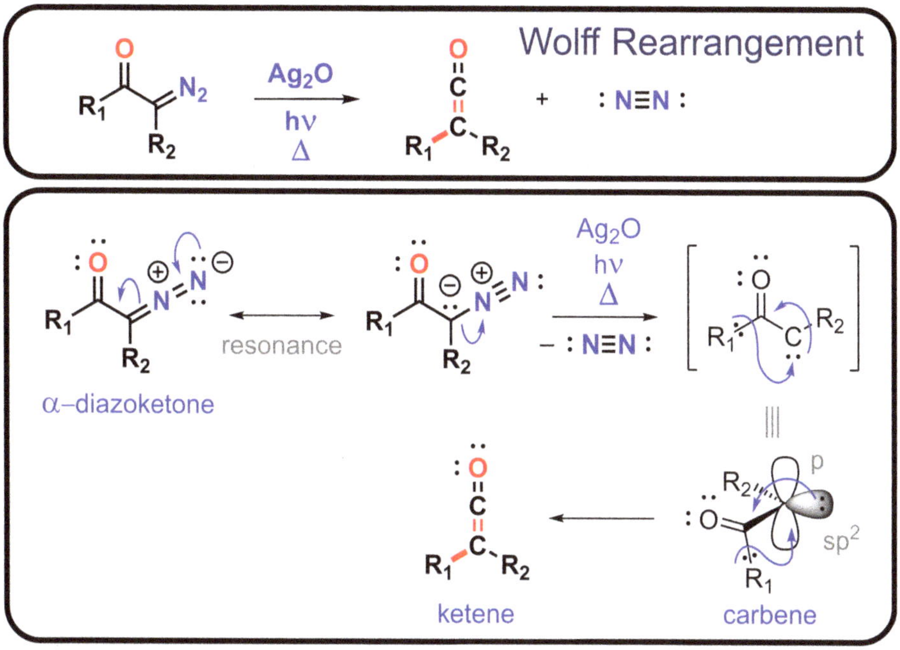

Fig. 10.4: The **Wolff** *rearrangement* mechanism.[37]

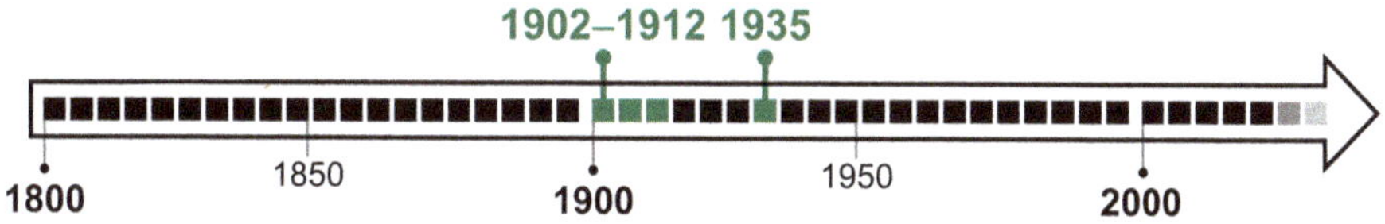

Fig. 10.5: The discovery of the **Arndt–Eistert** *synthesis*.[38]

---

37 The **Wolff** *rearrangement* (α-diazoketone) is part of the **Arndt–Eistert** *synthesis* mechanism [10a].
38 The related reaction was likely first described by Wolff between 1902 and 1912 [10a, 10b] and by Arndt and Eistert around 1935 [10c].

11 Baeyer–Villiger Oxidation

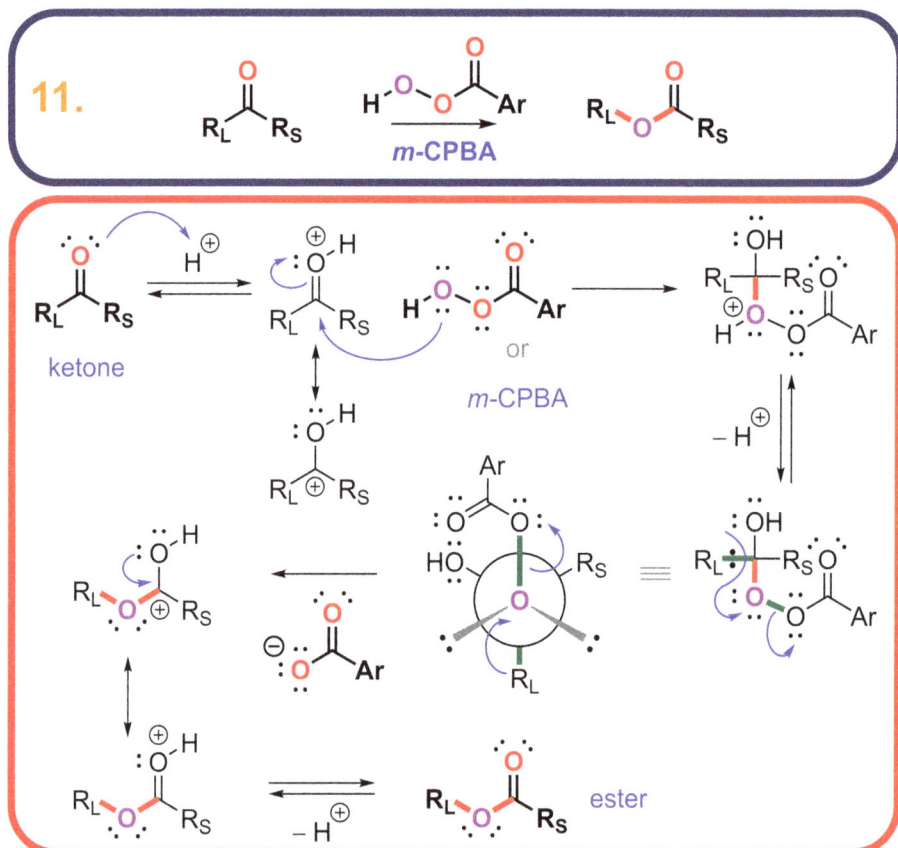

11.

Fig. 11.1: The **Baeyer–Villiger** oxidation mechanism.[39]

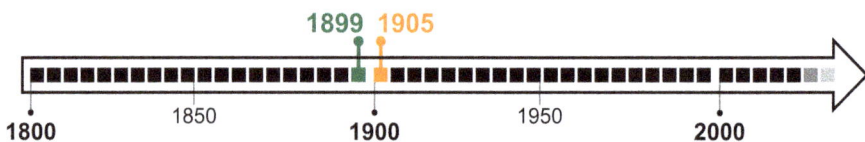

Fig. 11.2: The discovery of the **Baeyer–Villiger** oxidation.[40]

---

39 The **Baeyer–Villiger** oxidation is also called the **Baeyer–Villiger** rearrangement.
40 The reaction was likely first described around 1899 [11b]. In **1905**, Johann Friedrich Wilhelm Adolf von Baeyer received the Nobel Prize in Chemistry [11c].

https://doi.org/10.1515/9783110786835-011

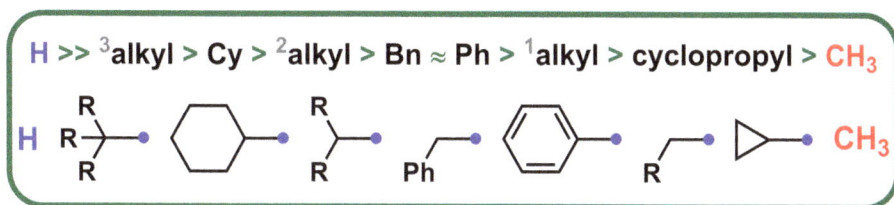

H >> ^{3}alkyl > Cy > ^{2}alkyl > Bn ≈ Ph > ^{1}alkyl > cyclopropyl > CH$_3$

Fig. 11.3: The order of group migration in the **Baeyer–Villiger** oxidation.[41]

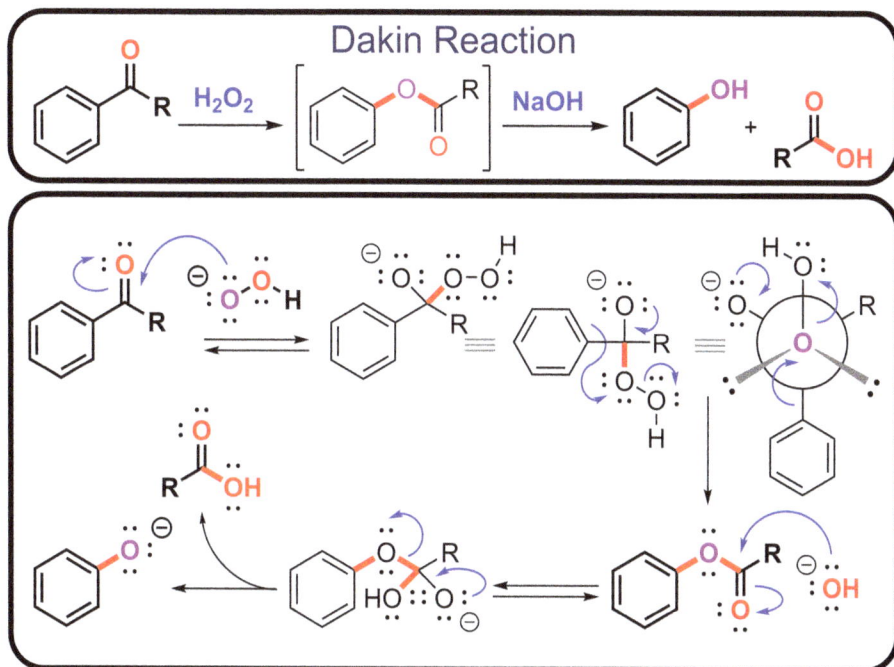

Dakin Reaction

Fig. 11.4: The **Dakin** reaction mechanism.[42]

---

41 The order of group migration is essential for the *asymmetrical* ketones. Please note that this preference for migration is a general empirical trend and not an absolute rule [1].

42 The **Dakin** reaction (oxidation) is closely related to the **Baeyer–Villiger** oxidation and it usually yields *ortho*-hydroxy or *para*-hydroxy phenols (or phenols with a strong *ortho*- or *para*- ERG) [11a].

12 Barton Decarboxylation

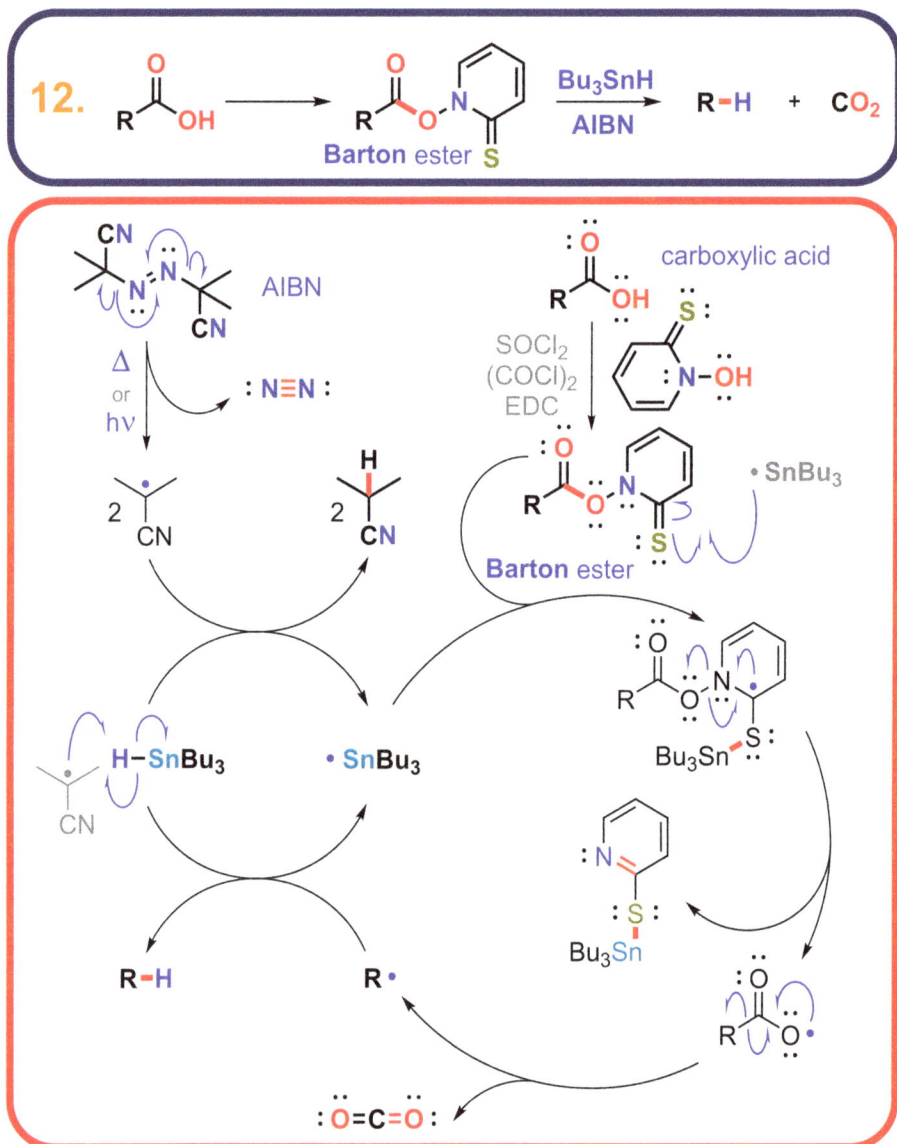

Fig. 12.1: The *Barton decarboxylation* mechanism.[43]

---

43 The ***Barton*** *decarboxylation* is a radical decarboxylation reaction of the ***Barton*** ester.

https://doi.org/10.1515/9783110786835-012

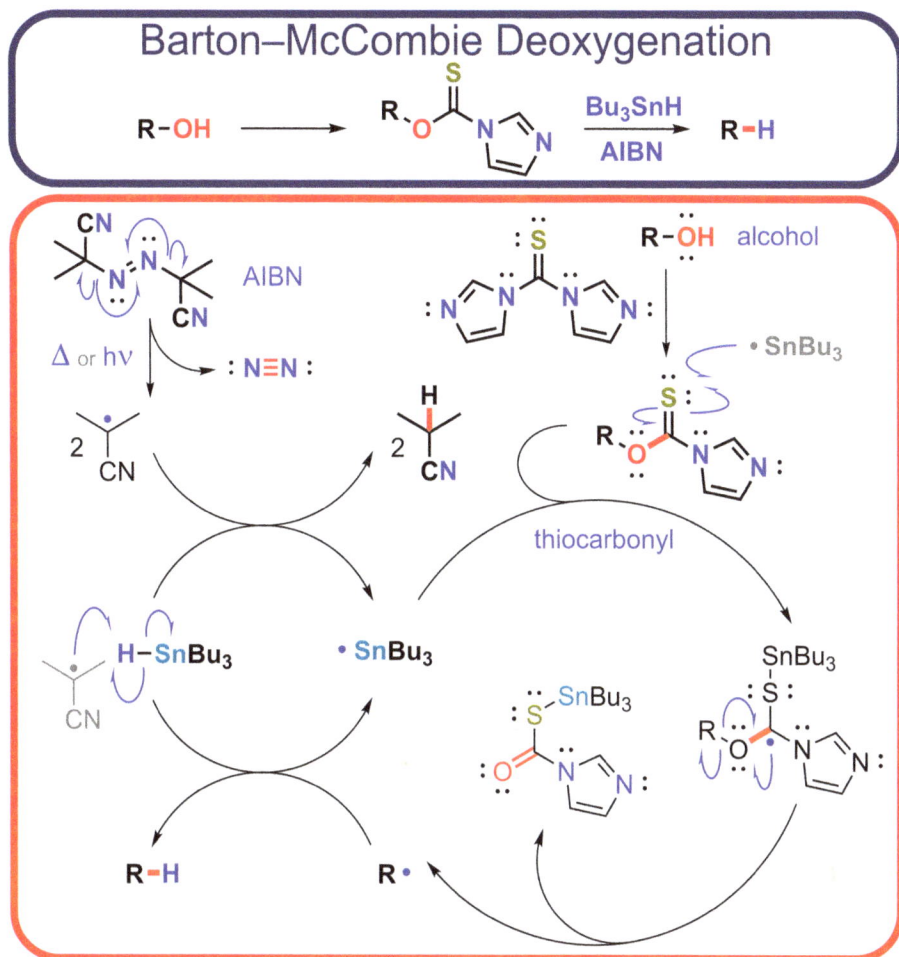

Fig. 12.2: The **Barton–McCombie** *deoxygenation* mechanism.[44]

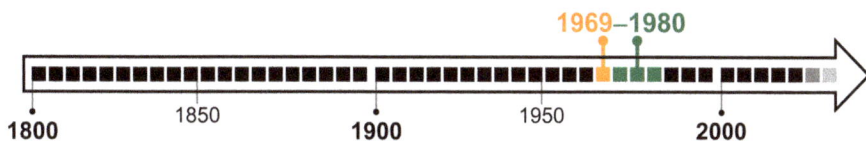

Fig. 12.3: The discovery of the **Barton** *decarboxylation*.[45]

---

44 The **Barton–McCombie** *deoxygenation* is a radical deoxygenation of a *thiocarbonyl*: *O,O*-thiocarbonate **R**OC(S)O**R**; *S,O*-dithiocarbonate = xanthate **R**OC(S)S**R**; or *O*-thiocarbamate **R**OC(S)N**R**$_2$.

45 The *decarboxylation* reaction was likely first described between 1980 and 1985 [12a, 12b], and the *deoxygenation* reaction was likely first described between 1975 and 1980 [12c, 12d]. In **1969**, Derek H. R. Barton (jointly with Odd Hassel) received the Nobel Prize in Chemistry [12e].

13 Baylis–Hillman Reaction

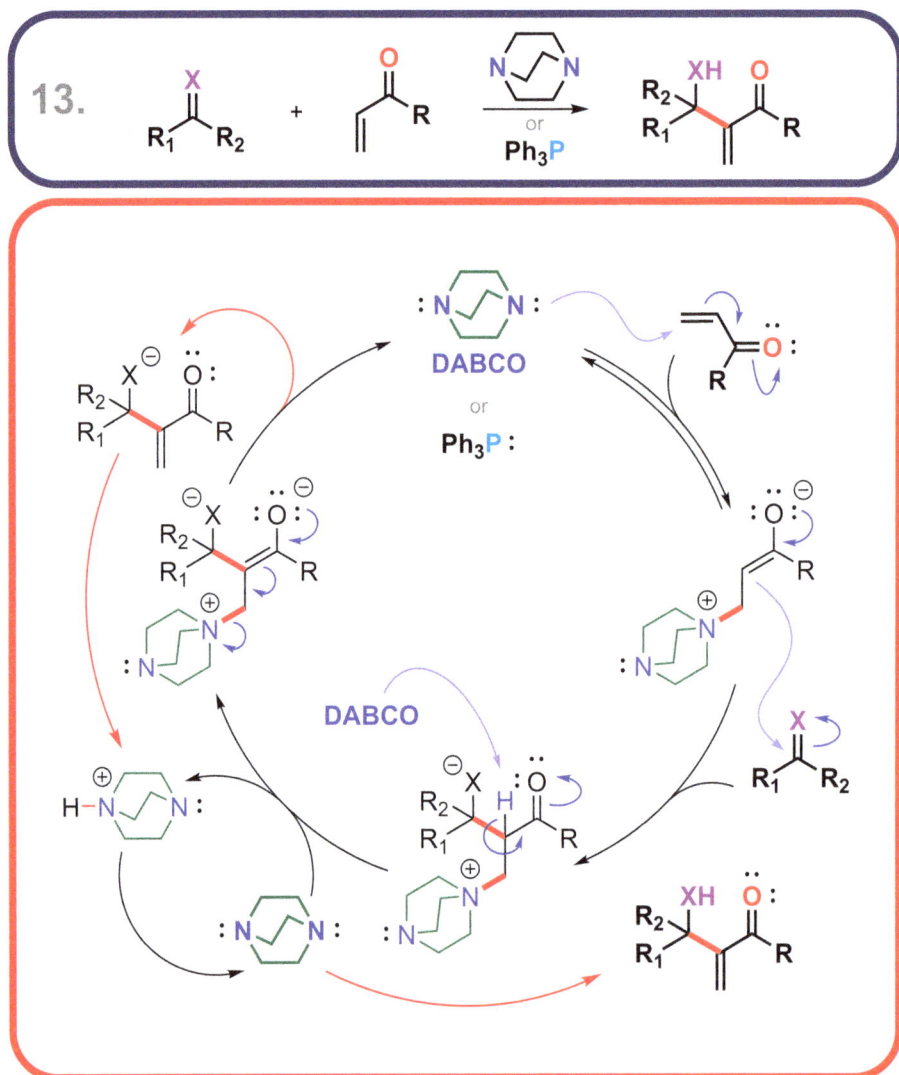

Fig. 13.1: The **Baylis–Hillman** reaction mechanism.[46]

---

46 The **Baylis–Hillman** reaction is also called the **Morita–Baylis–Hillman** reaction.

https://doi.org/10.1515/9783110786835-013

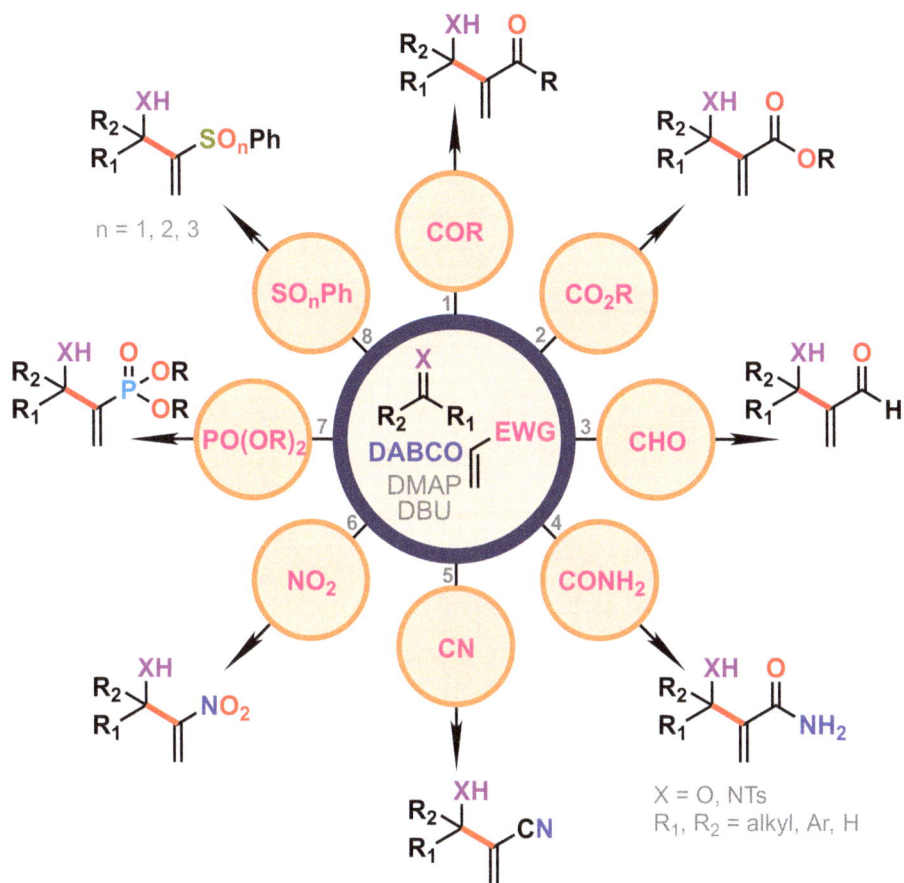

Fig. 13.2: The synthetic versatility of the *Baylis–Hillman reaction*.[47]

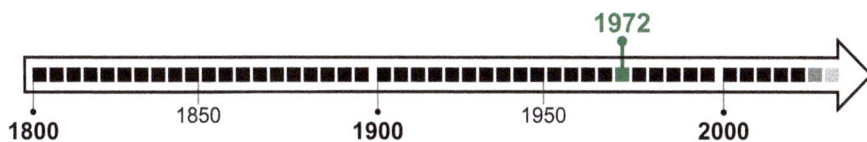

Fig. 13.3: The discovery of the *Baylis–Hillman reaction*.[48]

---

47 Many variations of the *Baylis–Hillman reaction* exist, depending on the nature of EWG (the *Michael* acceptor) and carbonyl compound (the electrophile). Please note that for X = NR, it is called the *aza-Baylis–Hillman reaction*.

48 The reaction was likely first described around 1972 [13].

14 Beckmann Rearrangement

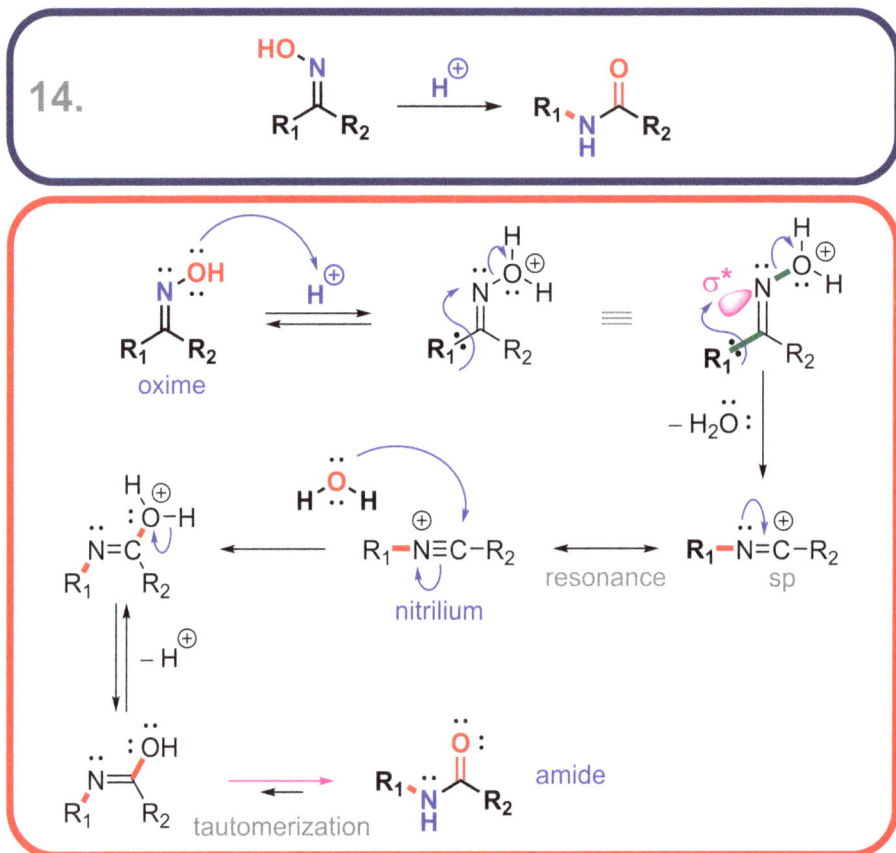

Fig. 14.1: The **Beckmann** *rearrangement* mechanism.[49]

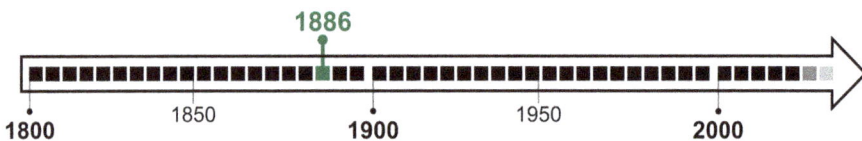

Fig. 14.2: The discovery of the **Beckmann** *rearrangement*.[50]

---

49 The **Beckmann** *rearrangement* is seldom called the **Beckmann** *oxime–amide rearrangement*.
50 The reaction was likely first described around 1886 [14a].

https://doi.org/10.1515/9783110786835-014

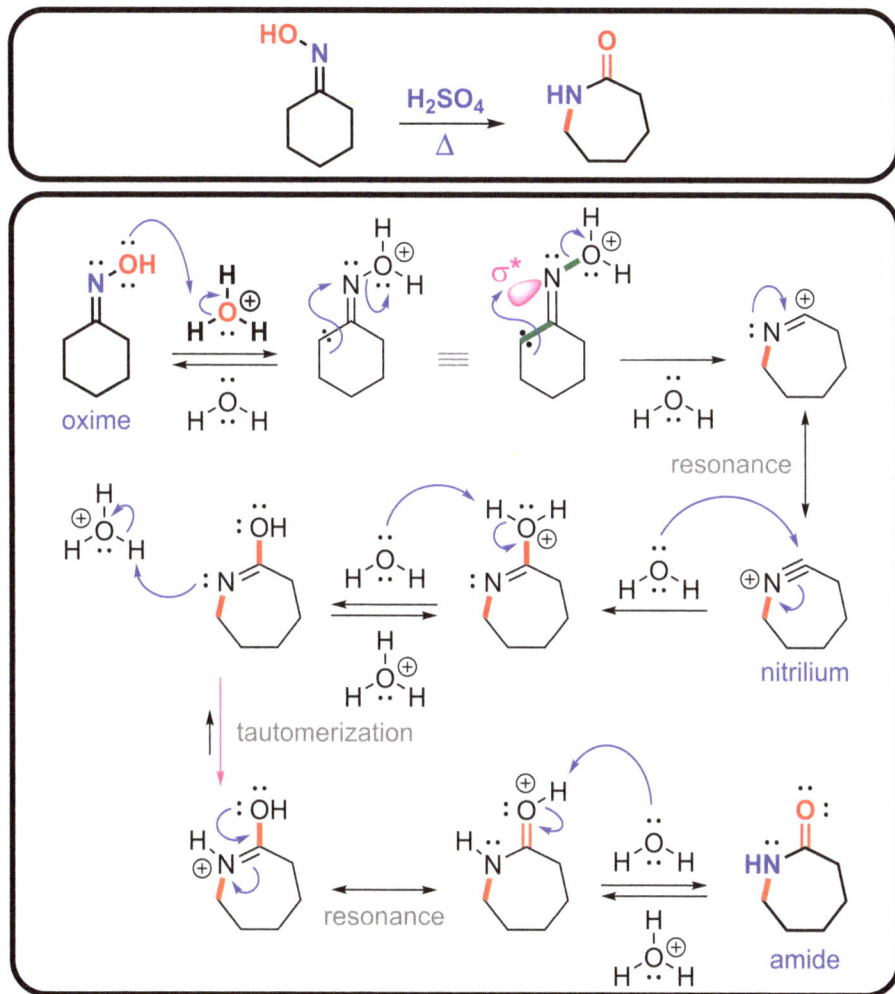

Fig. 14.3: The **Beckmann** *rearrangement* mechanism of cyclohexanone oxime.[51]

---

51 This is an example of the **Beckmann** *rearrangement* mechanism of cyclohexanone oxime to azepan-2-one (also known as **caprolactam**) [14b]. Several reactions are mechanistically related to the **Beckmann** *rearrangement*: the **Curtius** *rearrangement*, the **Schmidt** *reaction*, the **Hofmann** *rearrangement*, and the **Lossen** *rearrangement* (all covered in Chapter 31).

15 Benzoin Condensation

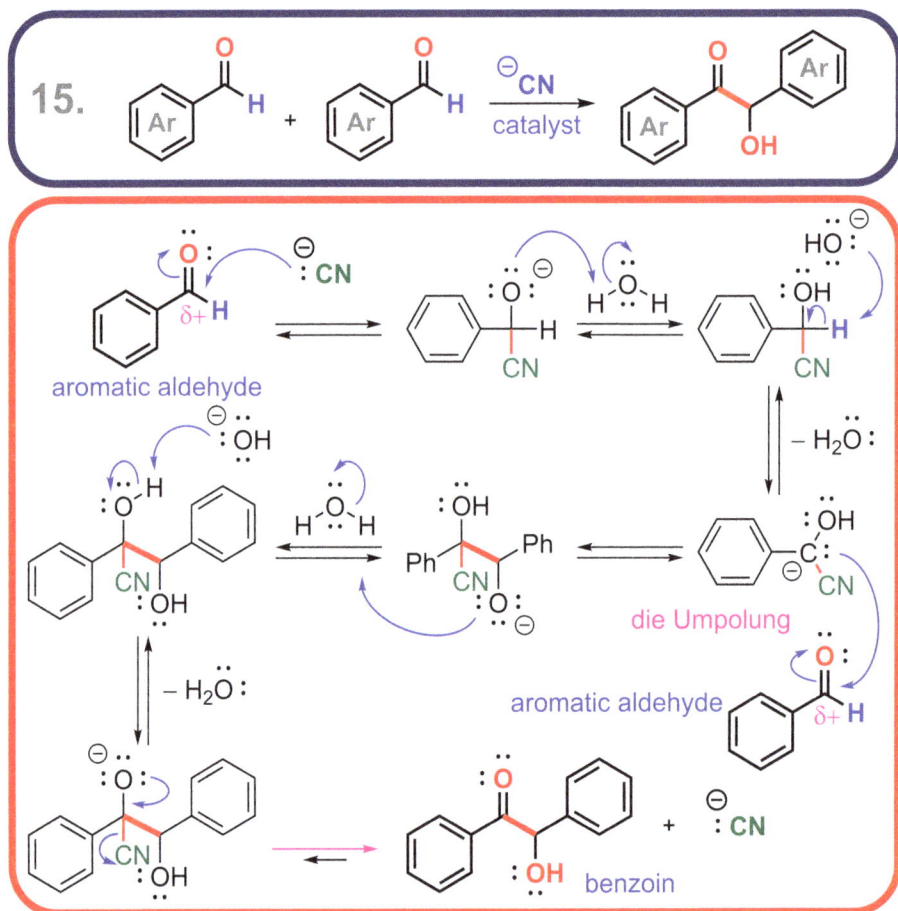

Fig. 15.1: The *benzoin condensation* mechanism.[52]

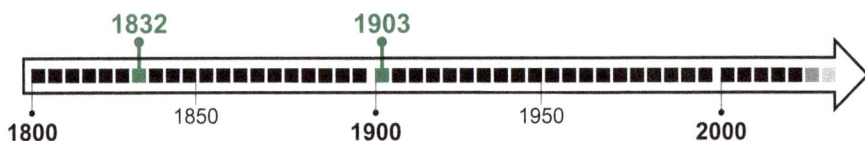

Fig. 15.2: The discovery of the *benzoin condensation*.[53]

---

52 The *benzoin condensation* is one of the oldest reactions in organic chemistry.
53 The reaction was likely first described around 1832 and the mechanism was proposed in 1903 [15c, 15d].

https://doi.org/10.1515/9783110786835-015

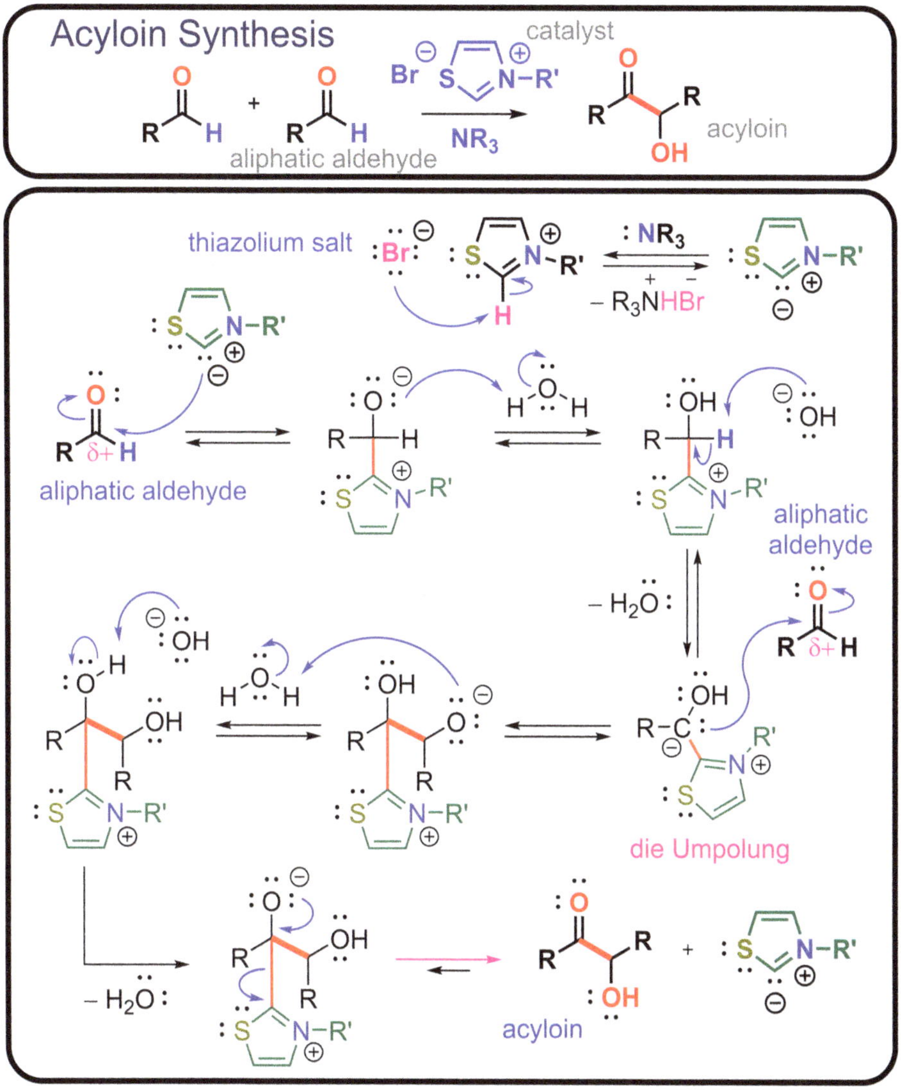

Fig. 15.3: The *acyloin synthesis* mechanism using thiazolium salts.[54]

---

[54] The *benzoin condensation* involves two *aromatic* aldehydes and is catalyzed by **cyanide ion** forming aromatic α-hydroxy ketones *(benzoins)*. The *acyloin synthesis* is a condensation of two *aliphatic* aldehydes, it is catalyzed by **thiazolium salts** [15a, 15b] and yields *aliphatic* (or mixed) α-hydroxy ketones *(acyloins)*. The *acyloin synthesis* should not be confused with the *acyloin condensation* (Chapter 7).

16 Benzyne Mechanism

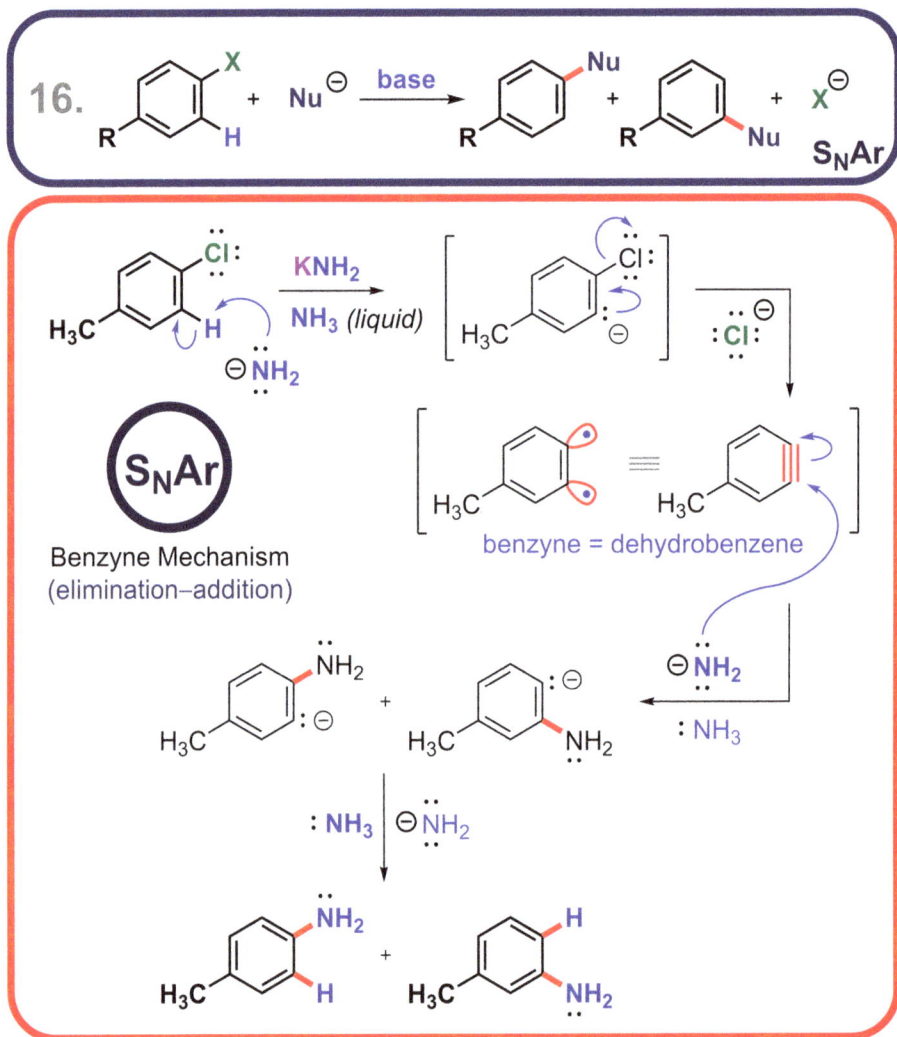

16.

Fig. 16.1: The *benzyne (elimination–addition)* mechanism.[55]

---

55 The *benzyne mechanism* is one of the fundamental **aromatic nucleophilic substitution** mechanisms; it is also called the *elimination–addition* mechanism, that is, the opposite of S_NAr (S_N2Ar), or the *addition–elimination* mechanism (covered in Chapter 4).

https://doi.org/10.1515/9783110786835-016

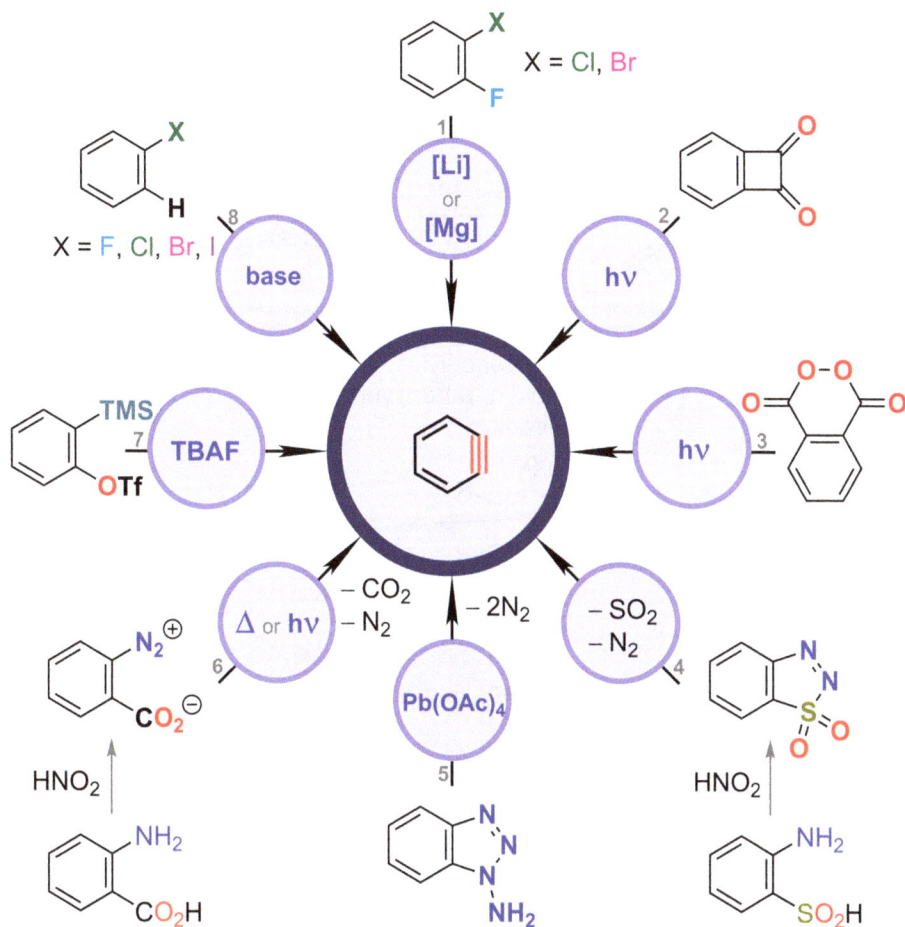

Fig. 16.2: Various synthetic methods leading to the formation of benzyne.[56]

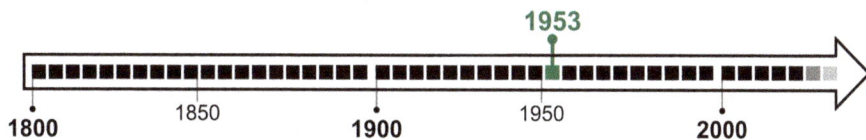

Fig. 16.3: The discovery of the *benzyne* mechanism.[57]

---

56 Since its first discovery, numerous methods evolved leading to the formation of the *benzyne* intermediate *(aryne)*. Note: *benzyne (aryne)* can also be called *dehydrobenzene (dehydroarene)* [16a, 16b].
57 The mechanism in its present form was likely first proposed around 1953 [16c].

17 Bergman Cyclization

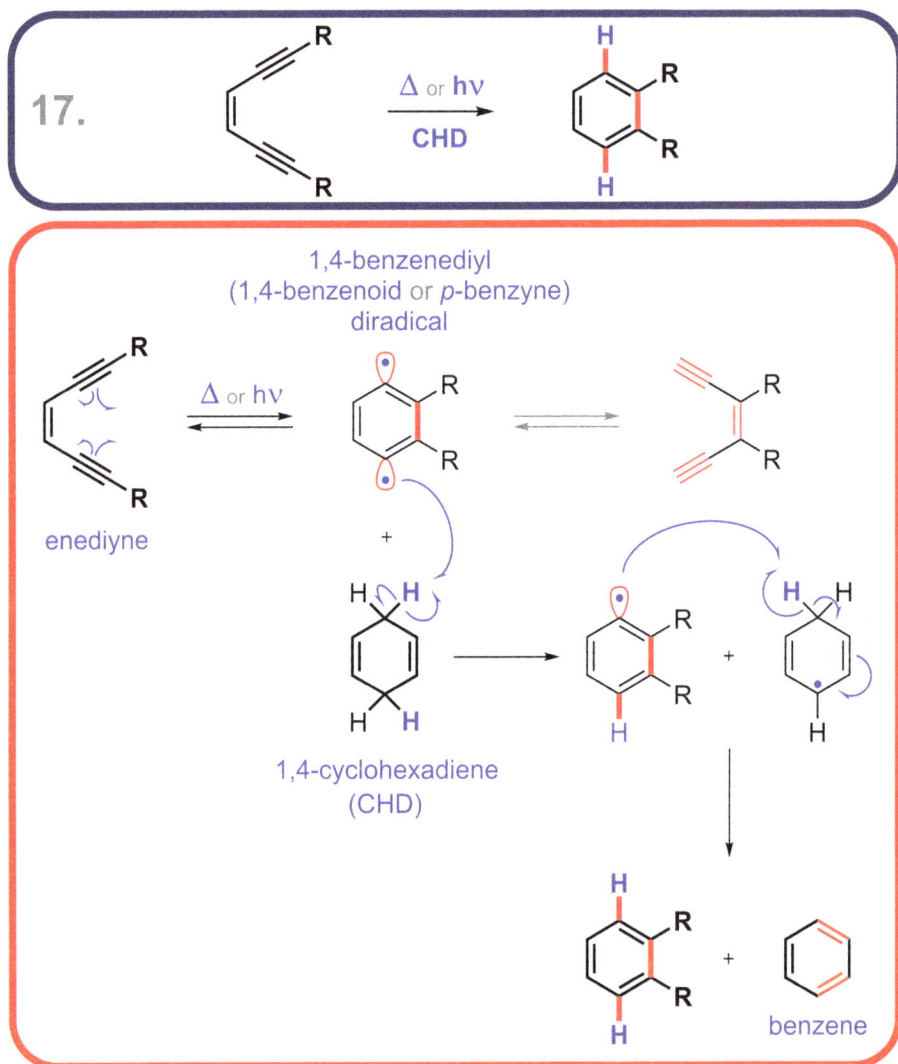

Fig. 17.1: The *Bergman* cyclization mechanism.[58]

---

58 The *Bergman* cyclization is also known as the *Bergman* reaction (isomerization or cycloaromatization).

https://doi.org/10.1515/9783110786835-017

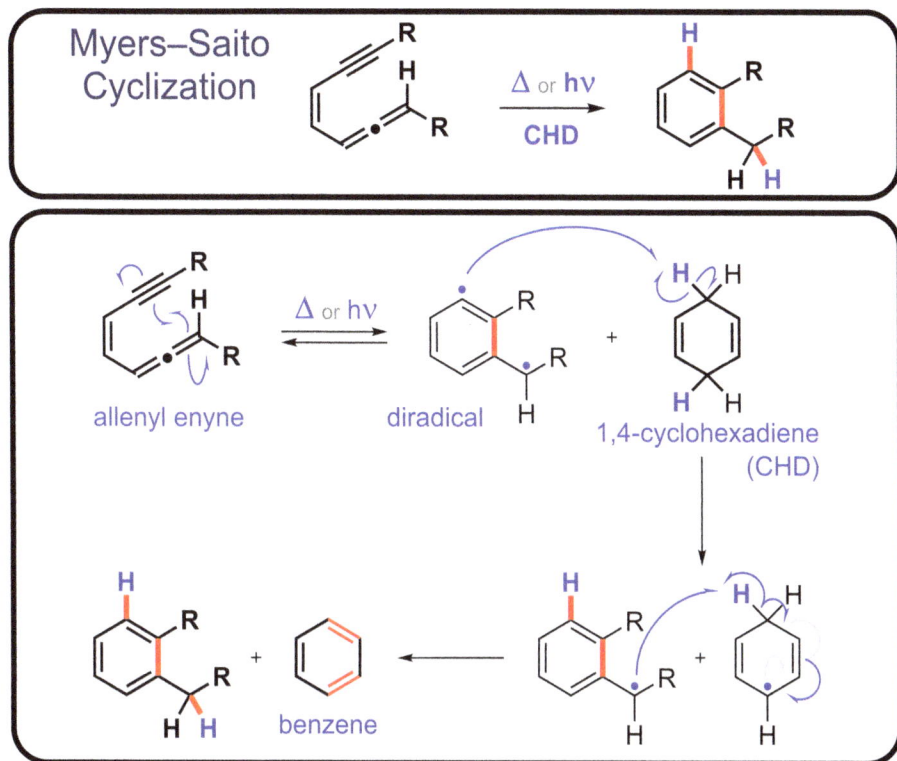

Fig. 17.2: The **Myers–Saito** *cyclization* mechanism.[59]

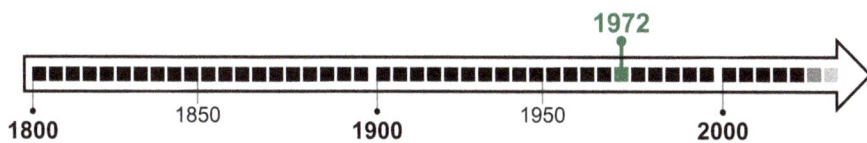

Fig. 17.3: The discovery of the **Bergman** *cyclization*.[60]

---

59 The **Myers–Saito** *cyclization* or *cycloaromatization* of enyne-allenes is related to the **Bergman** *cyclization* and the **Schmittel** *cyclization* (not shown) [17c].
60 The reaction was likely first described around 1972 [17a, 17b].

18 Birch Reduction

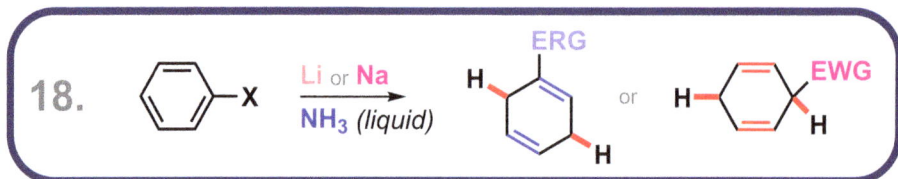

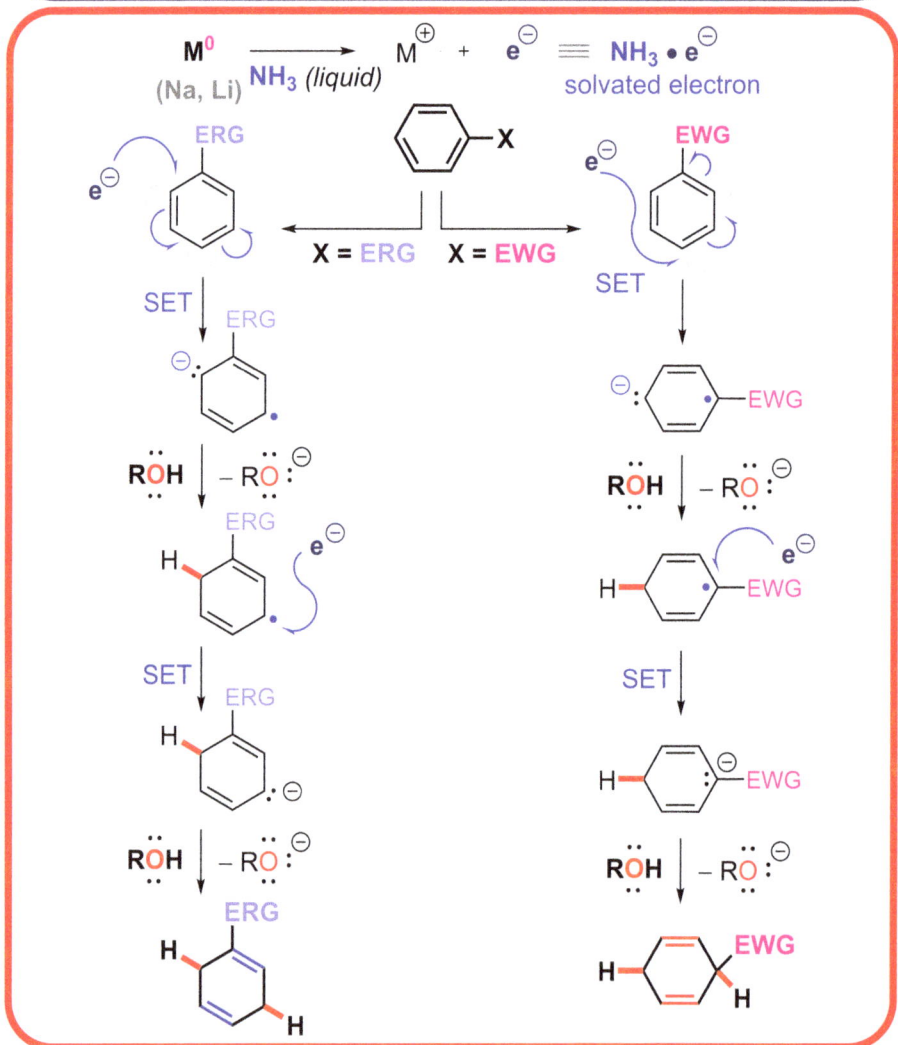

Fig. 18.1: The *Birch reduction* mechanism.[61]

https://doi.org/10.1515/9783110786835-018

Alkyne *trans*–Reduction

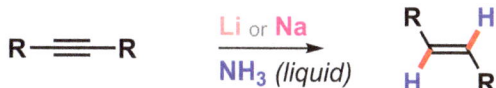

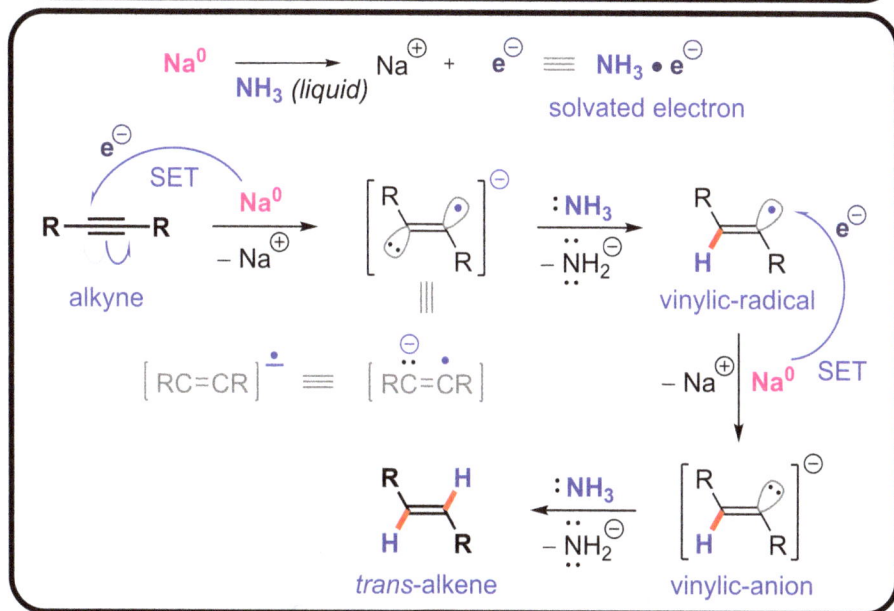

$$R \equiv R \xrightarrow[\text{NH}_3 \text{ (liquid)}]{\text{Li or Na}} \quad \begin{matrix} R \\ \diagup \\ H \end{matrix} = \begin{matrix} H \\ \diagup \\ R \end{matrix}$$

$$Na^0 \xrightarrow[\text{NH}_3 \text{ (liquid)}]{} Na^\oplus + e^\ominus \equiv NH_3 \cdot e^\ominus$$

solvated electron

$$[RC=CR]^{\overset{\bullet}{-}} \equiv [R\overset{\bullet\bullet}{C}=CR]$$

Fig. 18.2: The *alkyne trans-reduction* mechanism.[62]

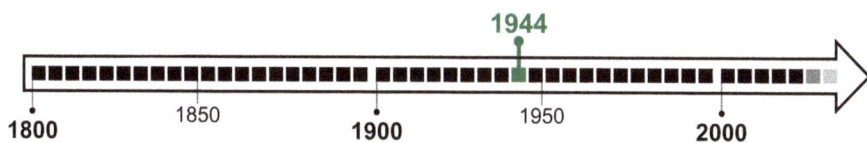

Fig. 18.3: The discovery of the **Birch** reduction.[63]

---

61 The first step in the **Birch** reduction mechanism is a *single electron transfer* (SET) (see Chapter 5). The regiochemistry of the formed products depends on the nature of the substitution (ERG versus EWG).

62 The *alkyne trans-reduction (alkyne metal reduction)* mechanism is much like the **Birch** reduction. Please note that under the **Birch** reduction conditions *alkynes* are reduced to ***trans*-alkenes** [18a, 18b]. Under Pd/C-catalyzed conditions, the ***cis*-alkene** is usually the major product.

63 The reaction was likely first described around 1944 [18c].

19 Bischler–Napieralski Cyclization

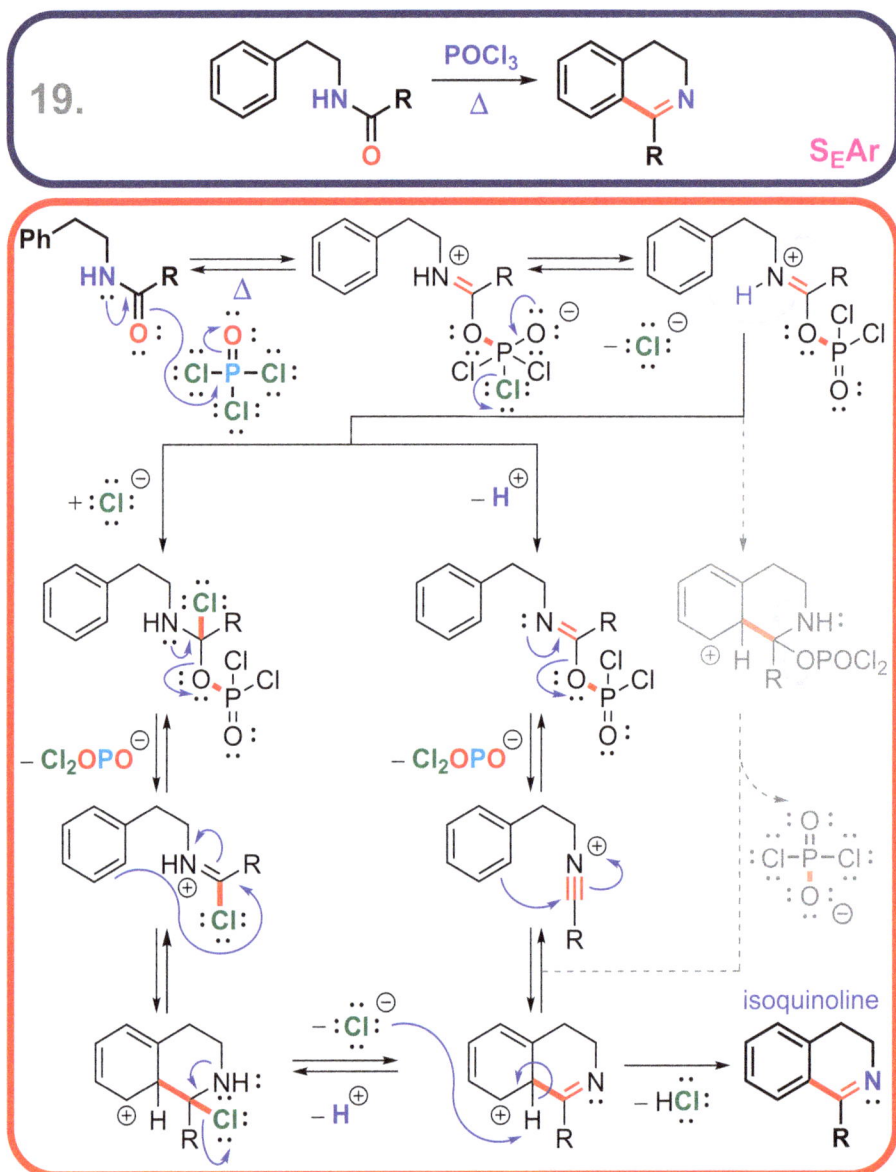

Fig. 19.1: The **Bischler–Napieralski** cyclization mechanism.[64]

---

[64] The **Bischler–Napieralski** cyclization (reaction) is a classic example of **aromatic electrophilic substitution** (the **arenium ion** mechanism or S_EAr, Chapter 3).

https://doi.org/10.1515/9783110786835-019

Pomeranz–Fritsch Reaction

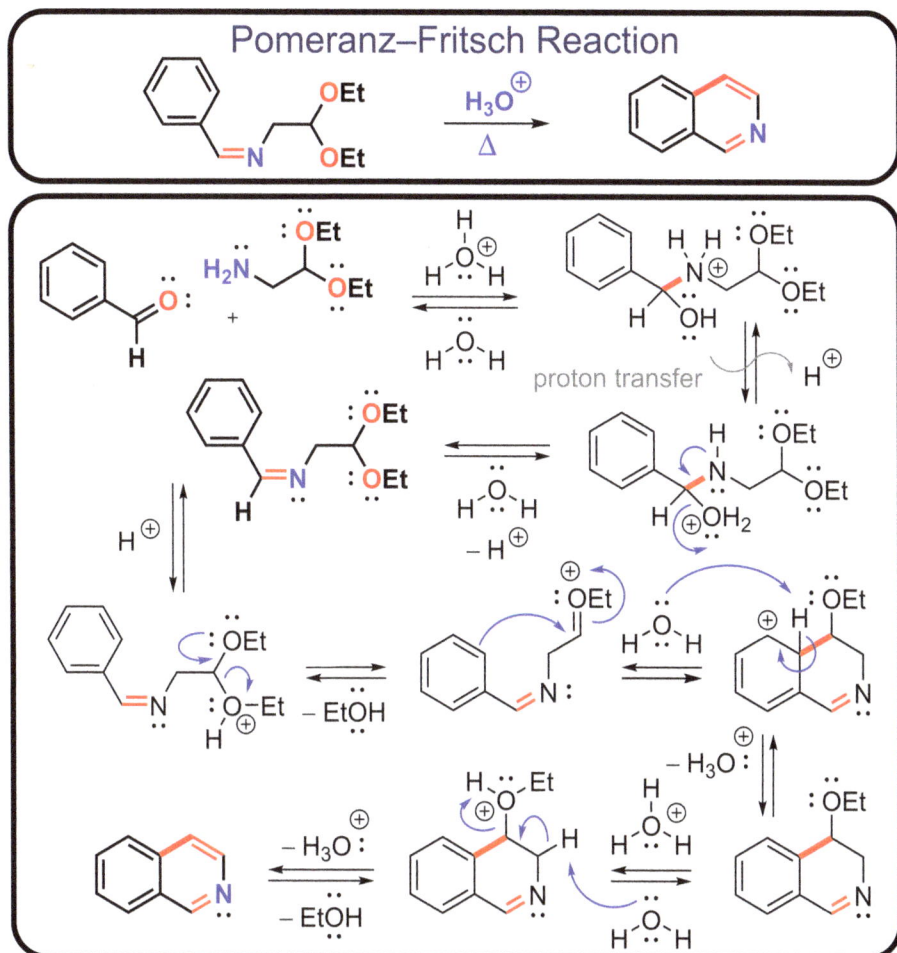

Fig. 19.2: The **Pomeranz–Fritsch** reaction mechanism.[65]

Fig. 19.3: The discovery of the **Bischler–Napieralski** cyclization.[66]

---

65 Several named reactions are related to the **Bischler–Napieralski** cyclization: the **Friedel–Crafts** acylation and alkylation (covered in Chapter 39), and the **Pomeranz–Fritsch** reaction, which is an alternative way to make isoquinolines [19a, 19b].

66 The reaction was likely first described around 1893 [19c].

20 Brown Hydroboration

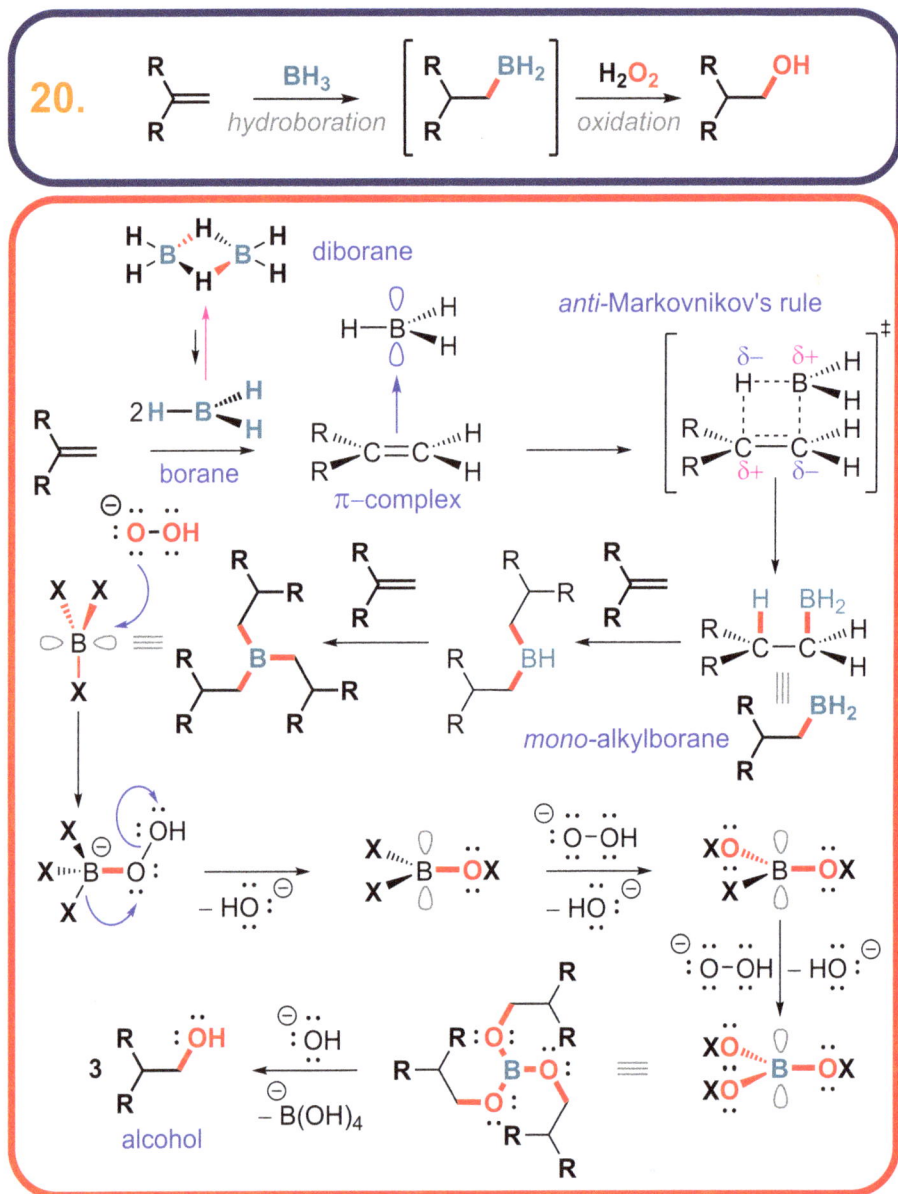

20. R₂C=CR₂ → (BH₃, hydroboration) → [R₂CH-BH₂] → (H₂O₂, oxidation) → R₂CH-OH

Fig. 20.1: The **Brown** *hydroboration* mechanism.[67]

https://doi.org/10.1515/9783110786835-020

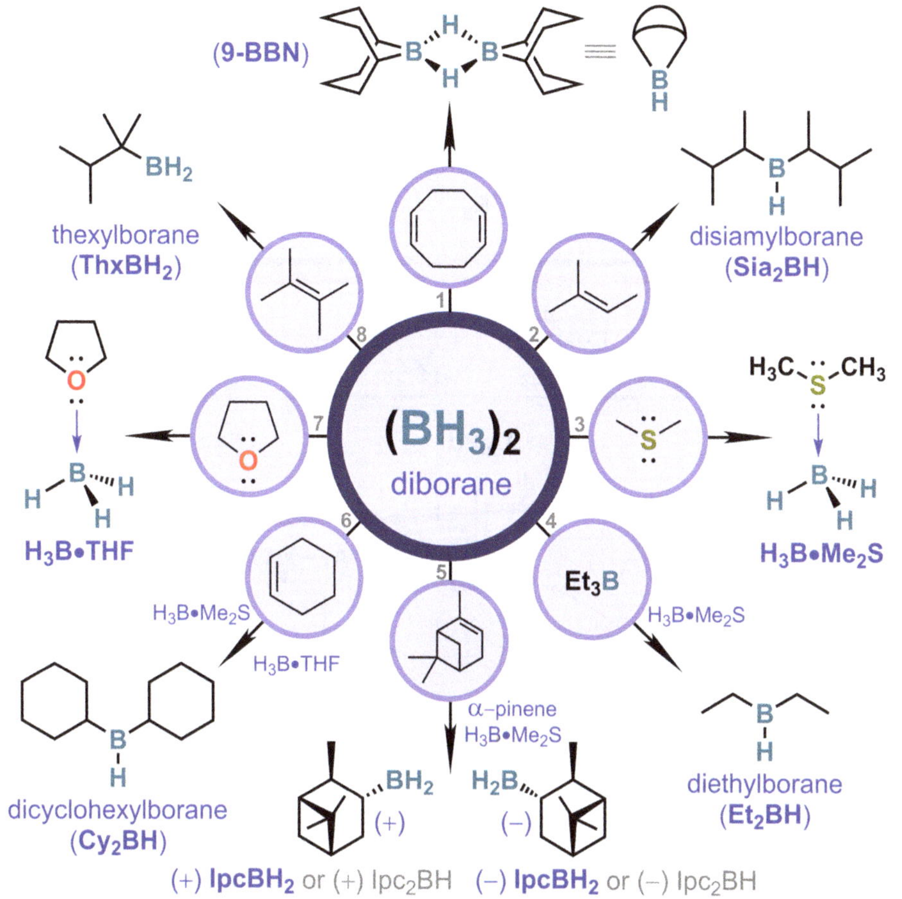

Fig. 20.2: Various borane derivatives formed from diborane.[68]

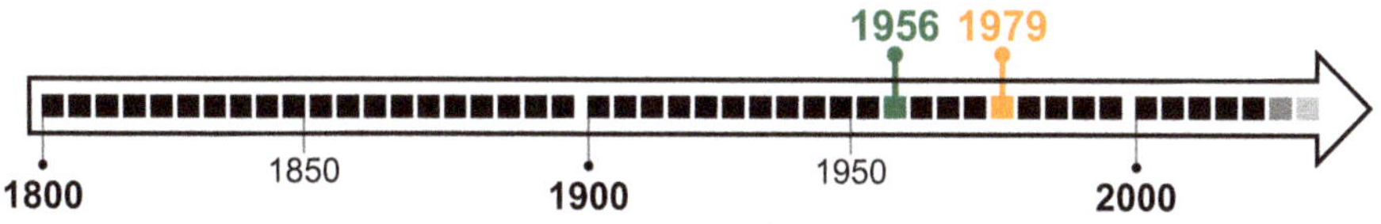

Fig. 20.3: The discovery of the **Brown** hydroboration.[69]

---

67 The **Brown** hydroboration is also known as the *hydroboration–oxidation*. The mechanism is believed to be concerted and **anti-Markovnikov's** product is usually formed. Compare to Chapter 52.

68 There are numerous examples of borane complexes (BH$_3$•X), the monoalkylborane (RBH$_2$), and dialkylborane (R$_2$BH) reagents, which can be prepared from the *diborane* (B$_2$H$_6$) via the *hydroboration reaction*: 9-BBN reagent is one of the most important among them [20a].

69 The reaction was likely first described around 1956 [20b]. In **1979**, Herbert C. Brown (jointly with Georg Wittig) received the Nobel Prize in Chemistry for the development of boron chemistry [20c].

21 Buchwald–Hartwig Cross-Coupling

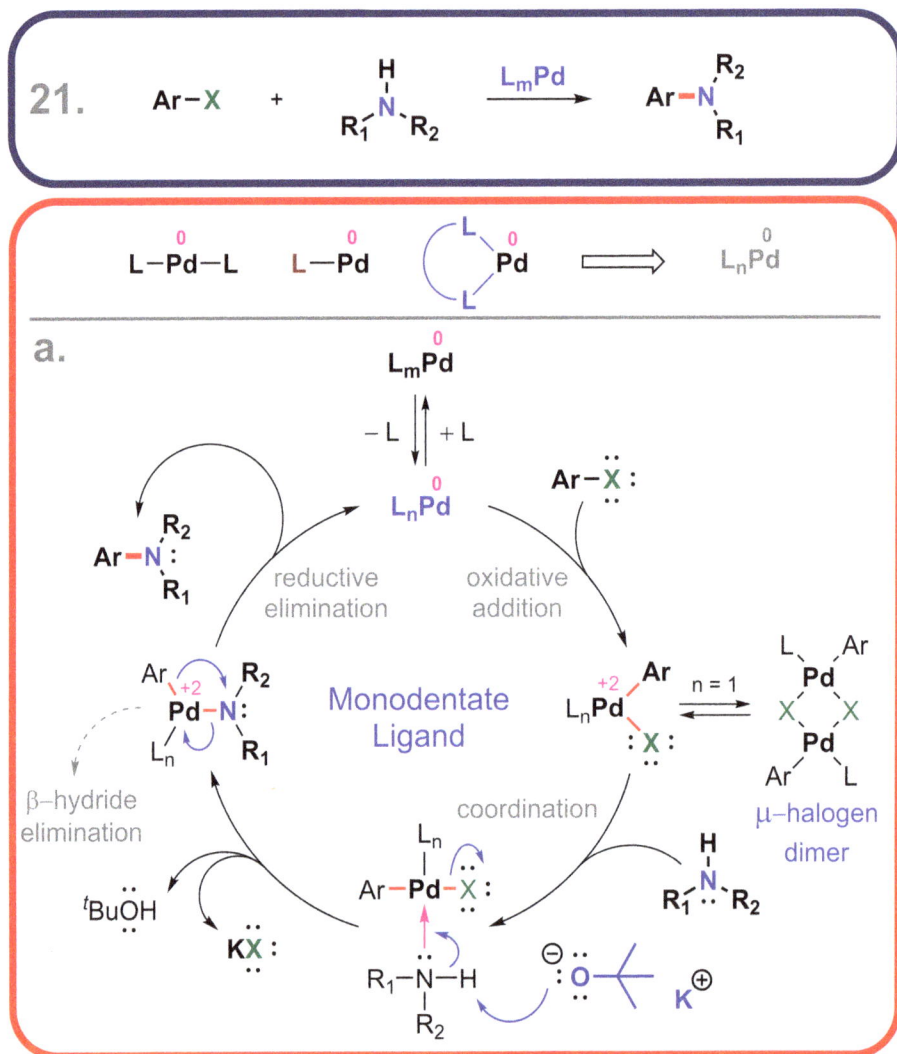

Fig. 21.1: The **Buchwald–Hartwig** cross-coupling mechanism (monodentate ligand).[70]

---

70 The **Buchwald–Hartwig** cross-coupling (amination) is a type of **Pd**-catalyzed cross-coupling reaction (C–N bond formation using aryl halides and amines). The mechanism varies and is usually substrate and ligand dependent. For teaching purposes, a simplified and general example is shown, which may take place in the presence of a monodentate ligand.

https://doi.org/10.1515/9783110786835-021

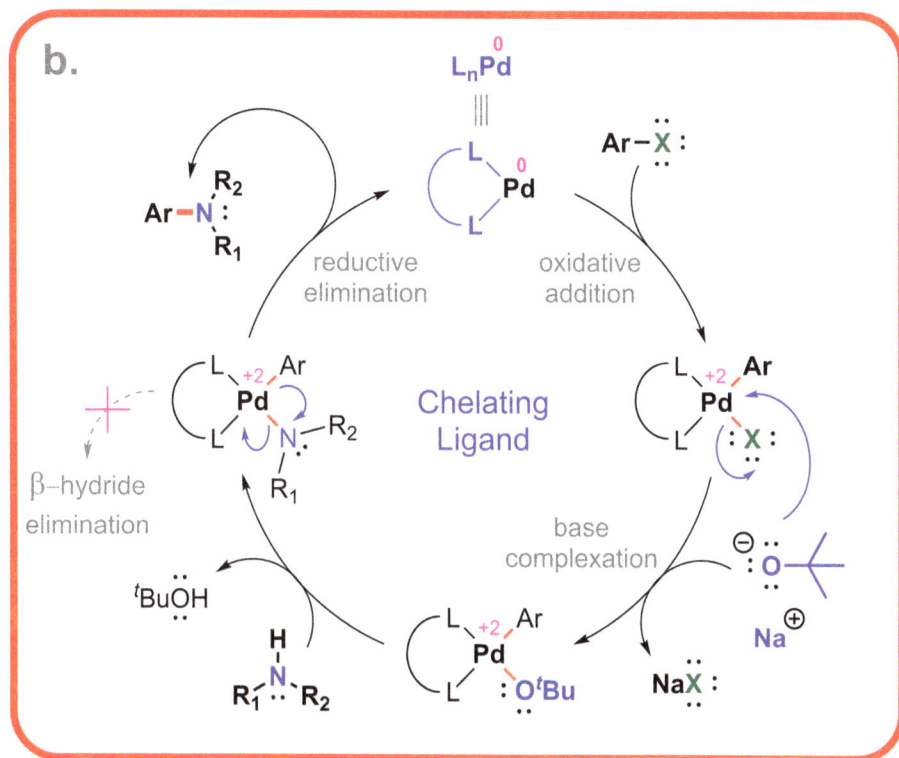

Fig. 21.2: The **Buchwald–Hartwig** cross-coupling mechanism (chelating ligand).[71]

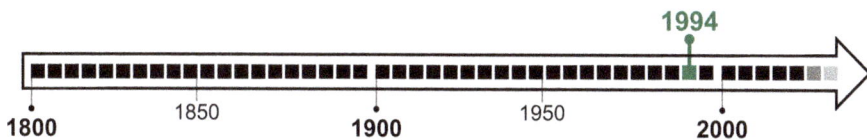

Fig. 21.3: The discovery of the **Buchwald–Hartwig** cross-coupling.[72]

---

71 For teaching purposes, a simplified and general example is shown, which may take place in the presence of a *chelating ligand*.

72 The reaction was likely first described around 1994 [21].

22 Cannizzaro Reaction

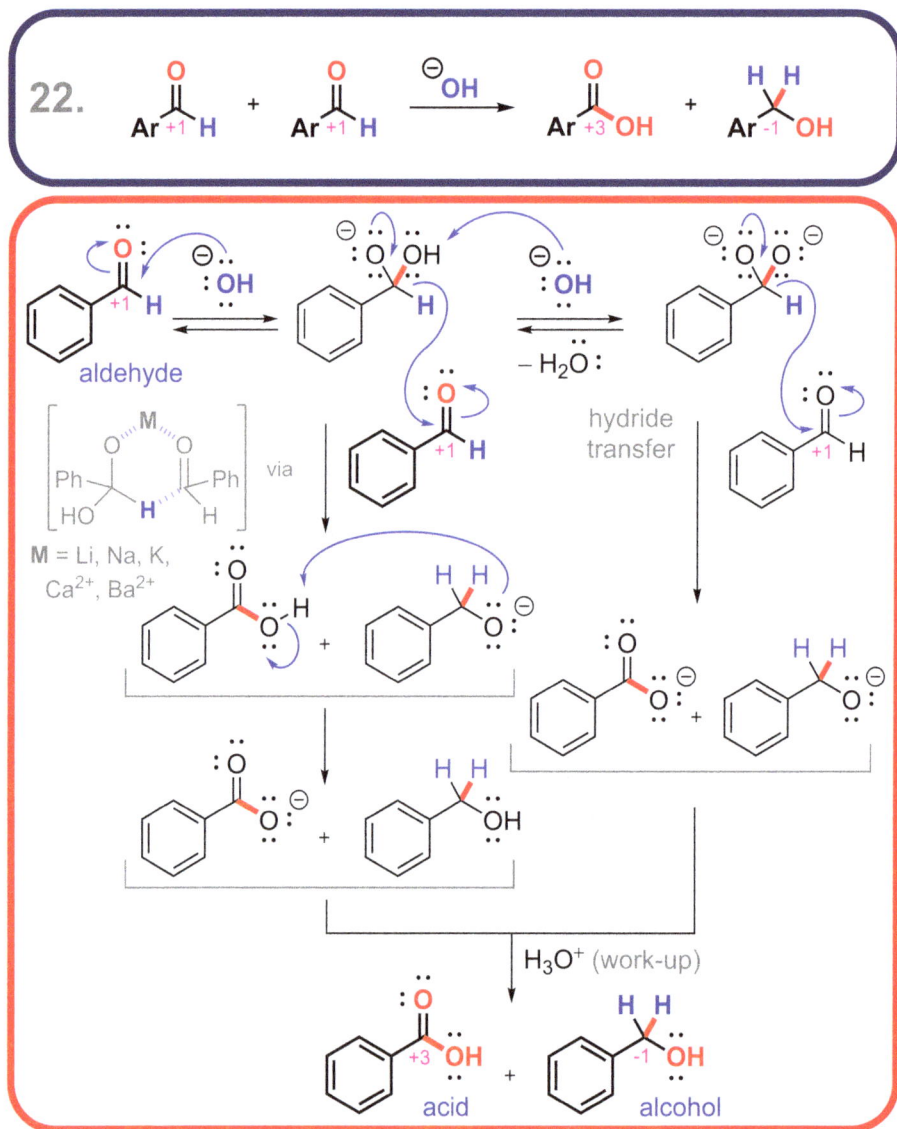

Fig. 22.1: The *Cannizzaro* reaction mechanism.[73]

---

[73] The *Cannizzaro* reaction is seldom called the *Cannizzaro* disproportionation *(RedOx) reaction*. It is one of the oldest reactions in organic chemistry.

https://doi.org/10.1515/9783110786835-022

Cannizzaro with Aromatic Aldehydes:

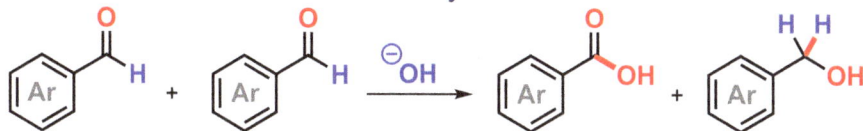

Cannizzaro with Aliphatic Aldehydes:

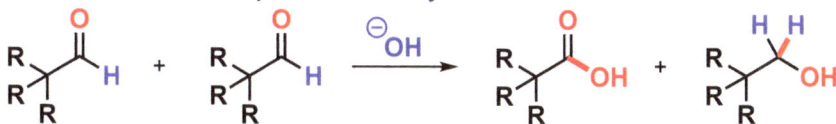

Cross–Cannizzaro Reaction:

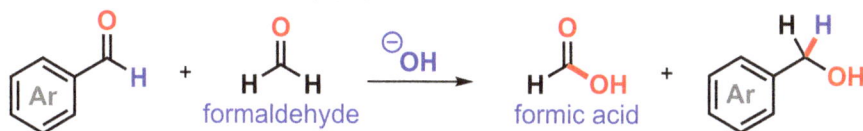

Intramolecular Cannizzaro Reaction:

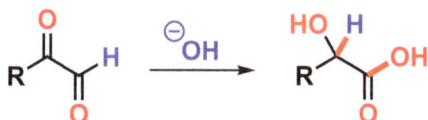

Fig. 22.2: Variations of the **Cannizzaro** reaction.[74]

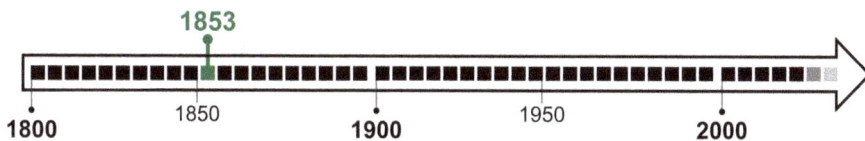

Fig. 22.3: The discovery of the **Cannizzaro** reaction.[75]

---

74 There are many variations of the **Cannizzaro** reaction: the **Cannizzaro** reaction with aromatic and aliphatic aldehydes containing no α-hydrogen atoms, and the cross-**Cannizzaro** reaction and the intramolecular **Cannizzaro** reaction [1].
75 The reaction was likely first described around 1853 [22].

23 Chan–Evans–Lam Cross-Coupling

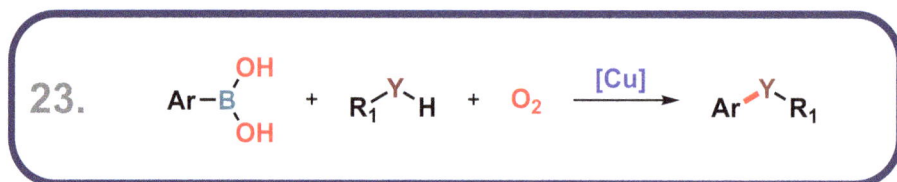

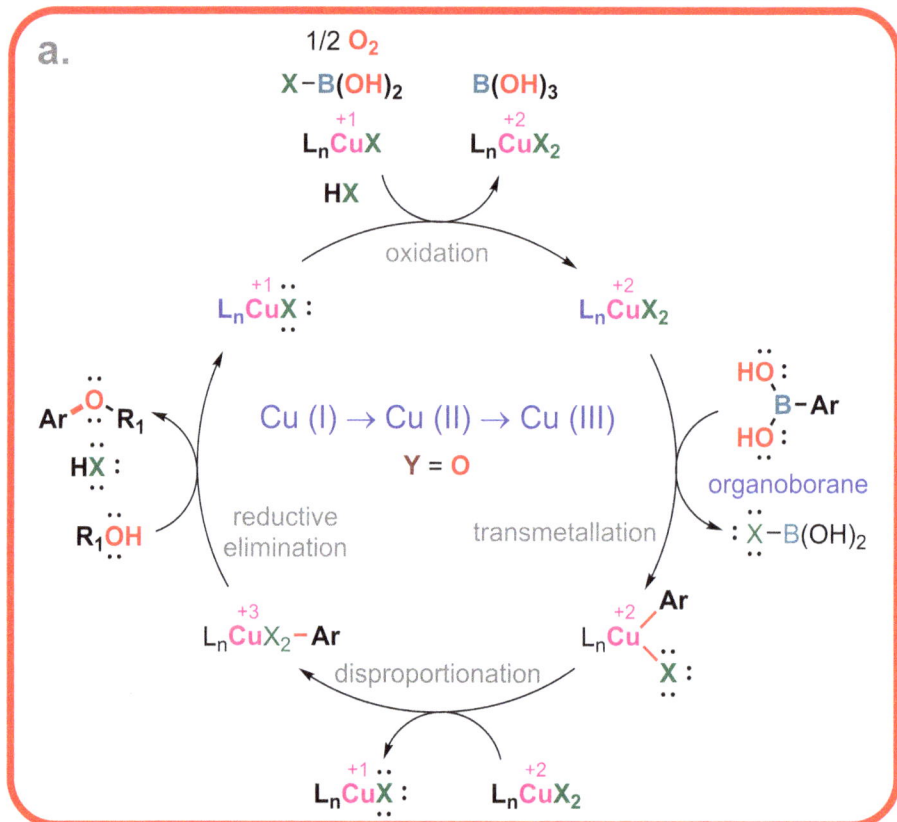

23. $Ar-B(OH)_2$ + R_1-Y-H + O_2 $\xrightarrow{\text{[Cu]}}$ $Ar-Y-R_1$

Fig. 23.1: The *Chan–Evans–Lam* cross-coupling mechanism (Y = O).[76]

---

76 The **Chan–Evans–Lam** cross-coupling (also simply called the **Chan–Lam** cross-coupling) is a type of **Cu**-catalyzed cross-coupling reaction (C–O and C–N bond formation using *aryl boronic acids* and *alcohols* or *amines*). The mechanism is not well-understood and is usually very substrate and ligand dependent. For teaching purposes, a simplified and general example is shown, which may take place in etherification reactions (C–O bond formation, Y = O) [23a, 23b].

https://doi.org/10.1515/9783110786835-023

b.

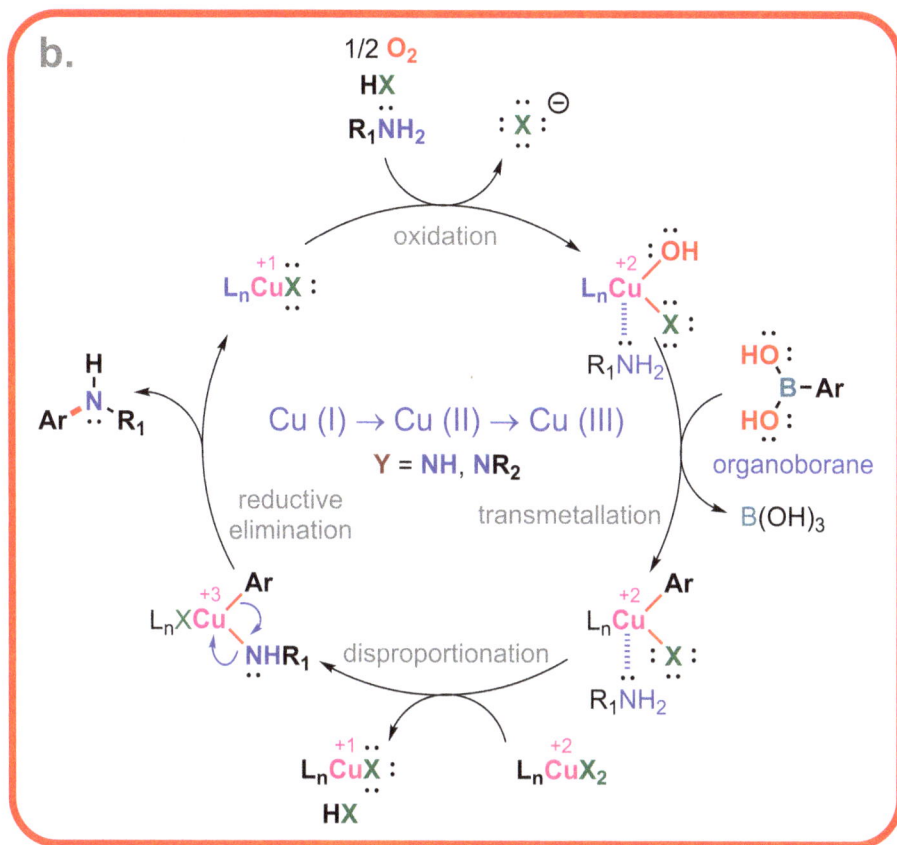

Fig. 23.2: The **Chan–Evans–Lam** cross-coupling mechanism (Y = NH, NR$_2$).[77]

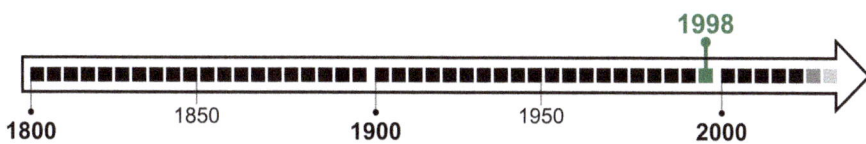

Fig. 23.3: The discovery of the **Chan–Evans–Lam** cross-coupling.[78]

---

77 The mechanism is not well-understood and is usually very substrate and ligand dependent. For teaching purposes, a simplified and general example is shown, which may take place in amination reactions (C–N bond formation, Y = NH, NR$_2$) [23c].

78 The reaction was likely first described around 1998 [23d, 23e, 23f].

24 Chichibabin Amination

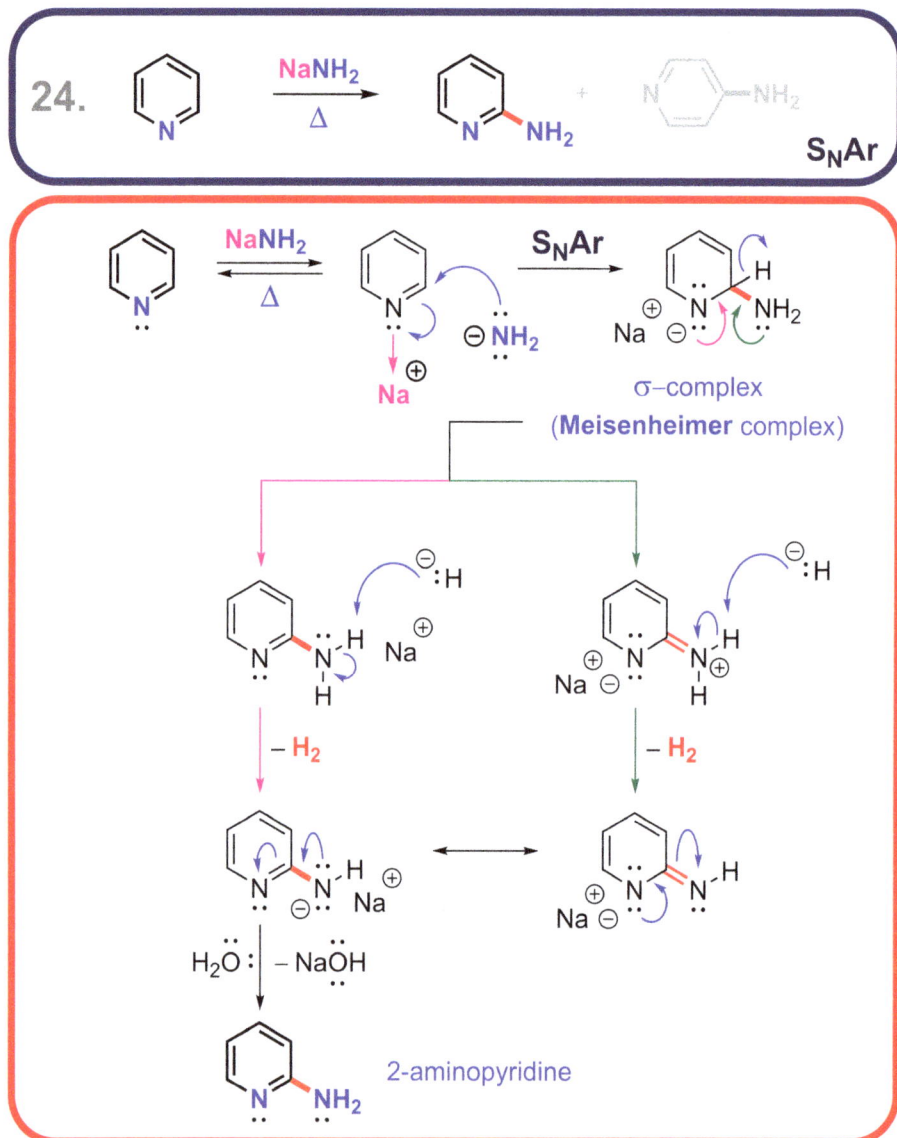

Fig. 24.1: The *Chichibabin amination* mechanism.[79]

https://doi.org/10.1515/9783110786835-024

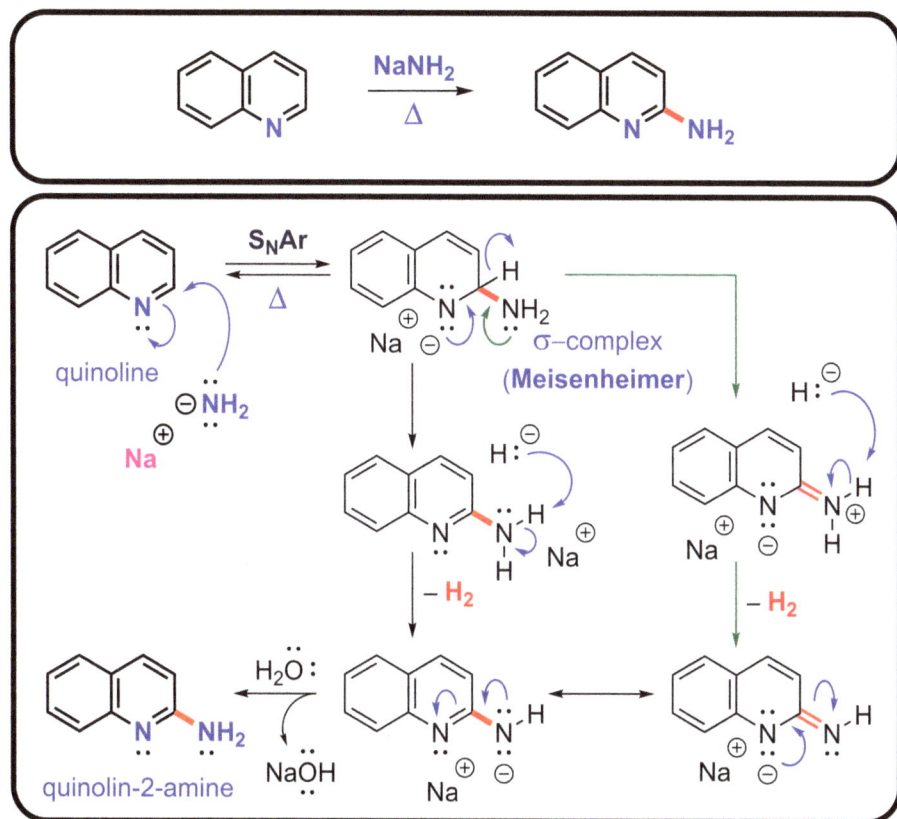

Fig. 24.2: The **Chichibabin** *amination* mechanism of quinoline.[80]

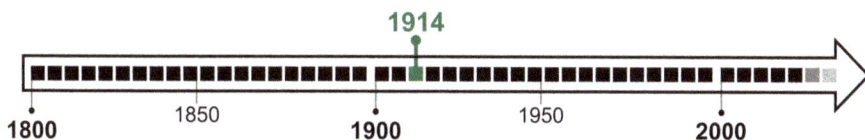

Fig. 24.3: The discovery of the **Chichibabin** *amination*.[81]

---

79 The **Chichibabin** *amination* (in Russian Чичибабин) is also called the **Chichibabin** *reaction*. It is a classic example of **aromatic nucleophilic substitution**. Specifically, it undergoes the *addition–elimination* mechanism: S_NAr (S_N2Ar), covered in Chapter 4.

80 An example of the **Chichibabin** *amination* of quinoline to yield 2-aminoquinoline [24c].

81 The reaction was likely first described around 1914 [24a, 24b].

25 Claisen Condensation

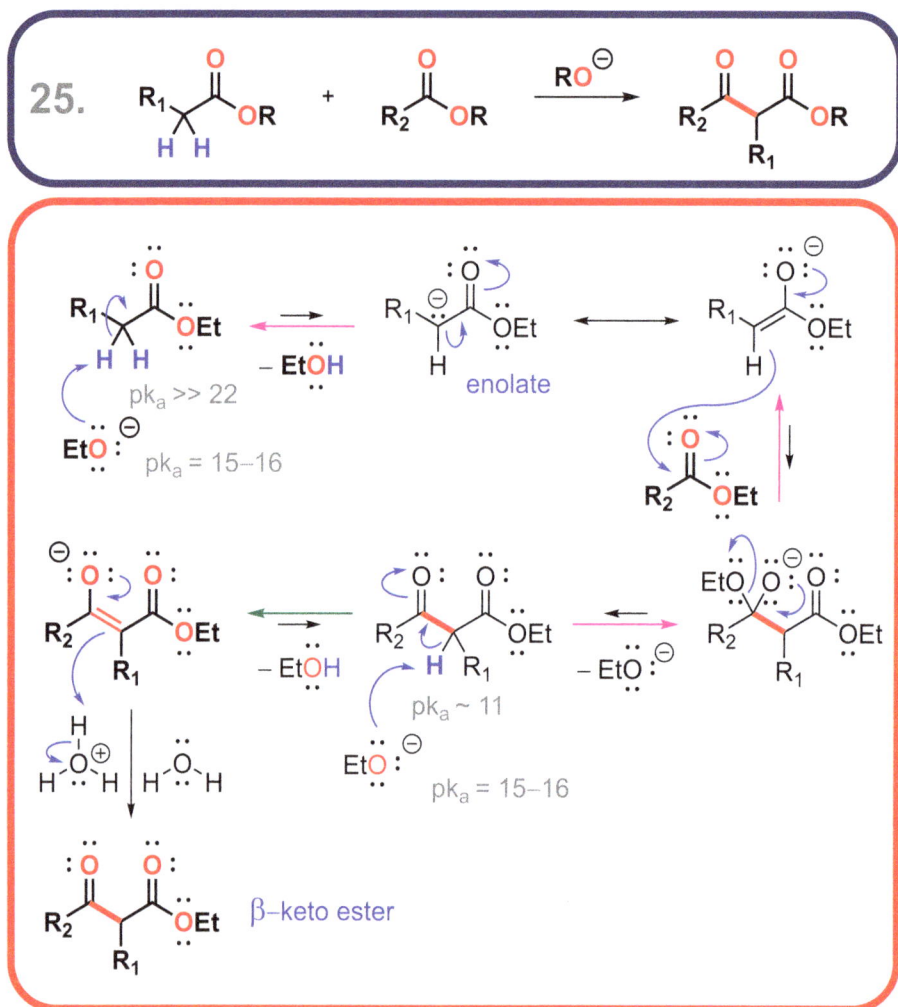

Fig. 25.1: The *Claisen condensation* mechanism.[82]

---

[82] The *Claisen condensation* is a condensation reaction between an *ester* and another carbonyl compound containing two enolizable H-atoms (α-hydrogen atoms).

https://doi.org/10.1515/9783110786835-025

Dieckmann Condensation

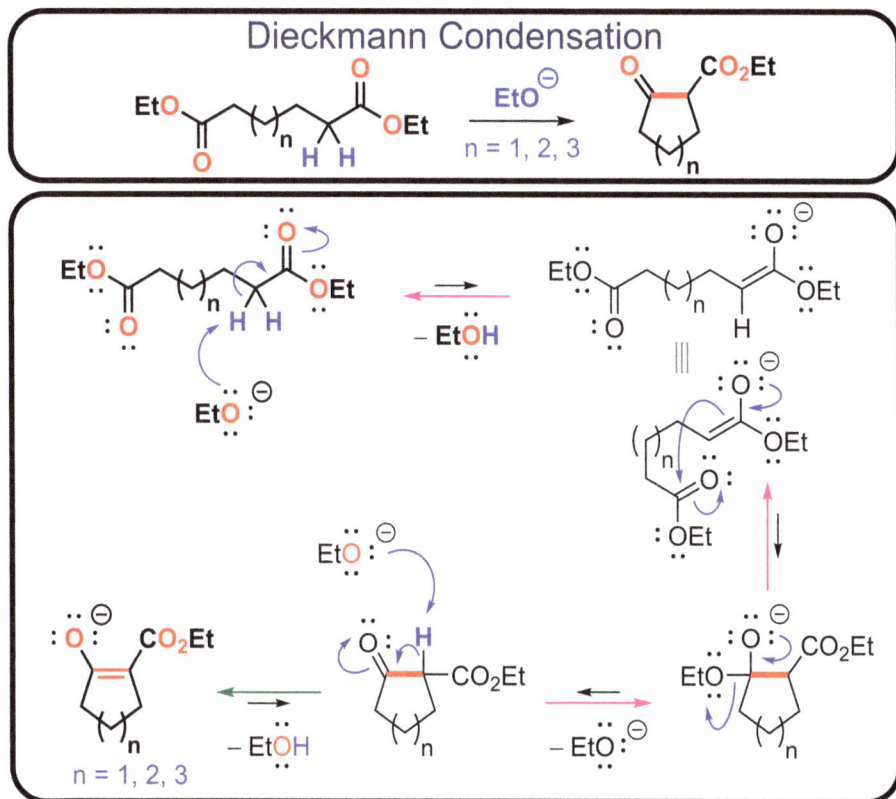

Fig. 25.2: The **Dieckmann** *condensation* mechanism.[83]

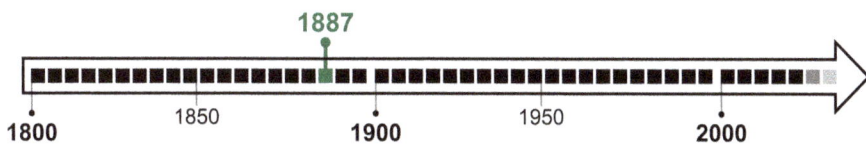

Fig. 25.3: The discovery of the **Claisen** *condensation*.[84]

---

83 The **Dieckmann** *condensation* is the *intramolecular* **Claisen** *condensation* and their mechanisms are almost identical. The **Dieckmann** *condensation* is ideal for the formation of five-, six-, and seven-membered rings [25a].
84 The reaction was likely first described around 1887 [25b].

26 Claisen Rearrangement

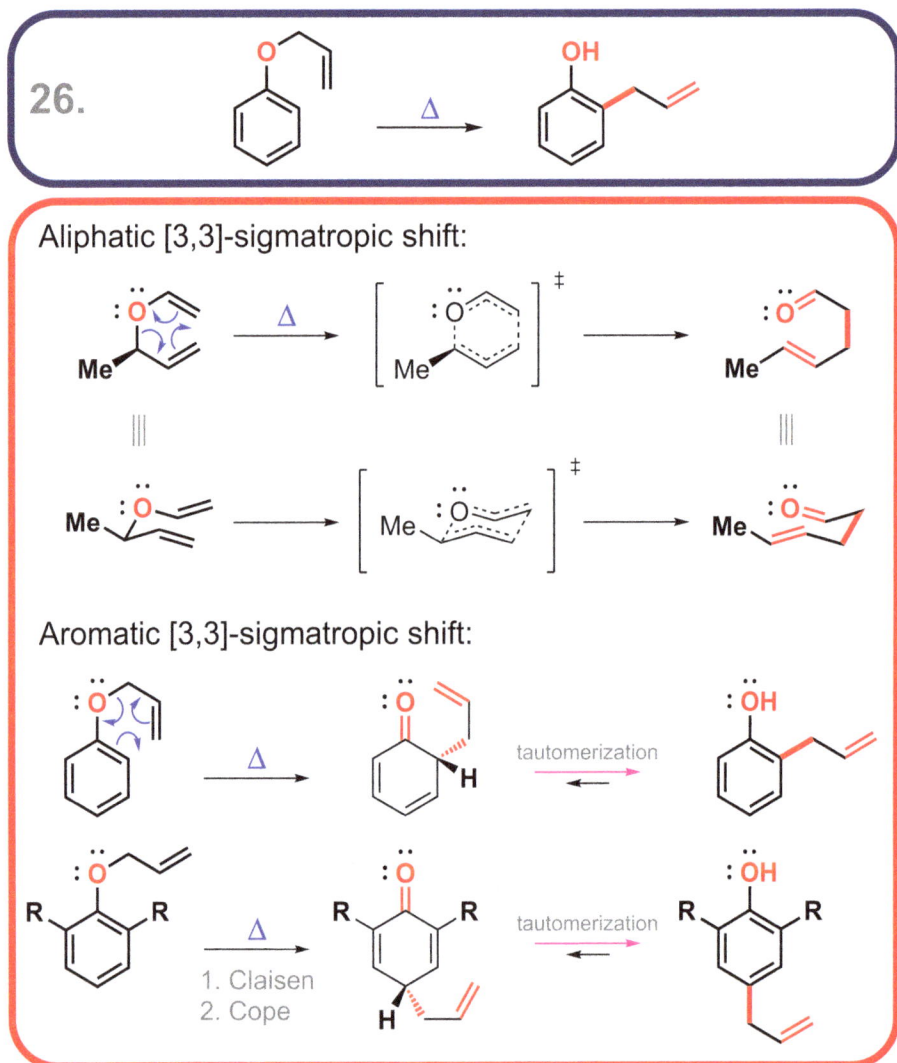

Aliphatic [3,3]-sigmatropic shift:

Aromatic [3,3]-sigmatropic shift:

Fig. 26.1: The *Claisen* rearrangement mechanism.[85]

---

[85] The *Claisen* rearrangement (different from the *Claisen* condensation and much like the *Cope* rearrangement, see Chapter 28) is a pericyclic reaction with a concerted mechanism. This is a classic example of a [3,3']-*sigmatropic rearrangement (shift)*.

https://doi.org/10.1515/9783110786835-026

Ireland–Claisen Rearrangement

Eschenmoser–Claisen Rearrangement

Johnson–Claisen Rearrangement

Aza–Claisen Rearrangement

Overman Rearrangement

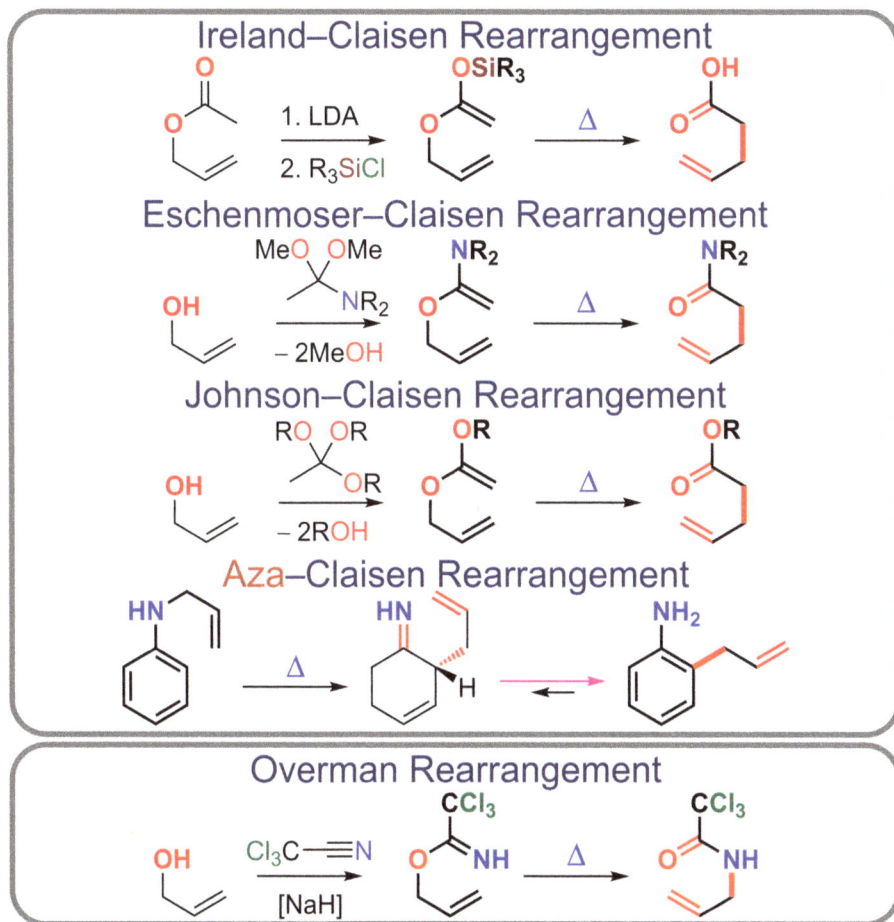

Fig. 26.2: Reactions related to the *Claisen rearrangement*.[86]

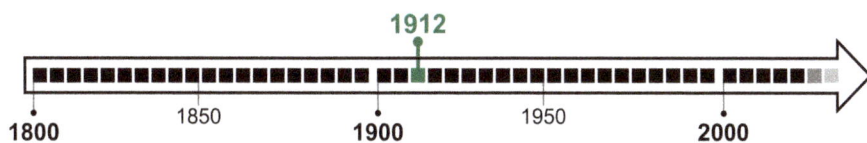

1912

1800   1850   1900   1950   2000

Fig. 26.3: The discovery of the *Claisen rearrangement*.[87]

---

86 There are numerous variations and modifications of the *Claisen rearrangement reaction*, to name a few: the *Ireland–Claisen* rearrangement, the *Eschenmoser–Claisen* rearrangement, the *Johnson–Claisen* rearrangement, the aza-*Claisen* (aza-*Cope*) rearrangement, the *Overman* rearrangement, and others [26a].
87 The reaction was likely first described around 1912 [26b].

27 Cope Elimination

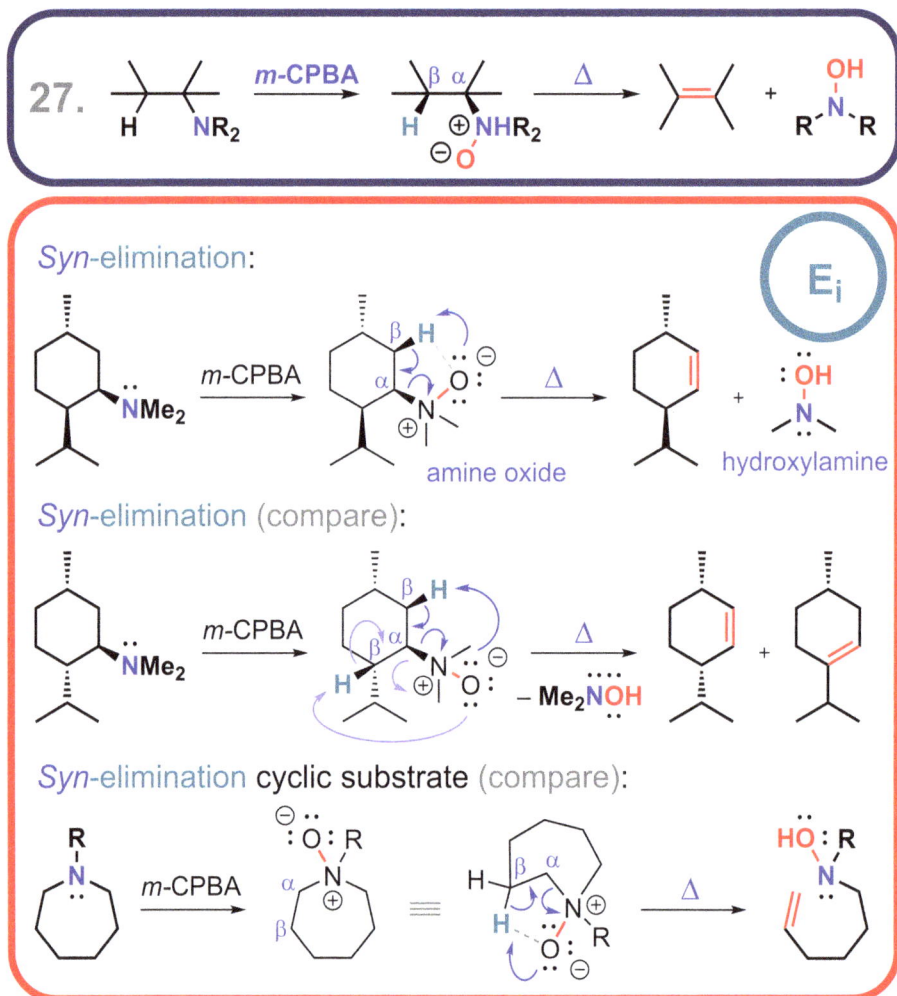

Fig. 27.1: The **Cope** *elimination* mechanism.[88]

---

88 The **Cope** *elimination* or the **Cope** *reaction* is an example of the *five-membered* **internal** or **intra-molecular β-elimination** reaction (**E$_i$**), mentioned in Chapter 6.

https://doi.org/10.1515/9783110786835-027

Hofmann Elimination

Selenoxide Elimination

Acetate Pyrolysis

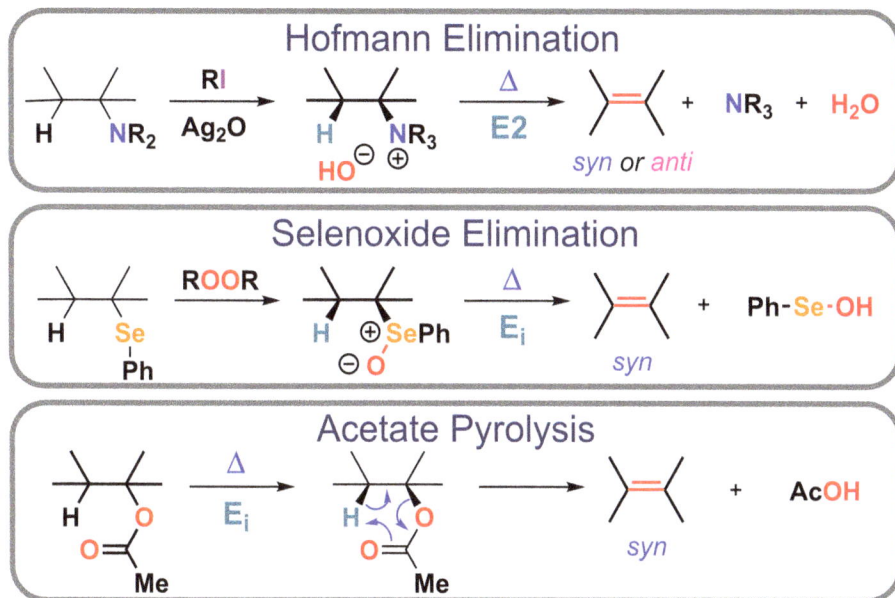

Fig. 27.2: Reactions related to the *Cope elimination*.[89]

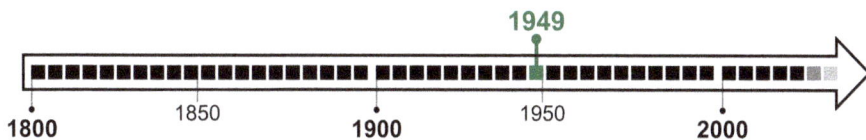

Fig. 27.3: The discovery of the *Cope elimination*.[90]

89 Several reactions are related to the *Cope elimination*: the *Hofmann* elimination (usually E2-type elimination, rarely E_i, covered in Chapter 49), the *selenoxide elimination* [27a, 27b], the *acetate pyrolysis* [1], and others (not mentioned here).
90 The reaction was likely first described around 1949 [27c].

28 Cope Rearrangement

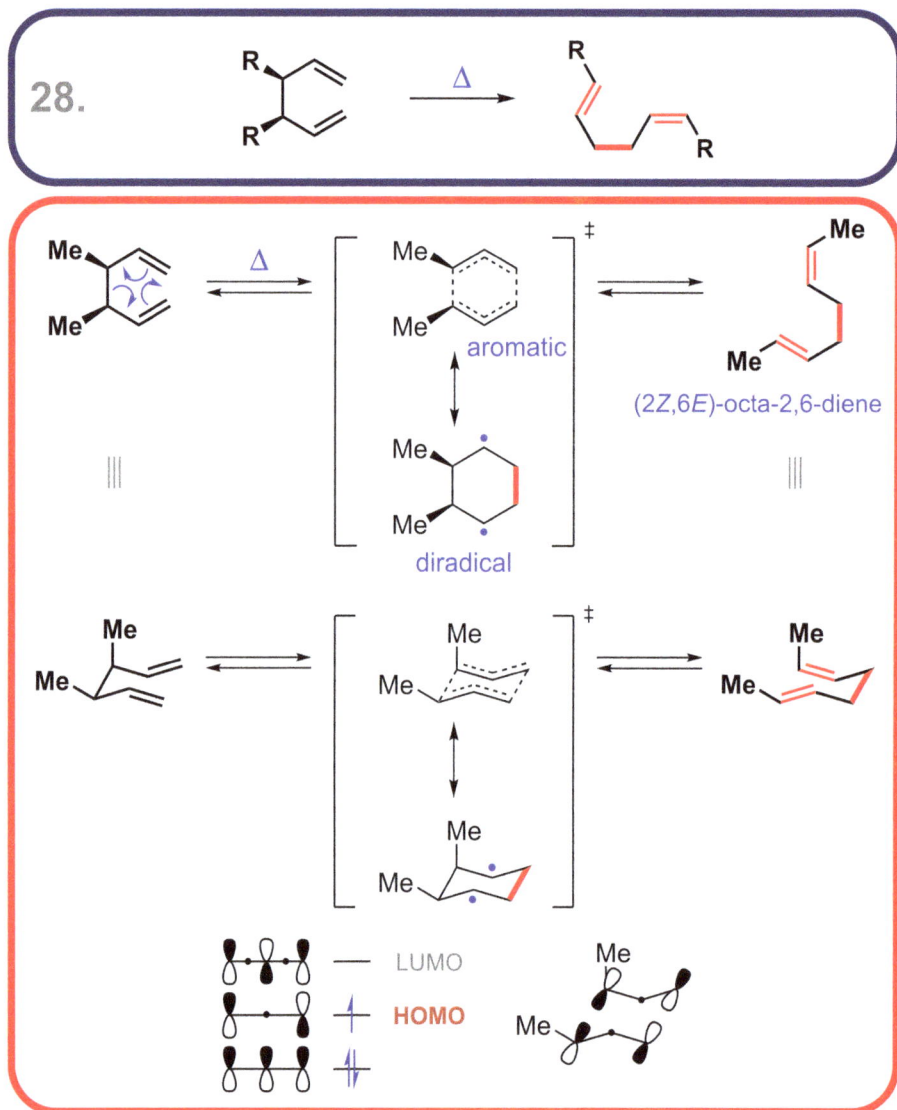

Fig. 28.1: The *Cope rearrangement* mechanism.[91]

---

91 The *Cope rearrangement* (different from the *Cope elimination* and much like the *Claisen rearrangement*, see Chapter 26) is a pericyclic reaction with a concerted mechanism. This is a classic example of a [3,3']-*sigmatropic rearrangement* (also referred to as [3,3']-*sigmatropic shift*).

https://doi.org/10.1515/9783110786835-028

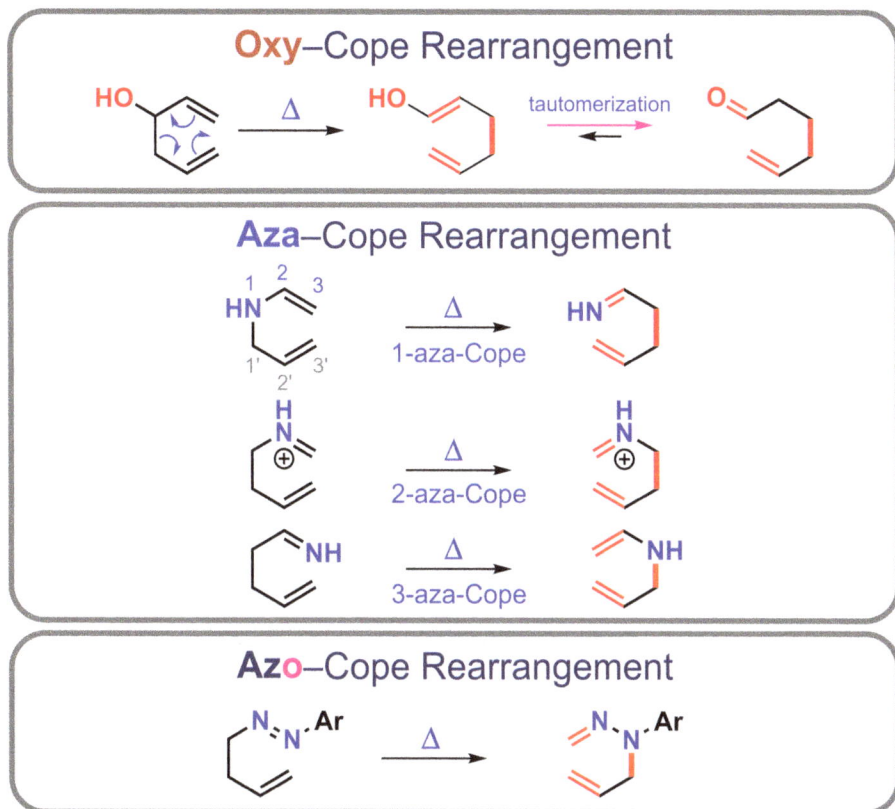

Fig. 28.2: Reactions related to the *Cope rearrangement*.[92]

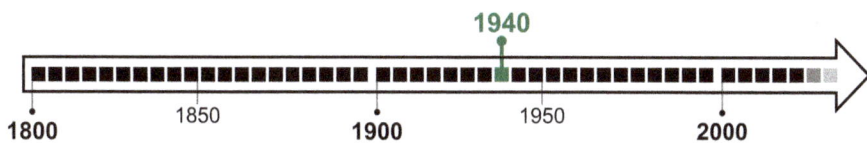

Fig. 28.3: The discovery of the *Cope rearrangement*.[93]

---

92 There are numerous variations of the *Cope rearrangement* [1], such as the *(anionic) oxy-Cope rearrangement*, the *aza-Cope* and/or *aza-Claisen rearrangement* (confusing), and the *azo-Cope rearrangement* [28a].
93 The reaction was likely first described around 1940 [28b].

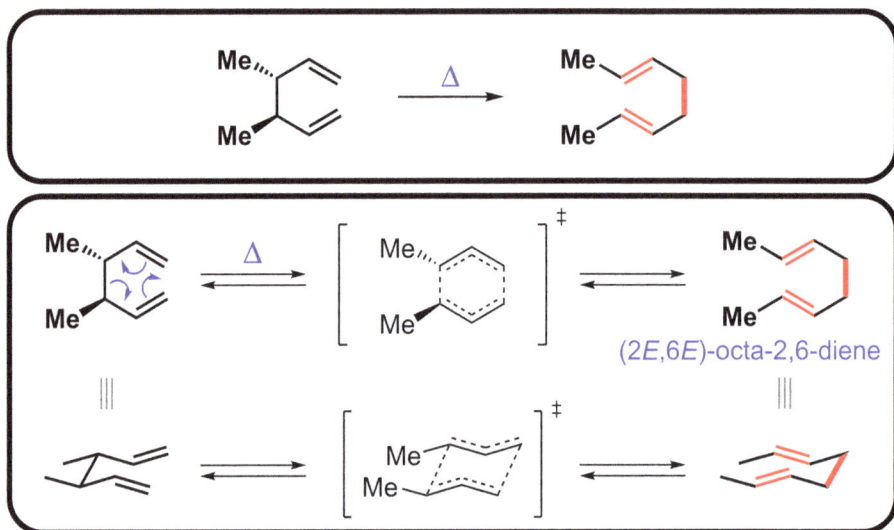

Fig. 28.4: The **Cope** rearrangement of (3R,4R)-3,4-dimethylhexa-1,5-diene.[94]

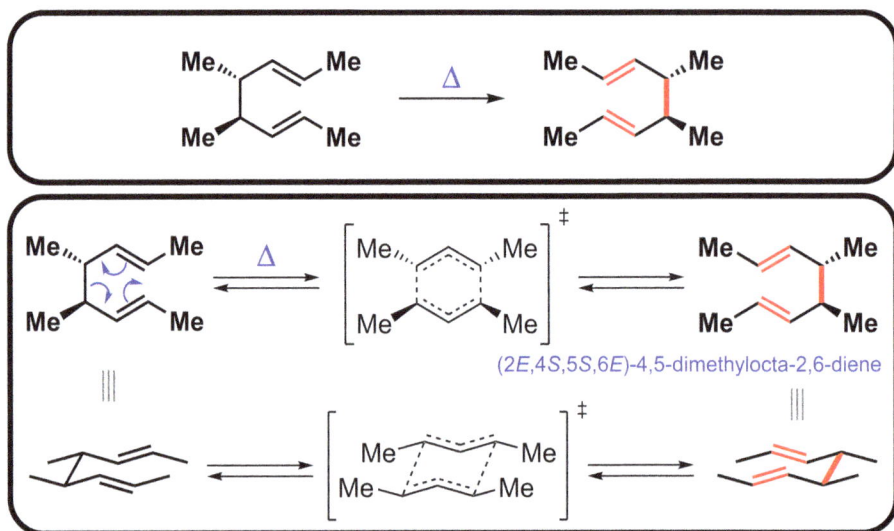

Fig. 28.5: The **Cope** rearrangement of (2E,4R,5R,6E)-4,5-dimethylocta-2,6-diene.[95]

---

94 The **Cope** rearrangement of (3R,4R)-3,4-dimethylhexa-1,5-diene to (2E,6E)-octa-2,6-diene.
95 The **Cope** rearrangement of (2E,4R,5R,6E)-4,5-dimethylocta-2,6-diene to (2E,4S,5S,6E)-4,5-dimethylocta-2,6-diene.

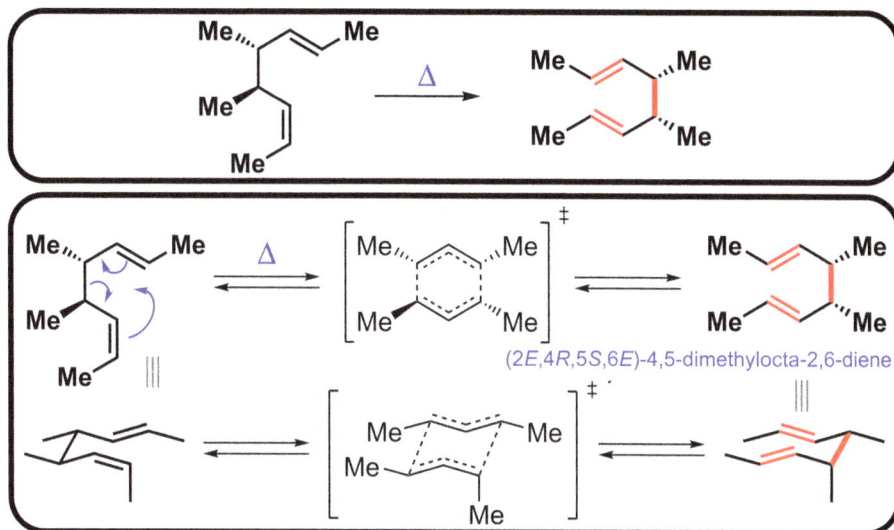

Fig. 28.6: The *Cope rearrangement* of (2Z,4R,5R,6E)-4,5-dimethylocta-2,6-diene.[96]

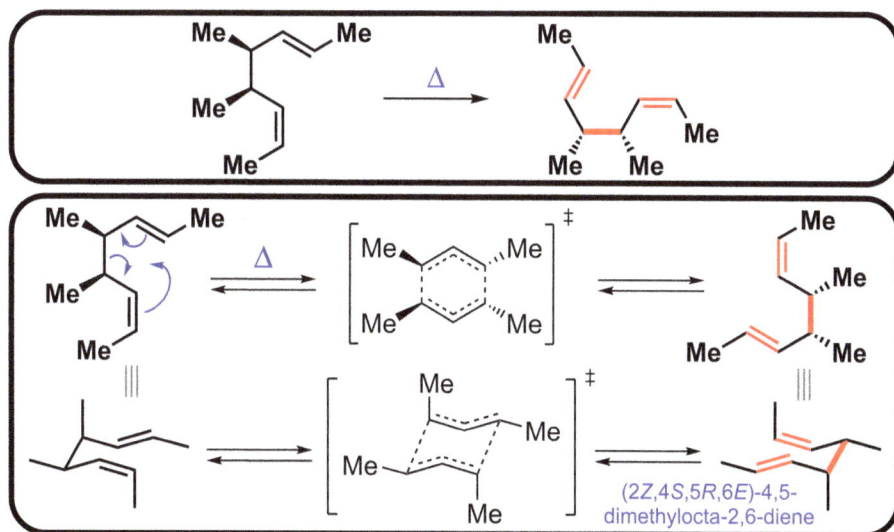

Fig. 28.7: The *Cope rearrangement* of (2Z,4R,5S,6E)-4,5-dimethylocta-2,6-diene.[97]

---

96 The *Cope rearrangement* of (2Z,4R,5R,6E)-4,5-dimethylocta-2,6-diene to (2E,4R,5S,6E)-4,5-dimethylocta-2,6-diene.

97 The *Cope rearrangement* of (2Z,4R,5S,6E)-4,5-dimethylocta-2,6-diene to (2Z,4S,5R,6E)-4,5-dimethylocta-2,6-diene.

29 Criegee and Malaprade Oxidation

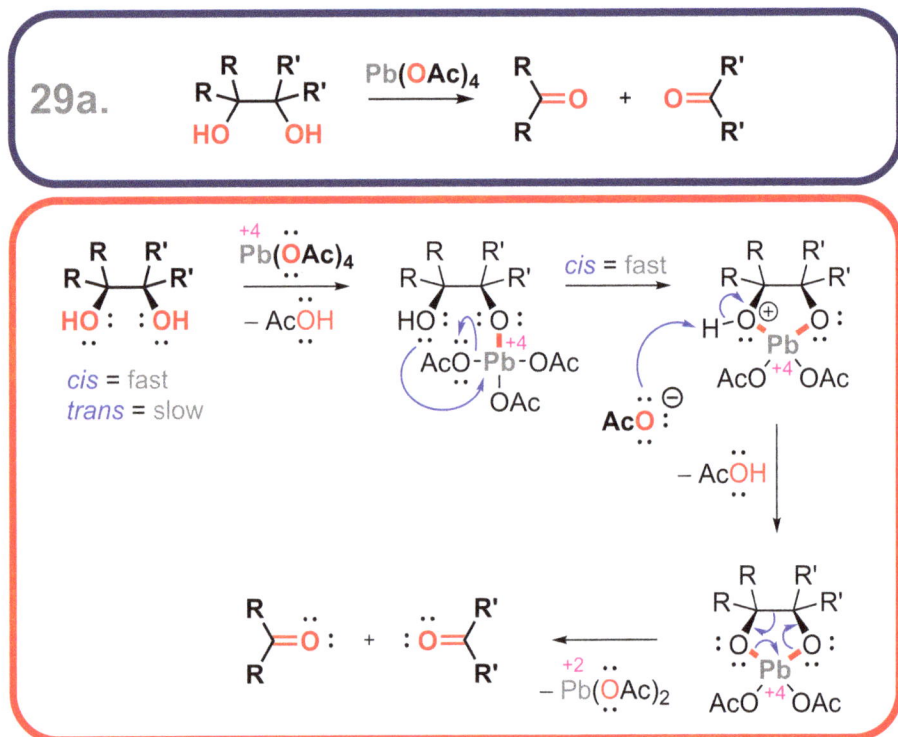

29a.

Fig. 29.1: The *Criegee* oxidation mechanism.[98]

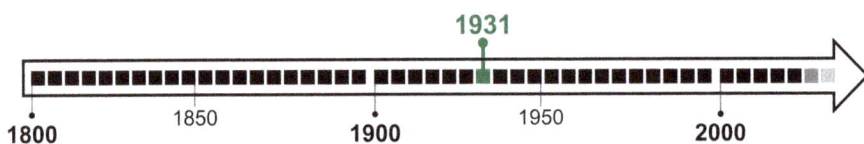

Fig. 29.2: The discovery of the *Criegee* oxidation.[99]

---

98 The *Criegee oxidation* or simply the *Criegee reaction* is different from the *Criegee mechanism* proposed for *ozonolysis* (covered in Chapter 70).
99 The reaction was likely first described around 1931 [29a].

https://doi.org/10.1515/9783110786835-029

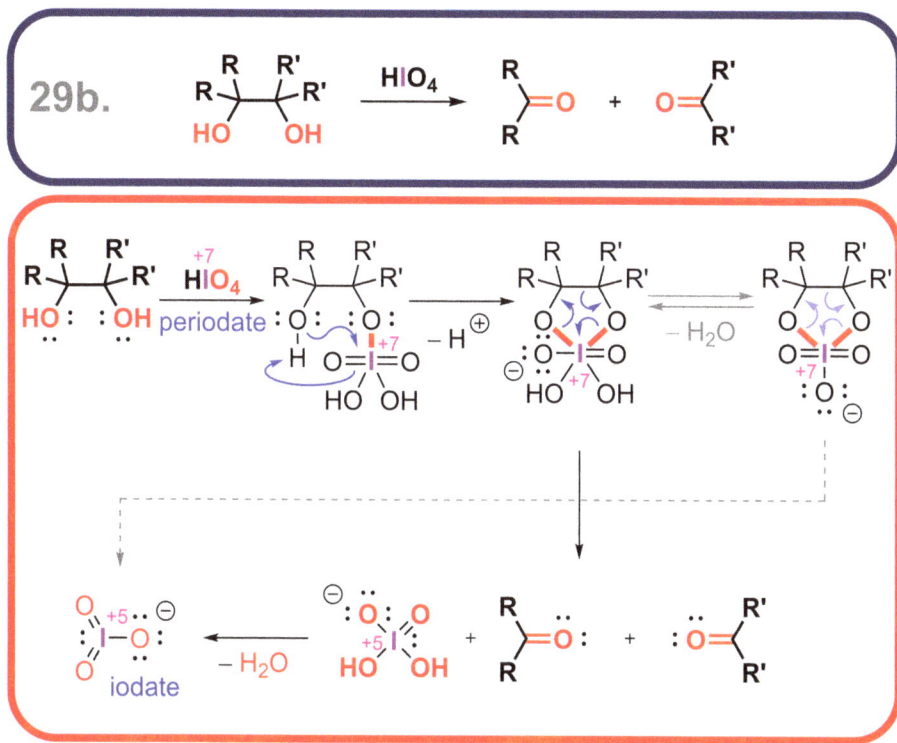

Fig. 29.3: The *Malaprade oxidation* mechanism.[100]

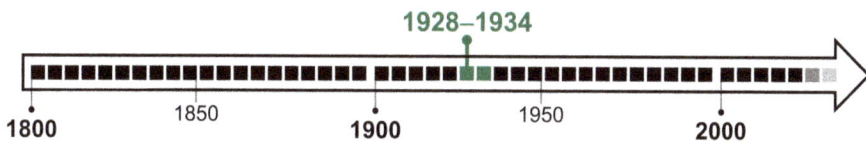

Fig. 29.4: The discovery of the *Malaprade oxidation*.[101]

---

100 The *Malaprade oxidation* is analogous to the *Criegee reaction*.
101 The reaction was likely first described between 1928 and 1934 [29b, 29c].

30. R₁—≡—H + ⊖N=N⊕=N—R₂ —[Cu]→ triazole product

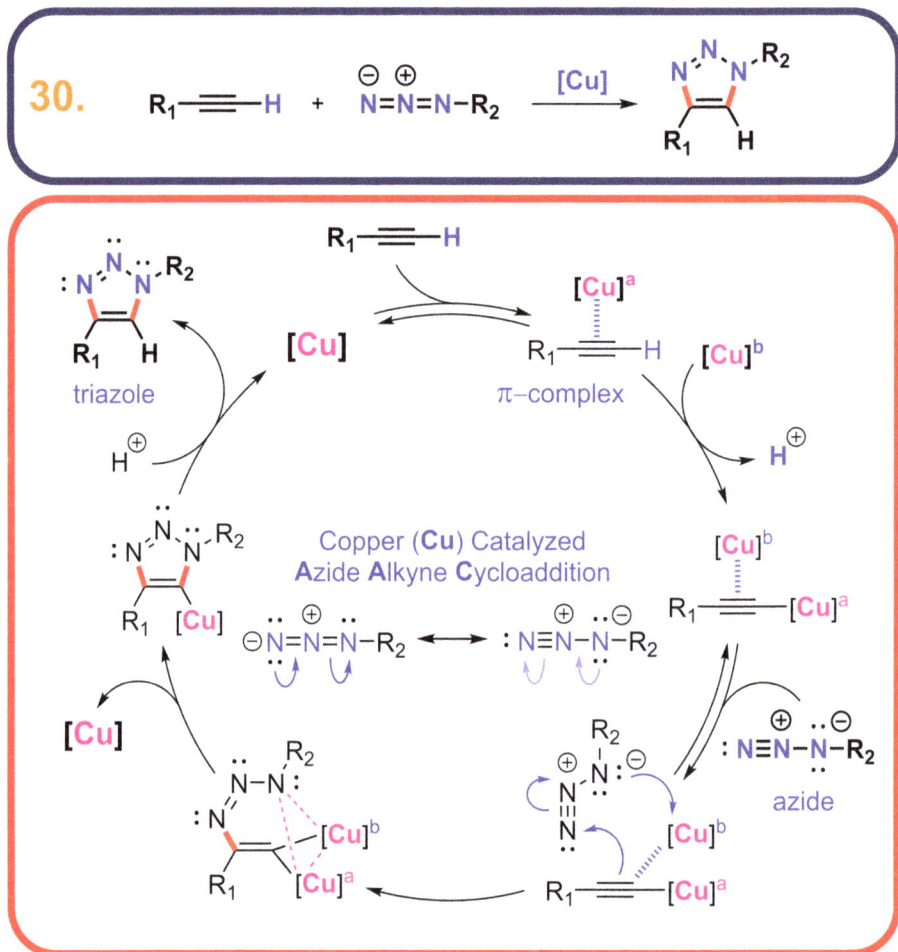

Fig. 30.1: The **CuAAC** mechanism.[102]

---

102 The acronym "**CuAAC**" stands for **Cu**-catalyzed **A**zide–**A**lkyne **C**ycloaddition (Copper(I)-catalyzed azide–alkyne cycloaddition). It is also often referred to as "**click chemistry.**" Formally, it is a *1,3-dipolar cycloaddition reaction* or a *(3 + 2)-cycloadditon reactions*. Please note that the notation (3 + 2) means the _atom count_ is used; the notation [4 + 2] means the _electron count_ involved in the reaction is used [30a]. IUPAC does *not* recommend mixed usage, but it is seen frequently in the literature: [3 + 2].

https://doi.org/10.1515/9783110786835-030

Huisgen 1,3-Dipolar Cycloaddition

$$MeO_2C-\!\!\!\equiv\!\!\!-CO_2Me \;+\; \overset{\ominus}{X}=\overset{\oplus}{X}=X-R_2 \;\xrightarrow{\Delta}\;$$

RuAAC

$$R_1-\!\!\!\equiv\!\!\!-H \;+\; N_3-R_2 \;\xrightarrow{[Ru]}\;$$

NiAAC

$$R_1-\!\!\!\equiv\!\!\!-H \;+\; N_3-R_2 \;\xrightarrow{[Ni]}\;$$

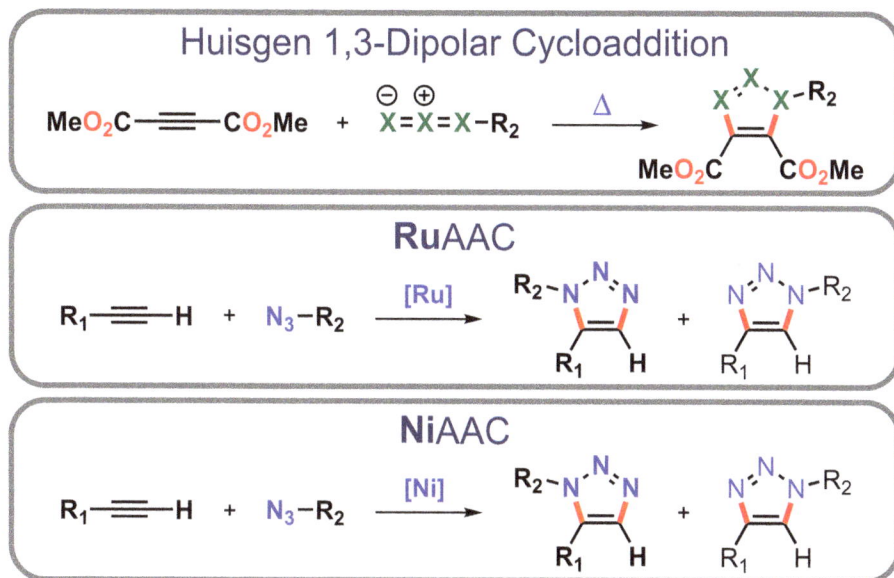

Fig. 30.2: Reactions related to the **CuAAC**.[103]

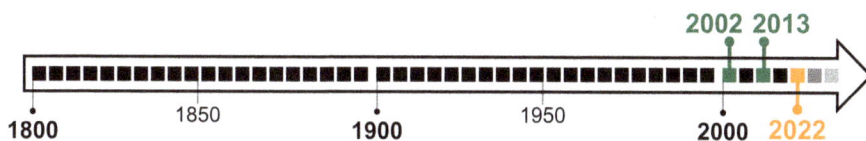

Fig. 30.3: The discovery of the **CuAAC**.[104]

---

103 The *Huisgen* cycloaddition [30b, 30c] is not catalytic but related to **CuAAC**. The *azide–alkyne cycloaddition* can be also catalyzed by **Ruthenium** (**RuAAC**) or Nickel (**NiAAC**); however, it undergoes a different mechanism (not shown).

104 The reaction was likely first described around 2002 [30d, 30e] and the mechanism, in its current form, proposed around 2013 [30f]. In **2022**, Carolyn R. Bertozzi, Morten Meldal, and K. Barry Sharpless received the Nobel Prize in Chemistry for the development of click chemistry and bioorthogonal chemistry [30g, 30h].

31 Curtius Rearrangement

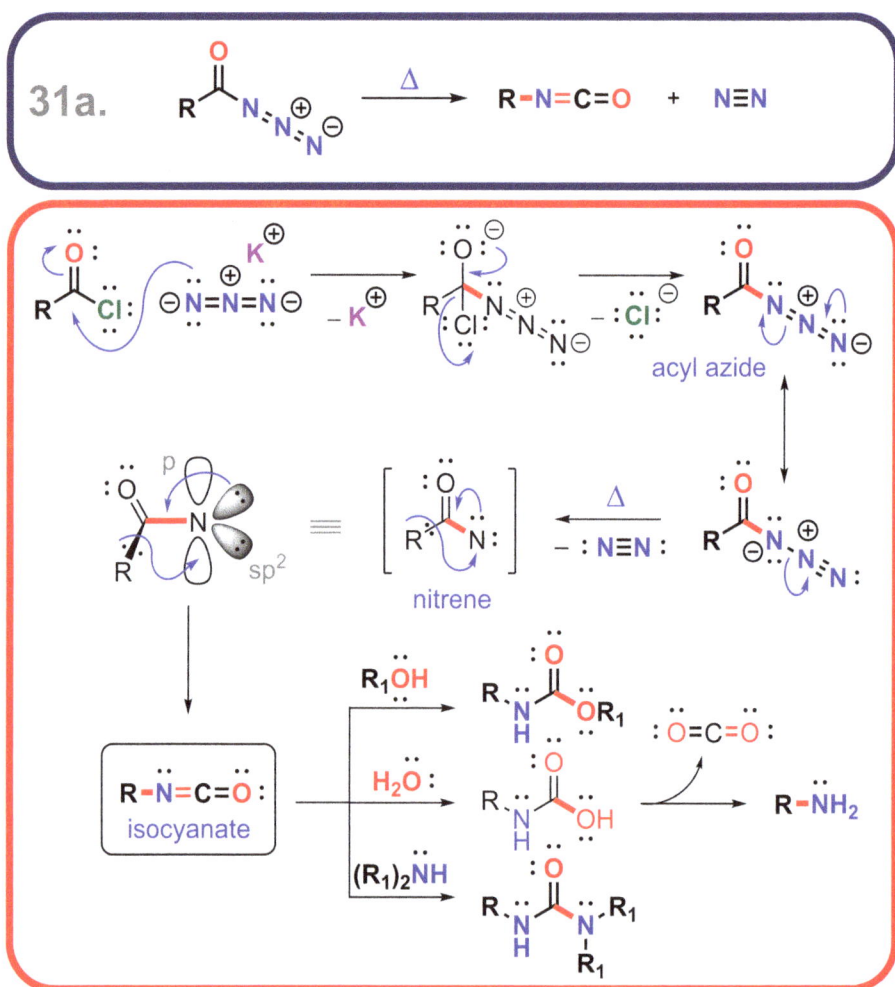

31a.

$$R-C(=O)-N=N^{\oplus}=N^{\ominus} \xrightarrow{\Delta} R-N=C=O + N\equiv N$$

Fig. 31.1: The **Curtius** *rearrangement* mechanism.[105]

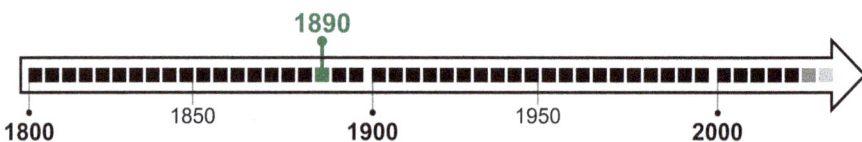

Fig. 31.2: The discovery of the **Curtius** *rearrangement*.[106]

---

105 The **Curtius** *rearrangement* is also called the **Curtius** *reaction*.

https://doi.org/10.1515/9783110786835-031

31b.

$$R-C(=O)-OH + HN_3 \xrightarrow{H^{\oplus}} R-N=C=O \xrightarrow{H_2O} R-NH_2$$

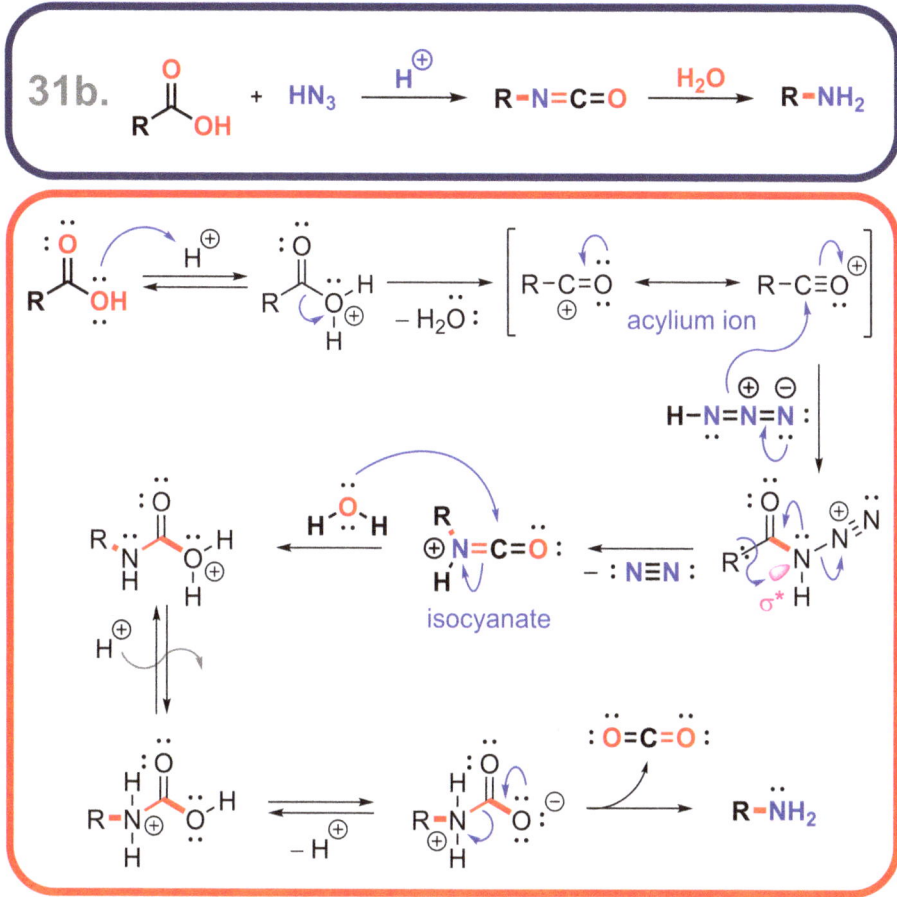

Fig. 31.3: The **Schmidt** reaction mechanism.[107]

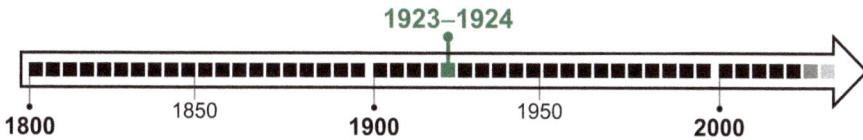

Fig. 31.4: The discovery of the **Schmidt** reaction mechanism.[108]

---

106 The reaction was likely first described around 1890 [31a, 31b].

107 The **Schmidt** reaction is also a rearrangement.

108 The reaction was likely first described between 1923 and 1924 [31c, 31d].

31c.
$$R-C(=O)-NH_2 \xrightarrow[\text{Br}_2]{\text{NaOH}} R-N=C=O \xrightarrow{H_2O} R-NH_2$$

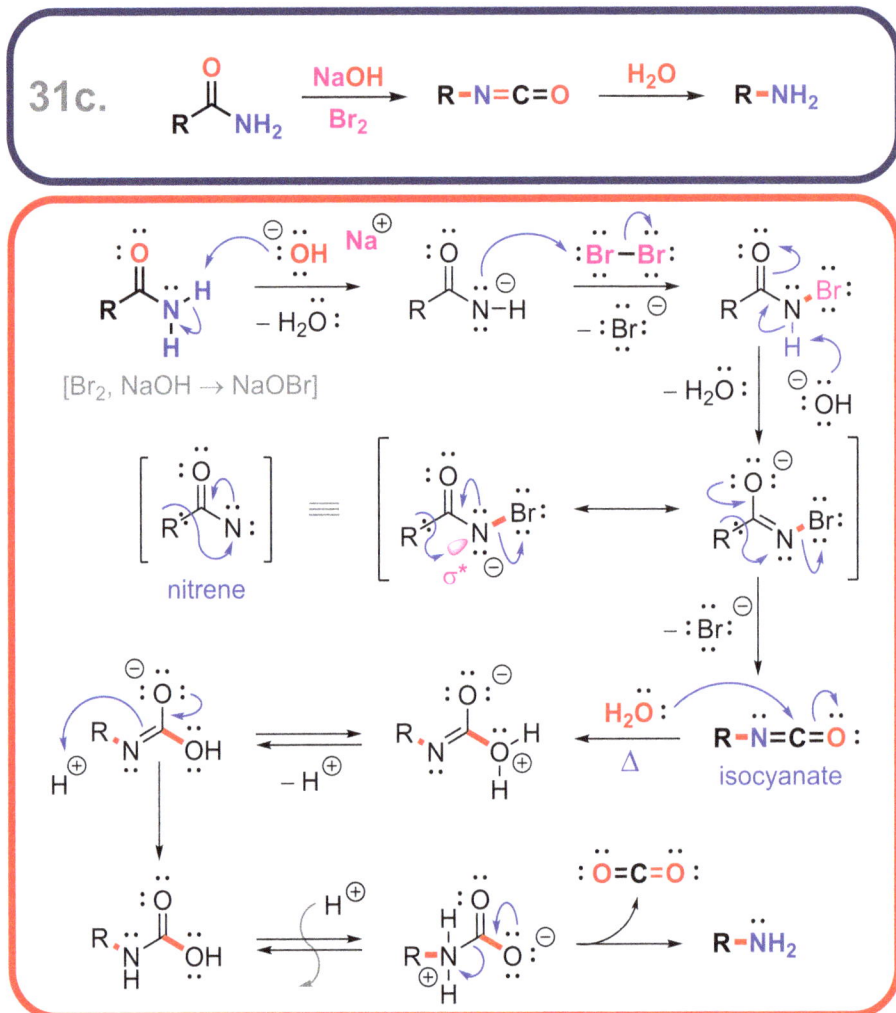

[Br$_2$, NaOH → NaOBr]

nitrene

isocyanate

Fig. 31.5: The *Hofmann rearrangement* mechanism.[109]

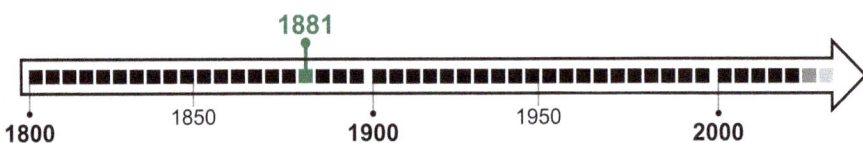

1881

1800   1850   1900   1950   2000

Fig. 31.6: The discovery of the *Hofmann rearrangement*.[110]

109 The *Hofmann rearrangement* is also known as the *Hofmann reaction*. It is completely different from the *Hofmann elimination* (see Chapter 49).
110 The reaction was likely first described around 1881 [31e].

31d.

$$R\text{-}C(=O)\text{-}N(H)\text{-}OH \xrightarrow[\text{NaOH}]{\text{AcCl}} R\text{-}N\text{=}C\text{=}O \xrightarrow{\text{H}_2\text{O}} R\text{-}NH_2$$

hydroxamic acid

X = Ac, SO₂Ar

isocyanate

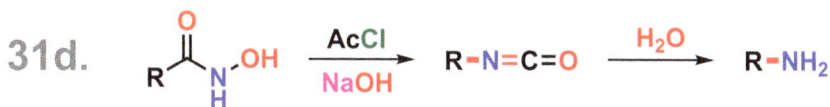

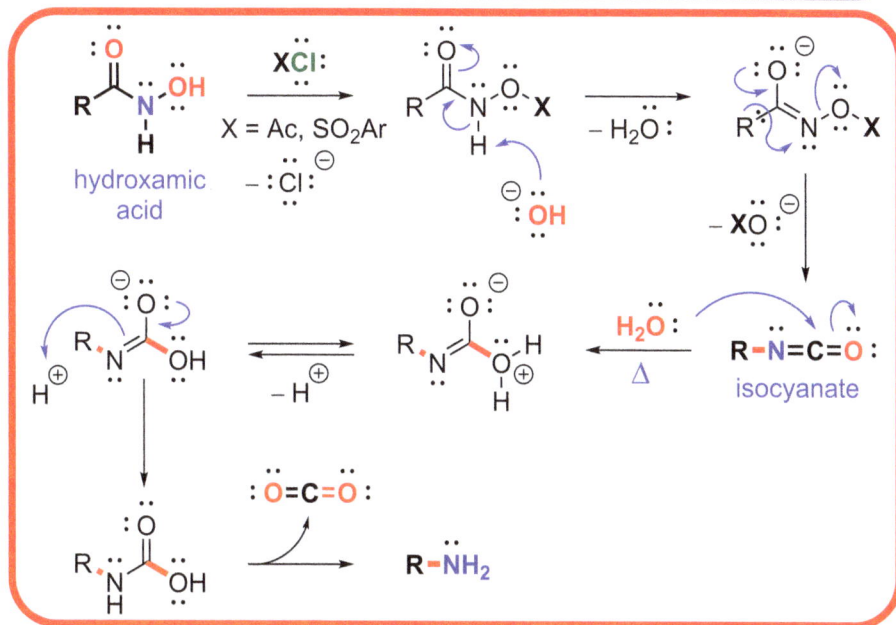

Fig. 31.7: The *Lossen rearrangement* mechanism.[111]

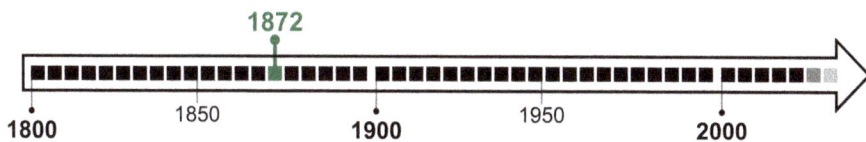

1872

1800   1850   1900   1950   2000

Fig. 31.8: The discovery of the *Lossen rearrangement*.[112]

---

111 The *Lossen rearrangement* is much like these reactions and is related to the *Beckmann rearrangement*, covered in Chapter 14.

112 The reaction was likely first described around 1872 [31f].

32 Darzens Condensation

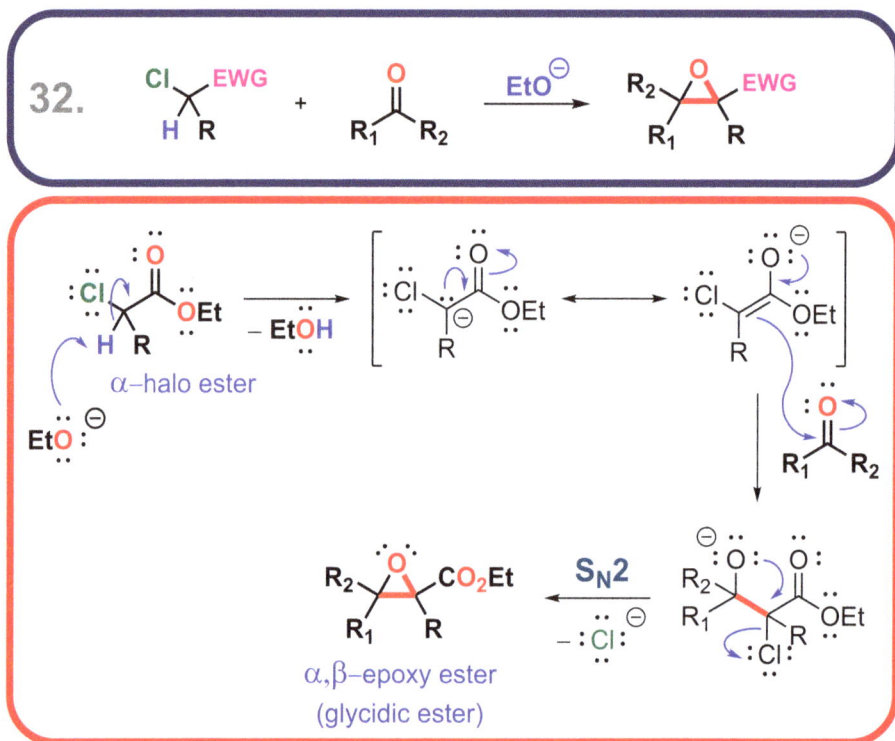

Fig. 32.1: The **Darzens** condensation mechanism.[113]

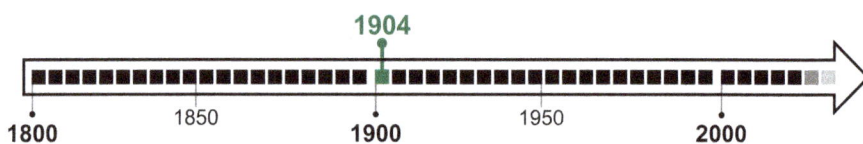

Fig. 32.2: The discovery of the **Darzens** condensation.[114]

---

113 The **Darzens** condensation is also called the **Darzens** glycidic ester condensation or the **Darzens** reaction. Please note that a glycidic ester is an α,β-epoxy ester.
114 The reaction was likely first described around 1904 [32c].

https://doi.org/10.1515/9783110786835-032

Corey–Chaykovsky Reaction

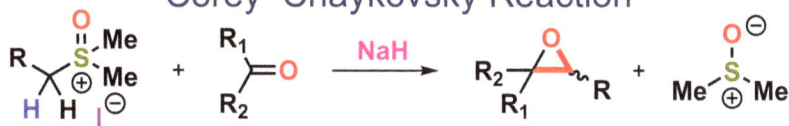

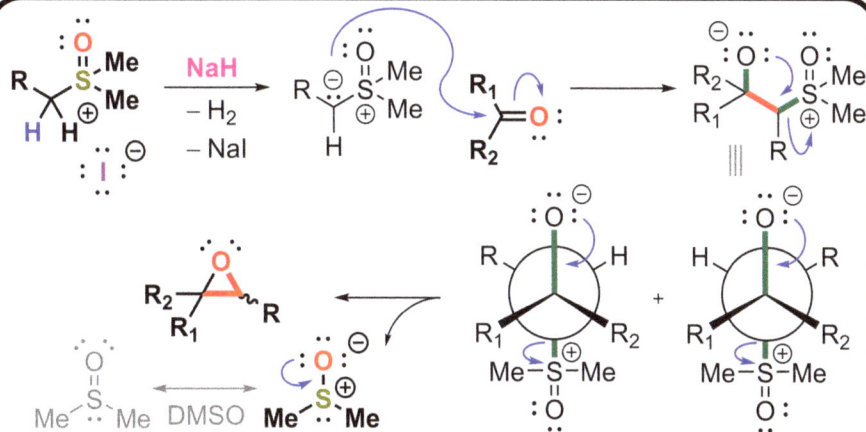

Fig. 32.3: The **Corey–Chaykovsky** *reaction* mechanism.[115]

Compare to Darzens Condensation:

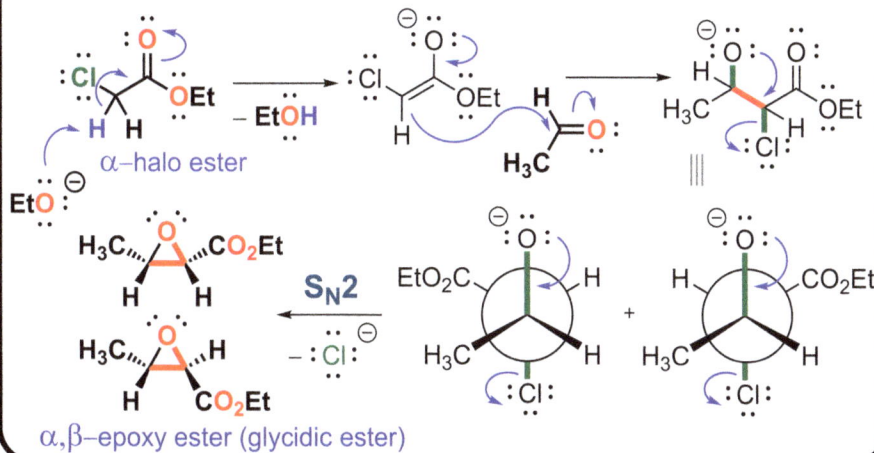

α–halo ester

α,β–epoxy ester (glycidic ester)

---

115 The **Corey–Chaykovsky** *reaction* (also known as the **Johnson–Corey–Chaykovsky** *reaction*) [32a, 32b] is related to both the **Darzens** condensation and the **Wittig** *reaction* (covered in Chapter 98).

33 Dess–Martin Oxidation

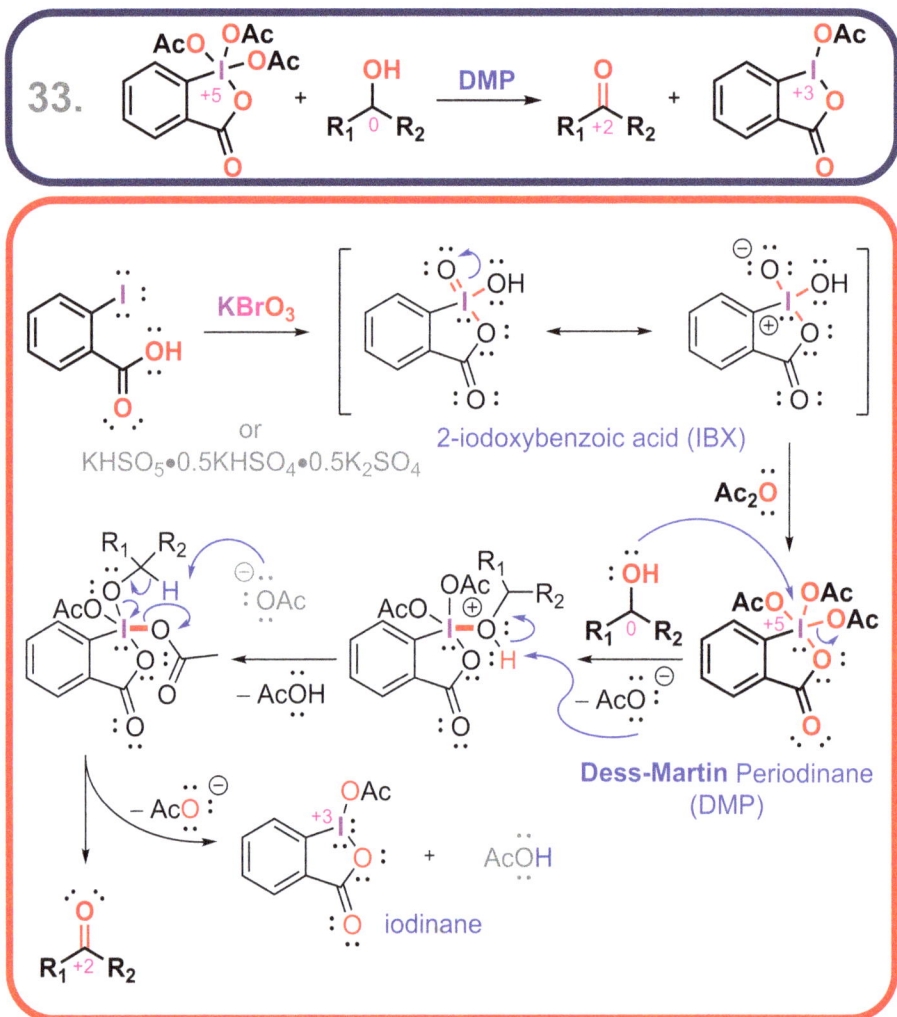

Fig. 33.1: The **Dess–Martin** oxidation mechanism.[116]

---

116 The **Dess–Martin** oxidation is based on the use of a named reagent: the **Dess–Martin** periodinane (**DMP**) [33a, 33b].

https://doi.org/10.1515/9783110786835-033

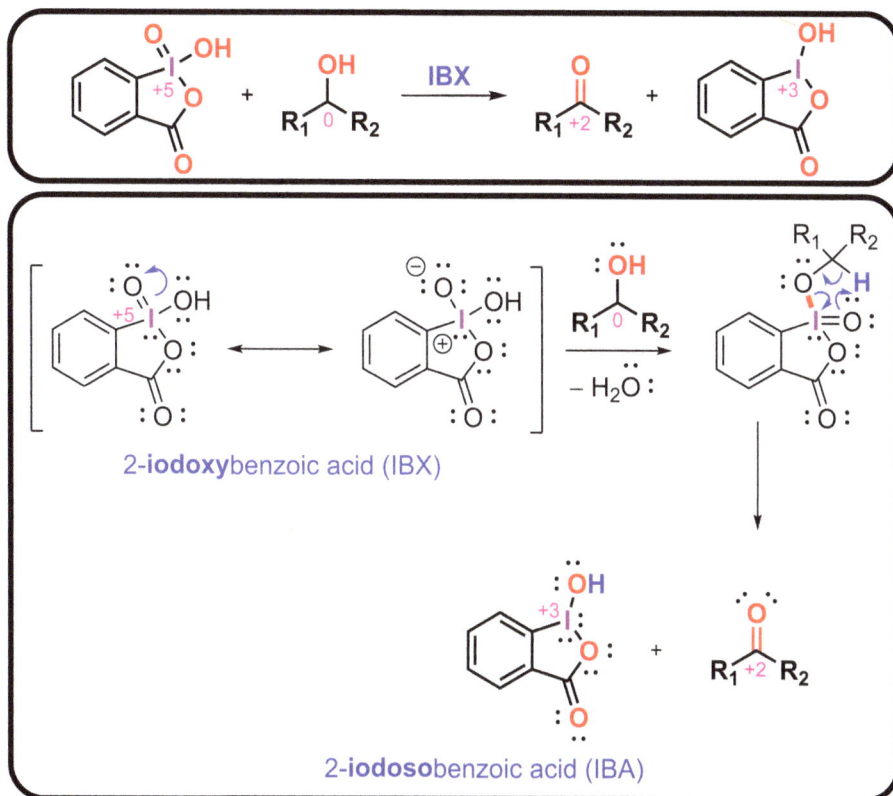

Fig. 33.2: The **IBX** *oxidation* mechanism.[117]

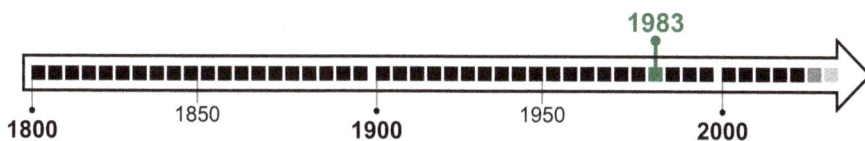

Fig. 33.3: The discovery of the ***Dess–Martin*** *oxidation.*[118]

---

117 2-Iodoxybenzoic acid (**IBX**) is a precursor for the preparation of the ***Dess–Martin*** *periodinane* (**DMP**). IBX can also be used as an oxidant.

118 The reaction was likely first described around 1983 [33c].

34 Diazotization (Diazonium Salt)

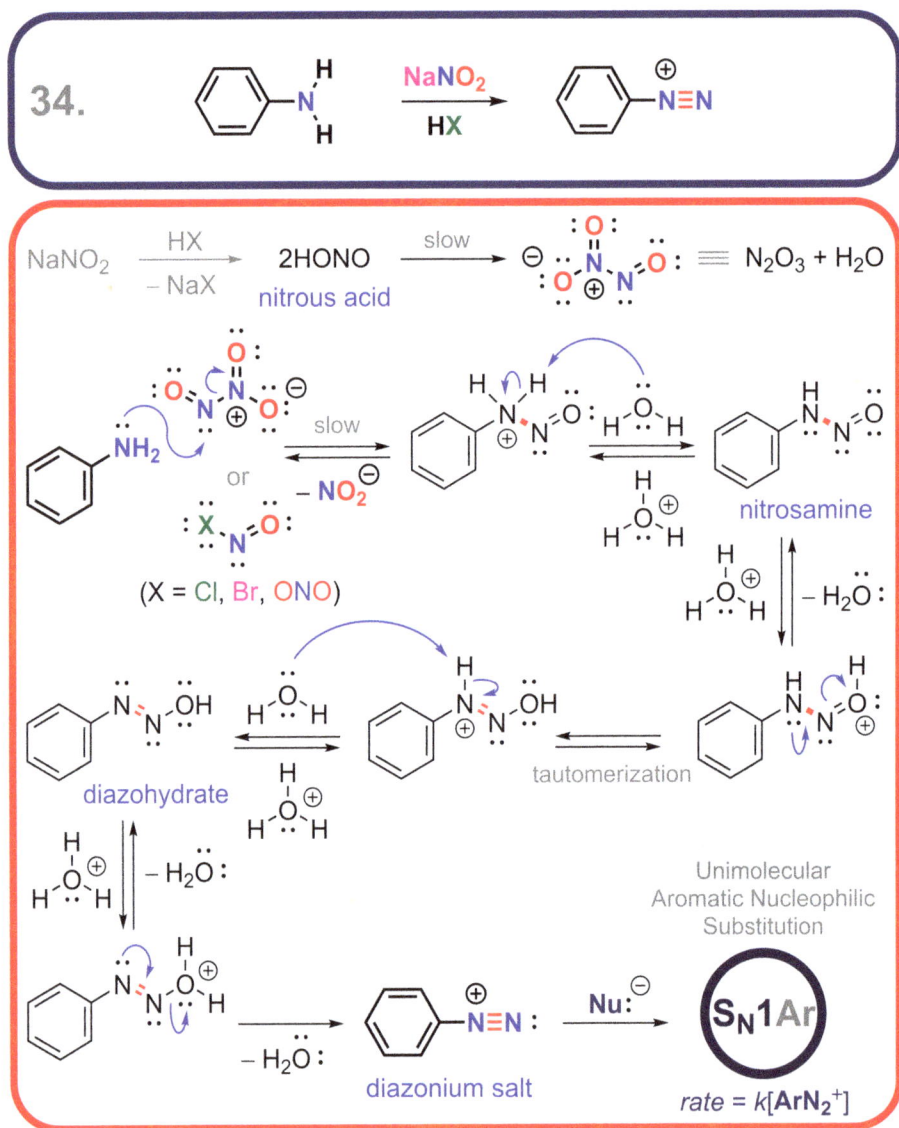

Fig. 34.1: The *diazonium salt formation (diazotization)* mechanism.[119]

---

119 The *diazonium salt formation reaction* is also known as the ***diazotization*** [1] (the term is also preferred in this book), or the *diazoniation* [1a], or the *diazotation* [34a].

https://doi.org/10.1515/9783110786835-034

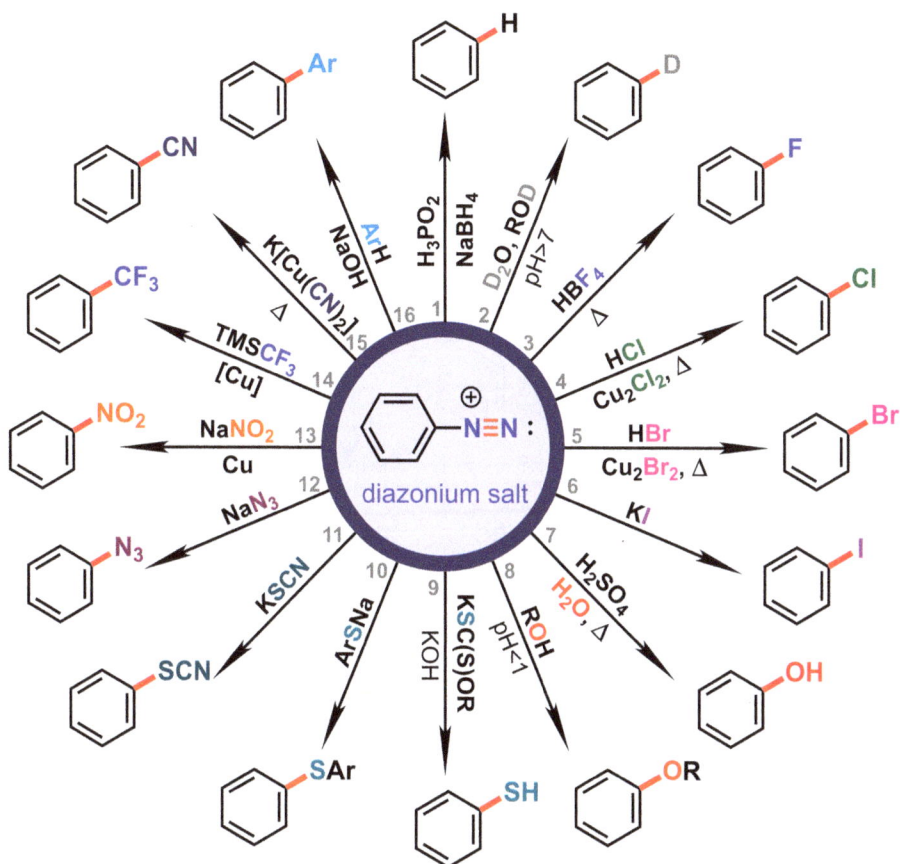

Fig. 34.2: Synthetic versatility of the diazonium salts.[120]

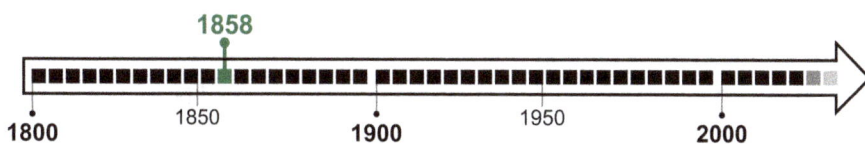

Fig. 34.3: The discovery of the *diazotization reaction*.[121]

---

120 The *diazonium salts* formed during the *diazotization* process have wide synthetic application and they can react with a variety of nucleophiles. These reactions go through the **aromatic nucleophilic substitution** mechanism (S_N1Ar or sometimes $S_{RN}1$). Symbol S_N1Ar stands for Substitution Nucleophilic Aromatic. It is a **Uni**molecular (1) reaction, that is, the rate of the reaction is first order, and the rate-determining step (the slow step) depends on the concentration of one reactant, the *diazonium salt* (ArN_2^+): *rate* $= k[ArN_2^+]^1$. This mechanism is different from the *addition–elimination* mechanism (S_NAr or S_N2Ar), covered in Chapter 4, because the first step is elimination and the formation of an *aryl cation*. It is also different from the *benzyne* mechanism (the *elimination–addition* mechanism, Chapter 16).
121 The reaction was likely first described around 1858 [34b].

35 Diels–Alder Cycloaddition

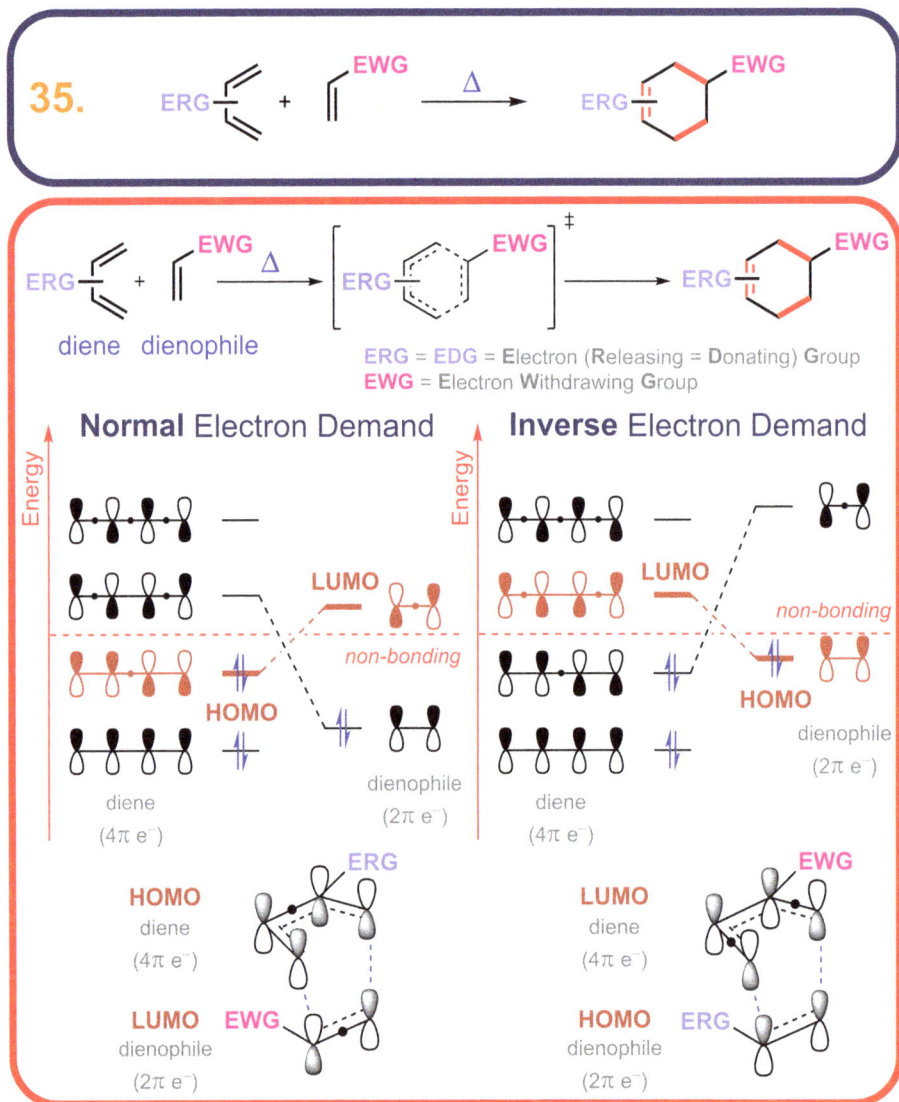

Fig. 35.1: The **Diels–Alder** cycloaddition mechanism.[122]

---

122 The **Diels–Alder** cycloaddition reaction or the **[4 + 2]**-cycloaddition reaction is a pericyclic reaction with a concerted mechanism. Note that the notation (4 + 2) means the *atom count* is used; the notation [4 + 2] means the *electron count* involved in the reaction is used [30a]. Compare with the *1,3-dipolar cycloaddition* (Chapter 30).

https://doi.org/10.1515/9783110786835-035

$[4_\pi+2_\pi]$ = Diels–Alder Cycloaddition

Homo–Diels–Alder Cycloaddition

homo-diene

dienophile

Retro–Diels–Alder Reaction

endo-

Δ

room temperature

Hetero–Diels–Alder Cycloaddition

X = N, *Aza*–Diels–Alder Reaction
X = O, *Oxa*–Diels–Alder Reaction

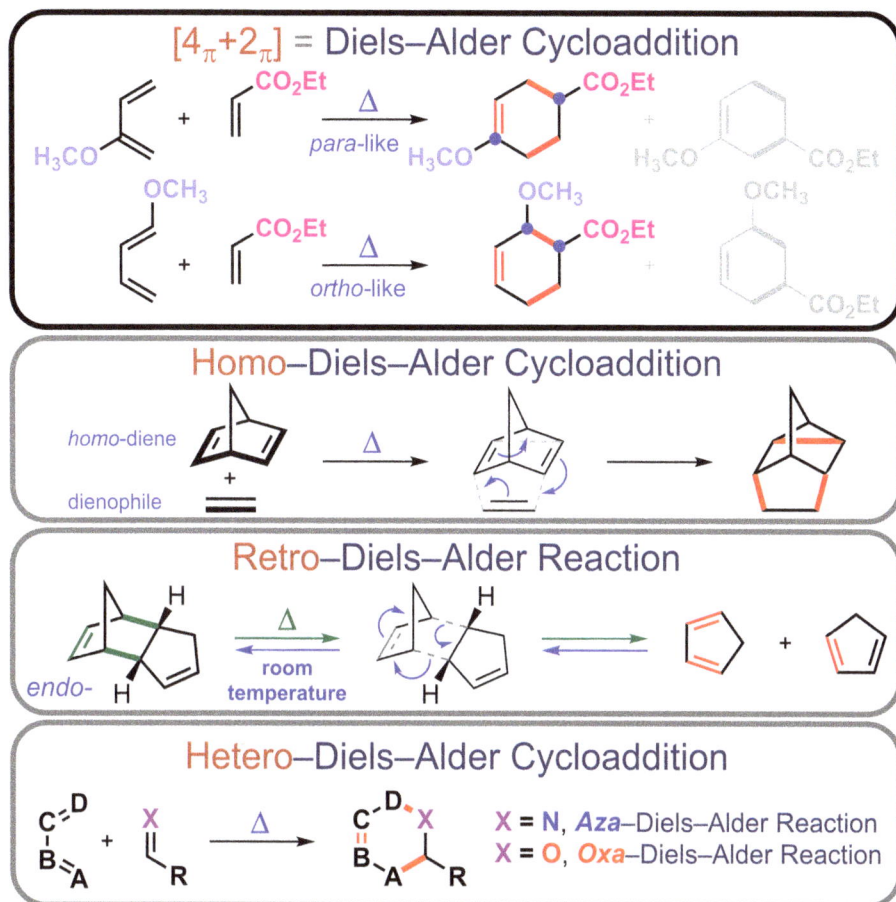

Fig. 35.2: Reactions related to the *Diels–Alder cycloaddition*.[123]

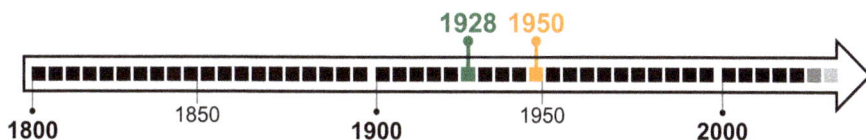

Fig. 35.3: The discovery of the *Diels–Alder cycloaddition*.[124]

---

123 There are numerous variations of this reaction: the *homo-Diels–Alder cycloaddition*, the *retro-Diels–Alder reaction*, the *hetero-Diels–Alder cycloaddition*, and many others (not shown). Please note the regiochemistry observed in the first case of the $[4_\pi + 2_\pi]$ = *Diels–Alder cycloaddition*.
124 The reaction was likely first described around 1928 [35a, 35b]. In **1950**, Otto Paul Hermann Diels and Kurt Alder received the Nobel Prize in Chemistry for the discovery of the diene synthesis [35c].

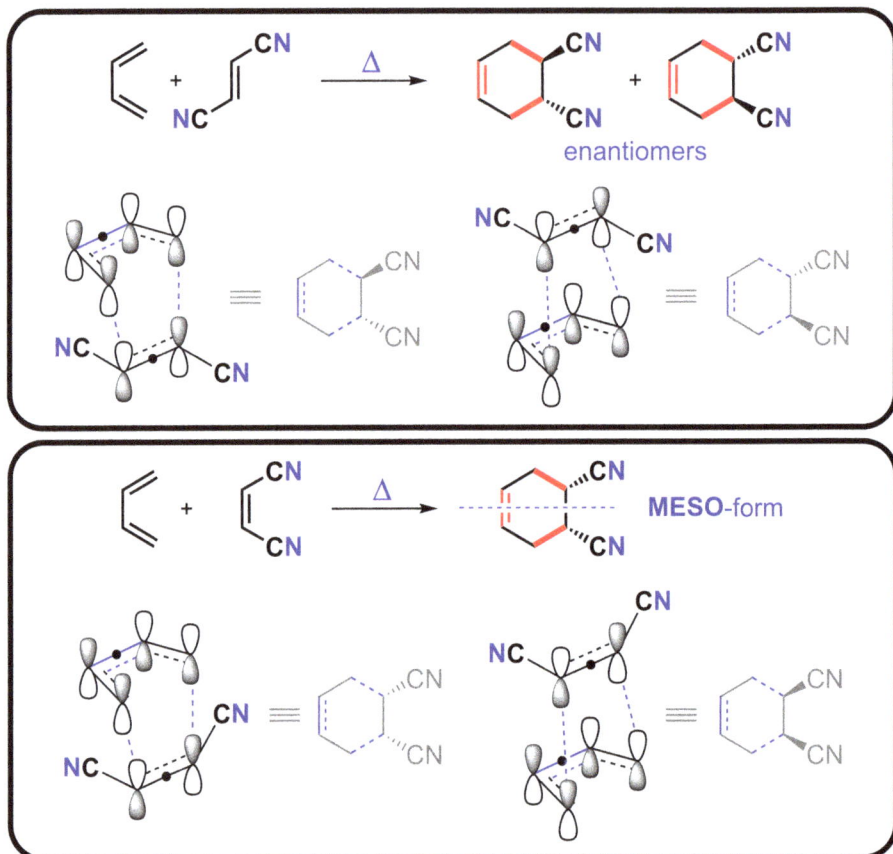

Fig. 35.4: The **Diels–Alder** cycloaddition using various dienophiles.[125]

---

125 The **Diels–Alder** cycloaddition reaction using buta-1,3-diene (diene) and either fumaronitrile or maleonitrile. Please note the difference in the reaction outcome.

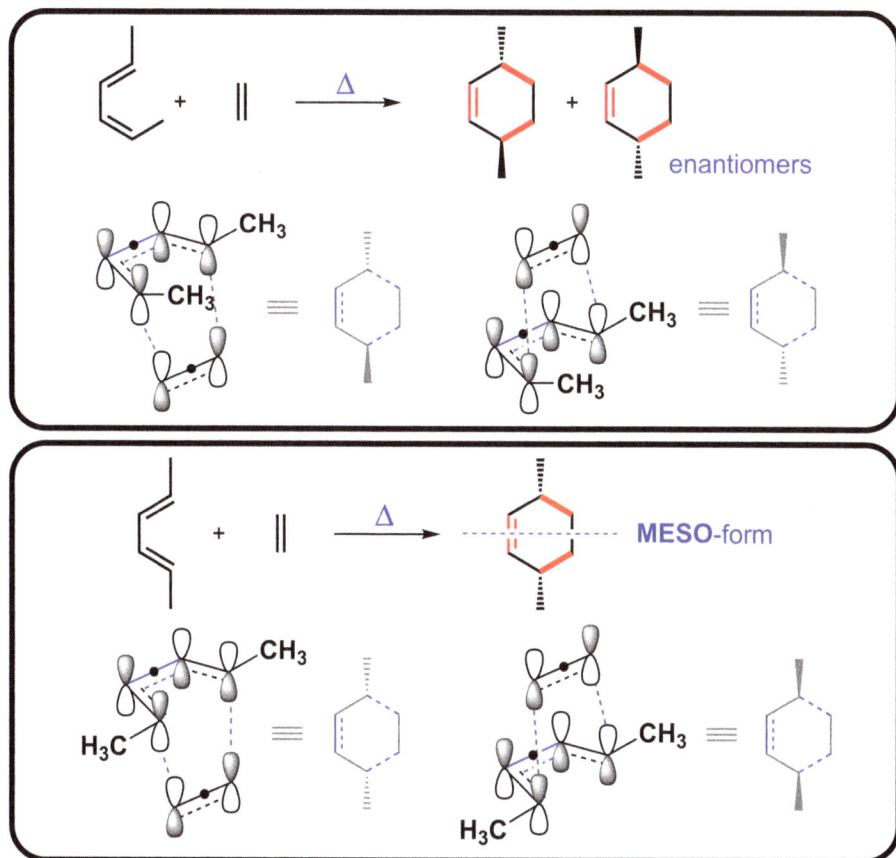

Fig. 35.5: The **Diels–Alder** cycloaddition using various dienes.[126]

---

[126] The **Diels–Alder** cycloaddition reaction using either (2Z,4E)-hexa-2,4-diene or (2E,4E)-hexa-2,4-diene and ethene (dienophile). Please note the difference in the reaction outcome.

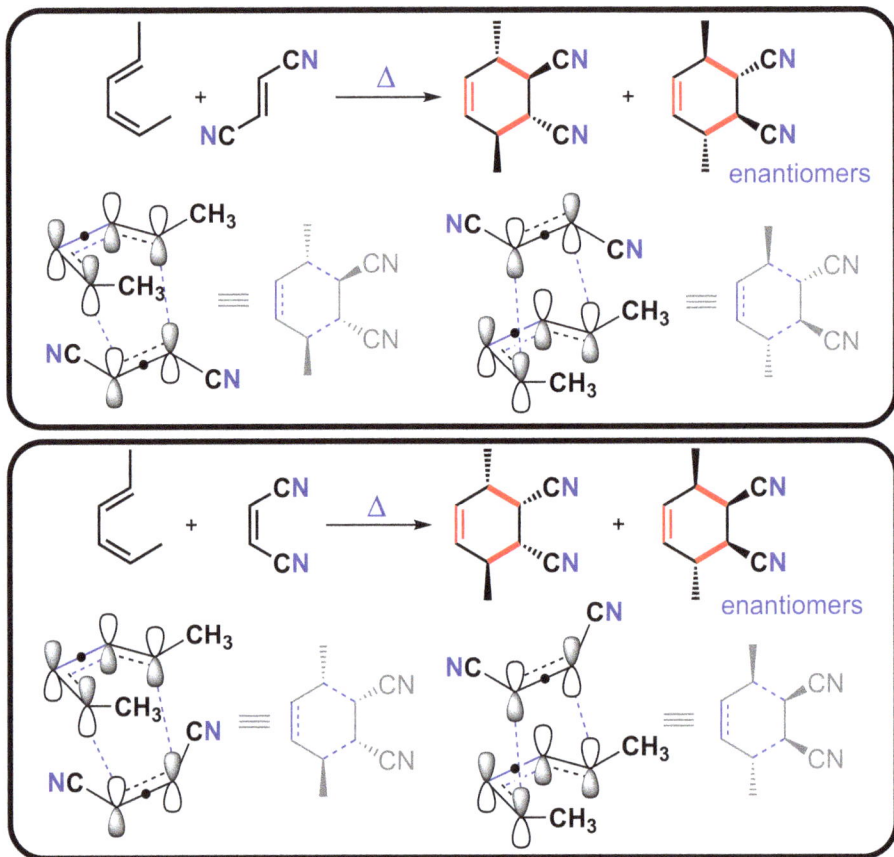

Fig. 35.6: The **Diels–Alder** *cycloaddition* using (2Z,4E)-hexa-2,4-diene and various dienophiles.[127]

---

127 The **Diels–Alder** *cycloaddition reaction* using (2Z,4E)-hexa-2,4-diene and either *fumaronitrile* or *maleonitrile*. Please note the difference in the reaction outcome.

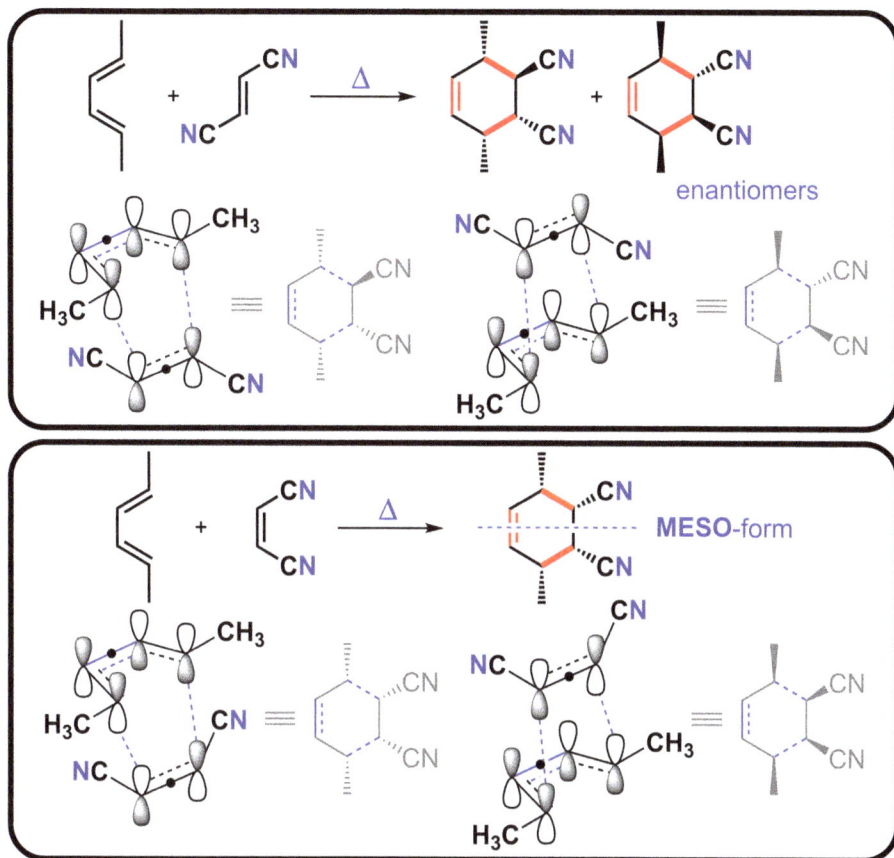

Fig. 35.7: The **_Diels–Alder_** _cycloaddition_ using (2E,4E)-hexa-2,4-diene and various dienophiles.[128]

---

[128] The **_Diels–Alder_** _cycloaddition reaction_ using (2E,4E)-hexa-2,4-diene and either _fumaronitrile_ or _maleonitrile_. Please note the difference in the reaction outcome.

36 Di-π-Methane Rearrangement

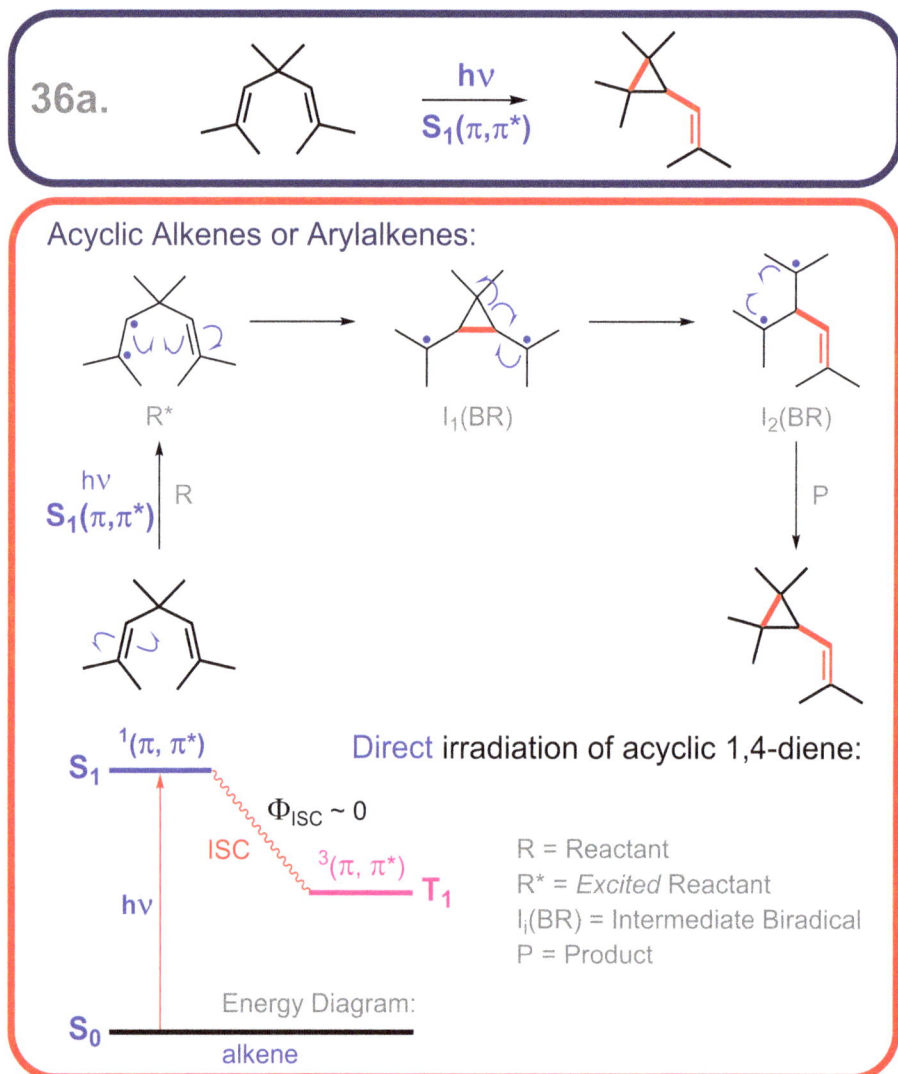

36a.

$$h\nu$$
$$S_1(\pi,\pi^*)$$

Acyclic Alkenes or Arylalkenes:

R*

I_1(BR)

I_2(BR)

$h\nu$ $\quad$ R
$S_1(\pi,\pi^*)$

P

Direct irradiation of acyclic 1,4-diene:

$^1(\pi, \pi^*)$
S_1

$\Phi_{ISC} \sim 0$

ISC $\quad ^3(\pi, \pi^*)$
$h\nu$ $\qquad\qquad$ T_1

R = Reactant
R* = *Excited* Reactant
I_i(BR) = Intermediate Biradical
P = Product

Energy Diagram:

S_0 ————————
alkene

Fig. 36.1: The *di-π-methane rearrangement* mechanism: direct irradiation.[129]

---

129 The *di-π-methane rearrangement* (**DPM**) is rarely called the **Zimmerman** *reaction*. If the reaction undergoes *direct irradiation*: the reaction occurs from the <u>*singlet*</u> excited state S_1, in this case $^1(\pi, \pi^*)$ [2b].

https://doi.org/10.1515/9783110786835-036

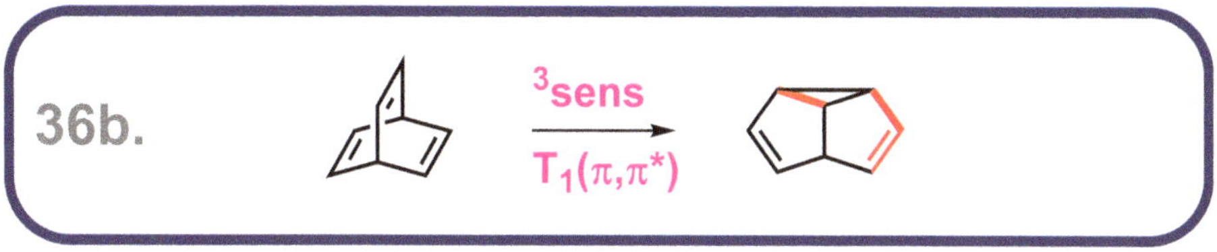
36b.
3sens
T1(π,π*)

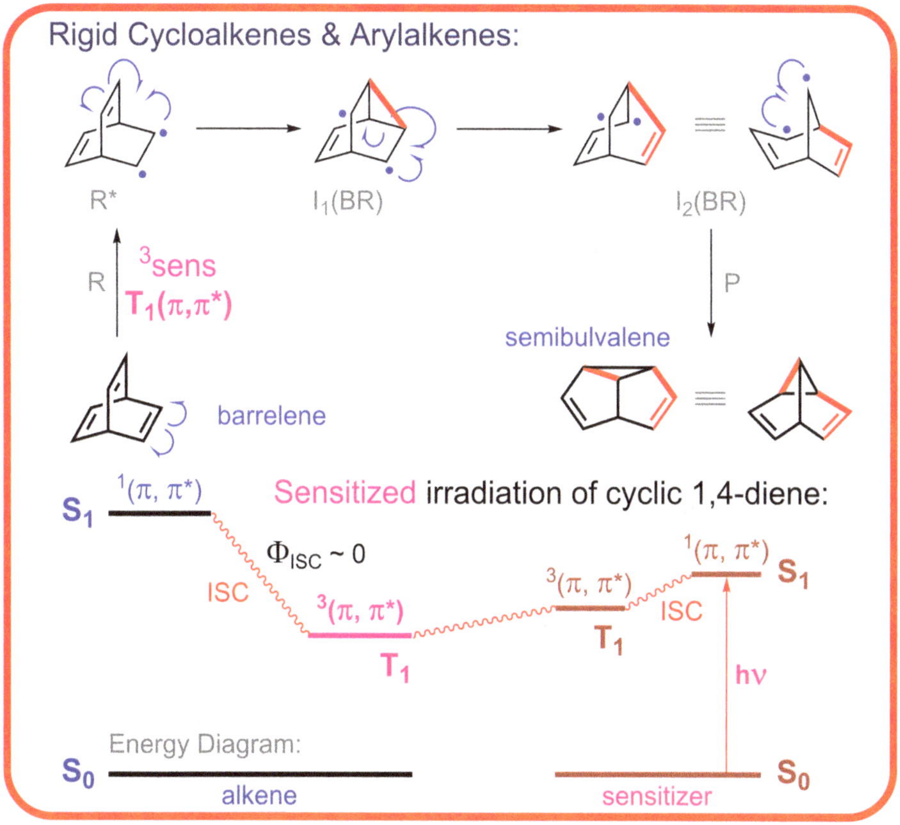
Rigid Cycloalkenes & Arylalkenes:
R*
I1(BR)
I2(BR)
R
3sens
T1(π,π*)
P
semibulvalene
barrelene
Sensitized irradiation of cyclic 1,4-diene:
S1
1(π, π*)
ΦISC ~ 0
ISC
3(π, π*)
T1
3(π, π*)
T1
1(π, π*)
S1
ISC
hν
Energy Diagram:
S0
alkene
sensitizer
S0

Fig. 36.2: The *di-π-methane rearrangement* mechanism: sensitized irradiation.[130]

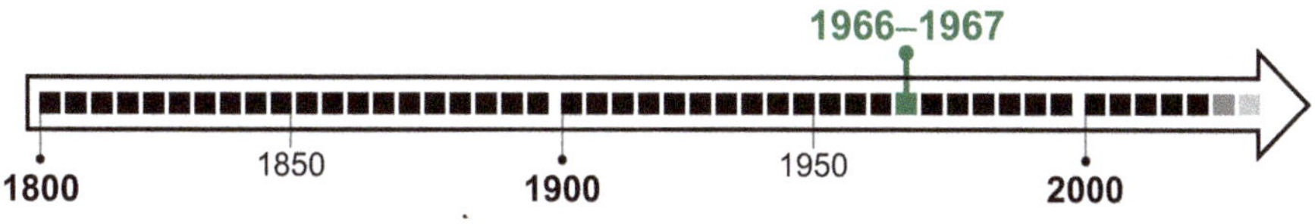
1966–1967
1800
1850
1900
1950
2000

Fig. 36.3: The discovery of the *di-π-methane rearrangement*.[131]

130 The *di-π-methane rearrangement* in the presence of a *photosensitizer*, that is, the reaction undergoes the *sensitized irradiation*: the product formation occurs from the *triplet* excited state $\mathbf{T_1}$, here $\mathbf{^3(\pi, \pi^*)}$ [2b].
131 The reaction was likely first described between 1966 and 1967 [36].

37 Favorskii Rearrangement

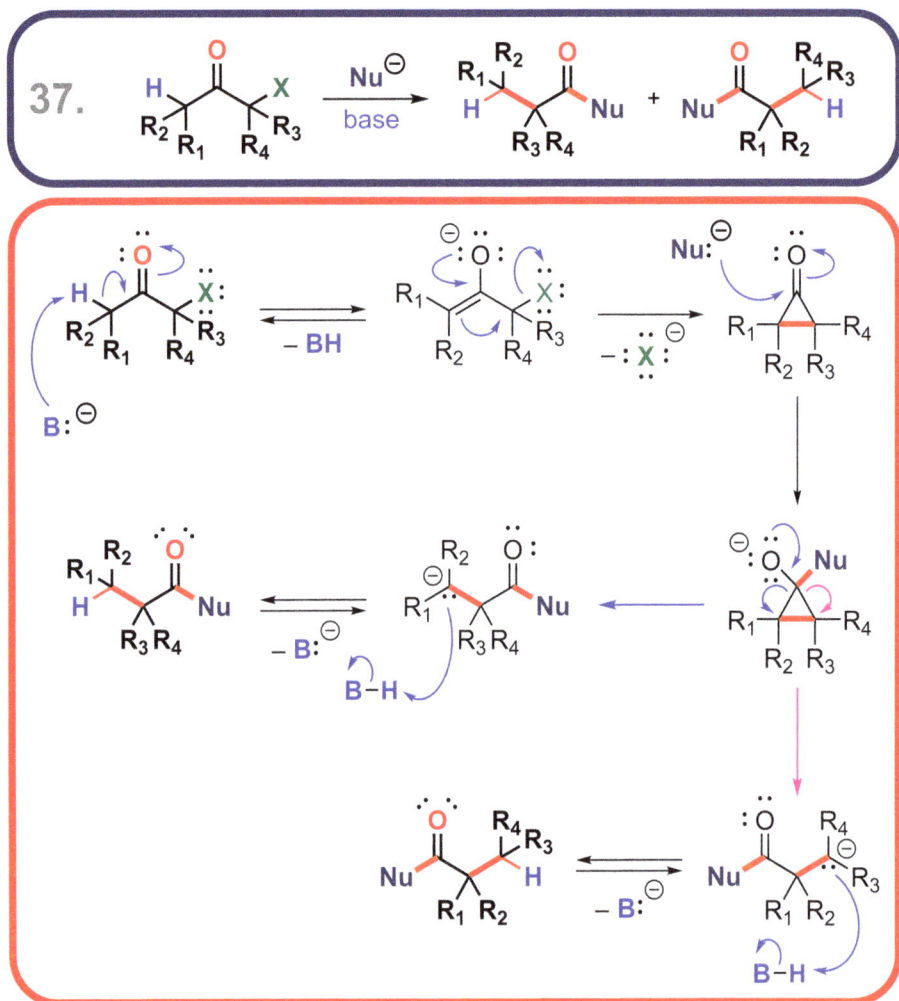

Fig. 37.1: The *Favorskii* rearrangement mechanism.[132]

---

132 The *Favorskii* rearrangement (also spelled Favorsky, in German transliteration Faworsky, and in Russian Алексей Евграфович Фаворский or А. Е. Фаворский) is different from the *Favorskii* reaction (not shown here).

https://doi.org/10.1515/9783110786835-037

Quasi–Favorskii Rearrangement

Semi-benzylic mechanism:

X = Cl, Br, I

X = Cl, Br, I

Homo–Favorskii Rearrangement

β–halo ketone

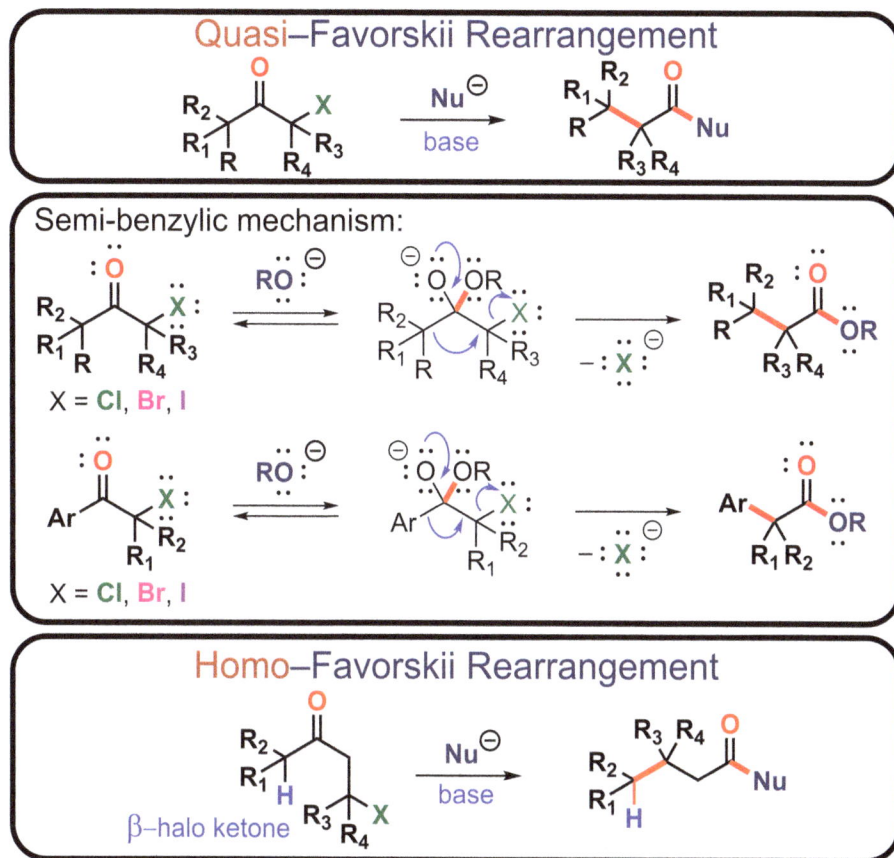

Fig. 37.2: The *quasi-Favorskii rearrangement* mechanism and related reactions.[133]

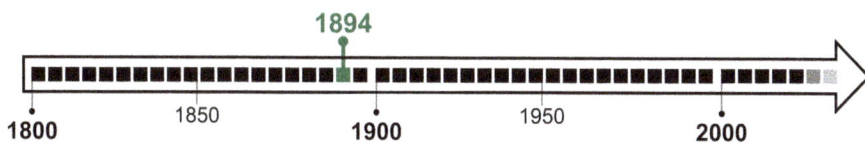

1894

1800   1850   1900   1950   2000

Fig. 37.3: The discovery of the *Favorskii rearrangement*.[134]

---

133 There are numerous variations of this reaction: for example, the *quasi-Favorskii rearrangement*, which undergoes a process similar to the *semi-benzylic* mechanism [37a, 37b], the *homo-Favorskii rearrangement*, and others (not shown).

134 The reaction was likely first described around 1894 [37c, 37d].

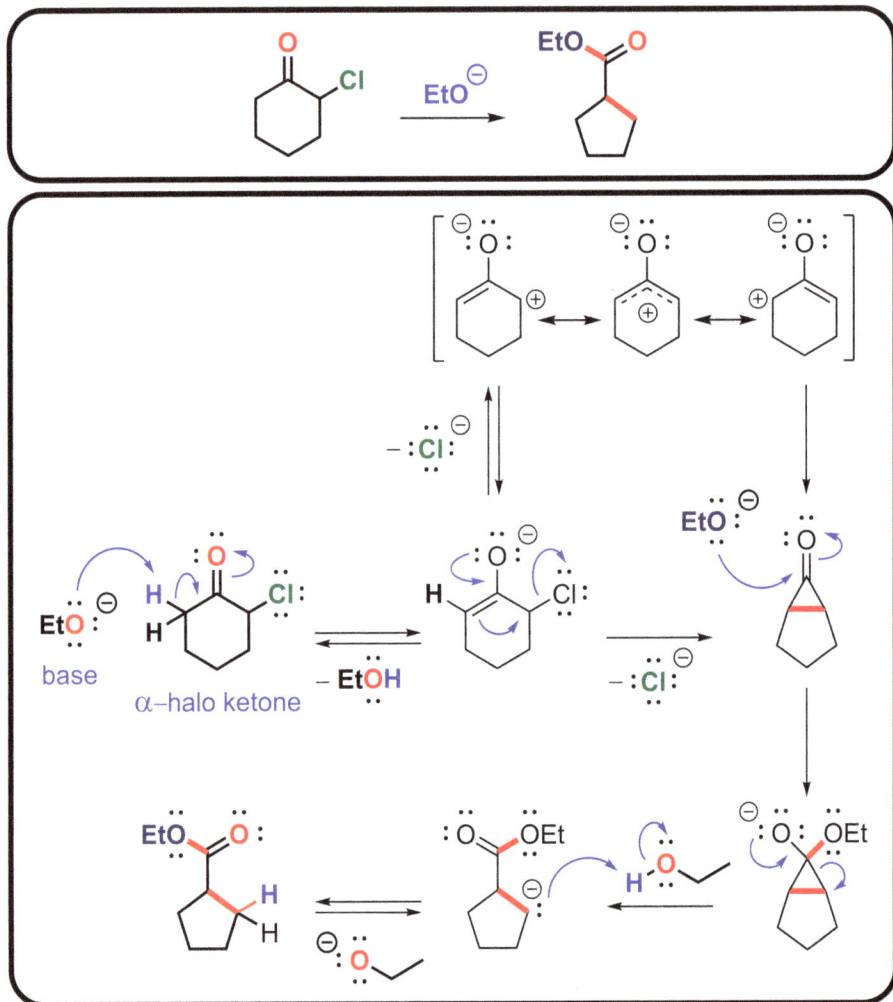

Fig. 37.4: The **Favorskii** *rearrangement* mechanism of 2-chlorocyclohexan-1-one.[135]

---

135 An example of the **Favorskii** *rearrangement* of 2-chlorocyclohexan-1-one yielding ring-contracted ethyl cyclopentanecarboxylate.

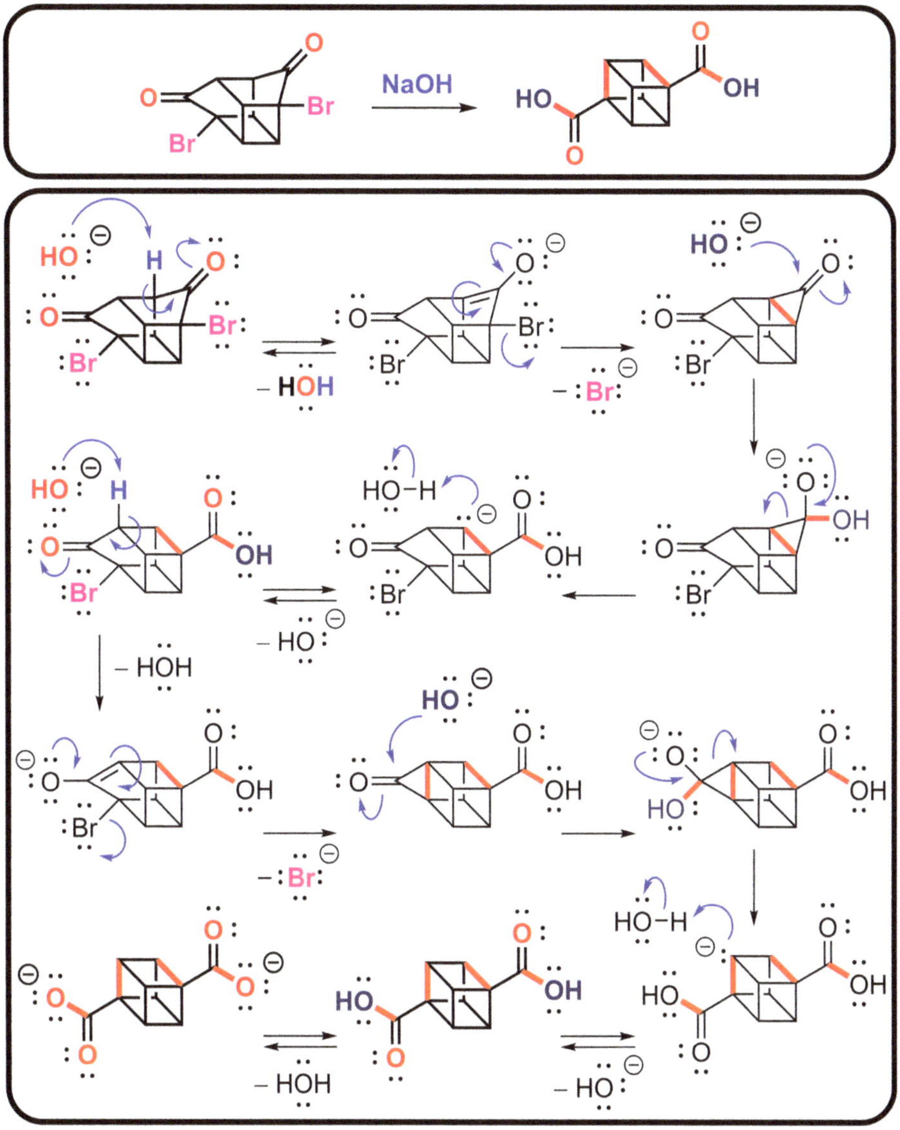

Fig. 37.5: Synthesis of cubane-1,4-dicarboxylic acid.[136]

---

136 The tandem *Favorskii rearrangement* plays the key role in the synthesis of cubane-1,4-dicarboxylic acid and other cubane derivatives [37e, 37f].

38 Fischer Indole Synthesis

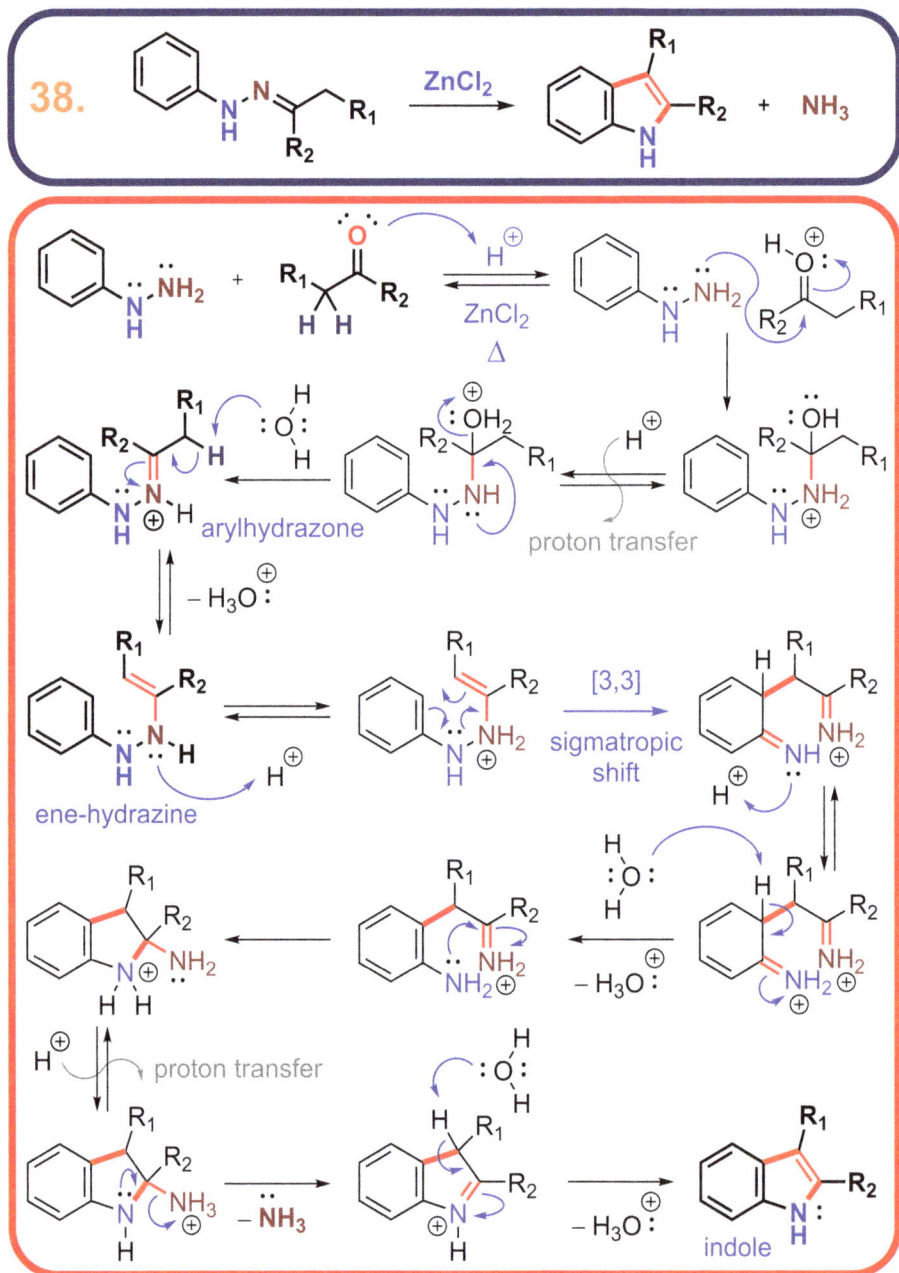

Fig. 38.1: The *Fischer indole synthesis* mechanism.[137]

https://doi.org/10.1515/9783110786835-038

Benzidine Rearrangement

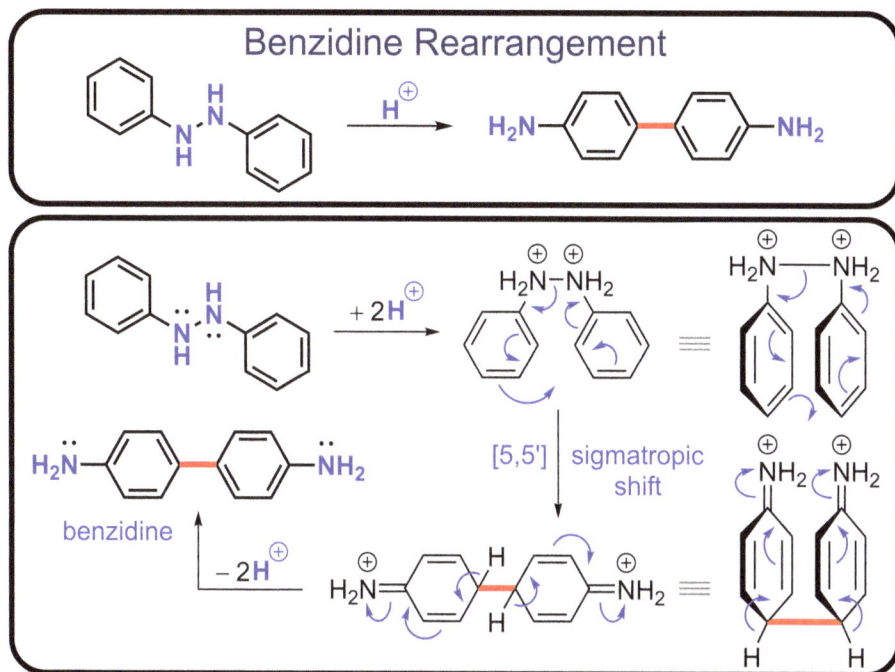

Fig. 38.2: The **benzidine** *rearrangement* mechanism.[138]

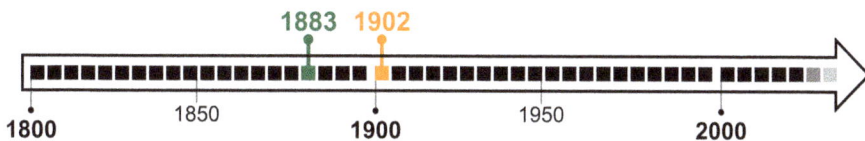

Fig. 38.3: The discovery of the **Fischer** *indole synthesis*.[139]

---

137 The **Fischer** *indole synthesis* (different from the **Fischer** *esterification*) is one of the most important reactions in organic chemistry. The key mechanistic step is the [3,3']-*sigmatropic shift (rearrangement)*.

138 The key mechanistic step in the **Fischer** *indole synthesis* is related to the **Cope** *rearrangement*, the *aza-***Cope**, and/or *aza-***Claisen** *rearrangement* (Chapter 28). Other related reactions include the **benzidine** *rearrangement* (its mechanism is not well-understood) [1, 38a].

139 The reaction was likely first described around 1883 [38b, 38c]. In **1902**, Emil Fischer received the Nobel Prize in Chemistry [38d].

39 Friedel–Crafts Acylation and Alkylation

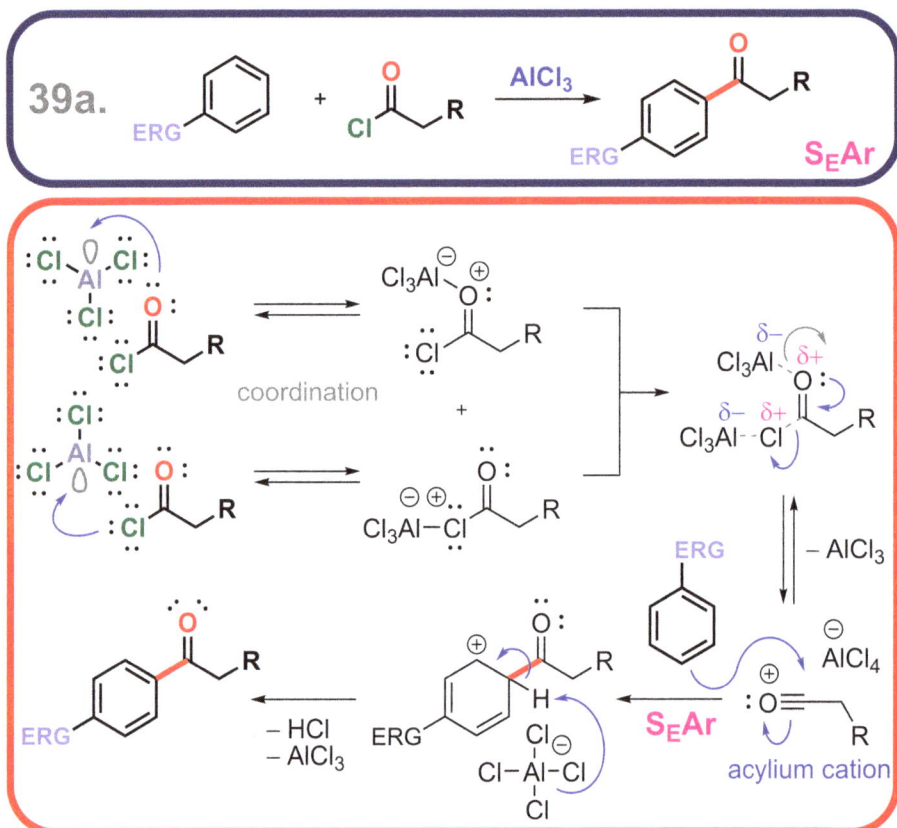

Fig. 39.1: The *Friedel–Crafts* acylation mechanism.[140]

Fig. 39.2: The discovery of the *Friedel–Crafts* acylation.[141]

---

140 The *Friedel–Crafts* acylation mechanism is an example of the **aromatic electrophilic substitution** (the *arenium ion* mechanism or S_EAr, covered in Chapter 3). The linear *acyl halides* react via *acylium cation* and form *aryl ketones* with *linear* alkyl chains.
141 The reaction was likely first described around 1877 [39a].

https://doi.org/10.1515/9783110786835-039

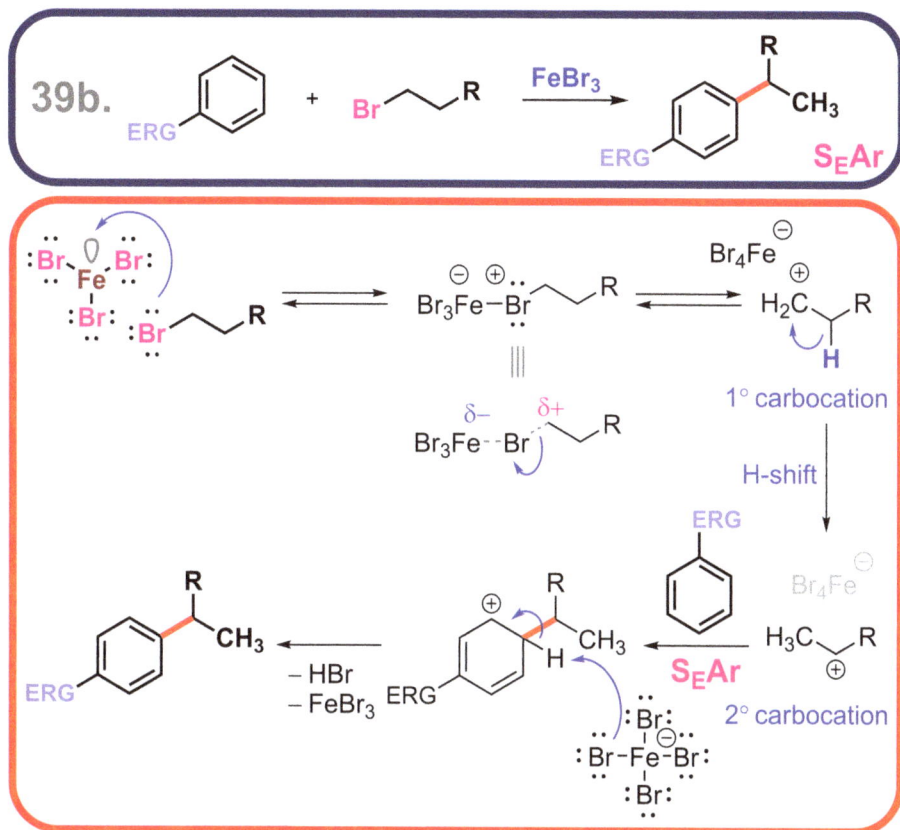

Fig. 39.3: The *Friedel–Crafts* alkylation mechanism.[142]

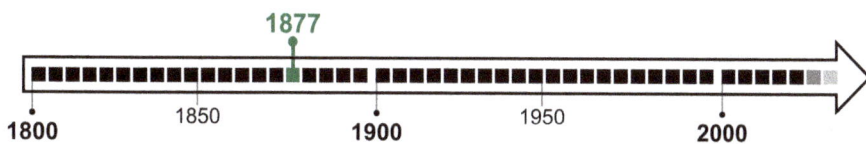

Fig. 39.4: The discovery of the *Friedel–Crafts* alkylation.[143]

142 The *Friedel–Crafts* alkylation is also the **aromatic electrophilic substitution**. The linear *alkyl halides* undergo the *carbocation rearrangement* (also called the **Wagner–Meerwein** *rearrangement* covered in Chapter 96) and always produce *branched* products.
143 The reaction was likely first described around 1877 [39b].

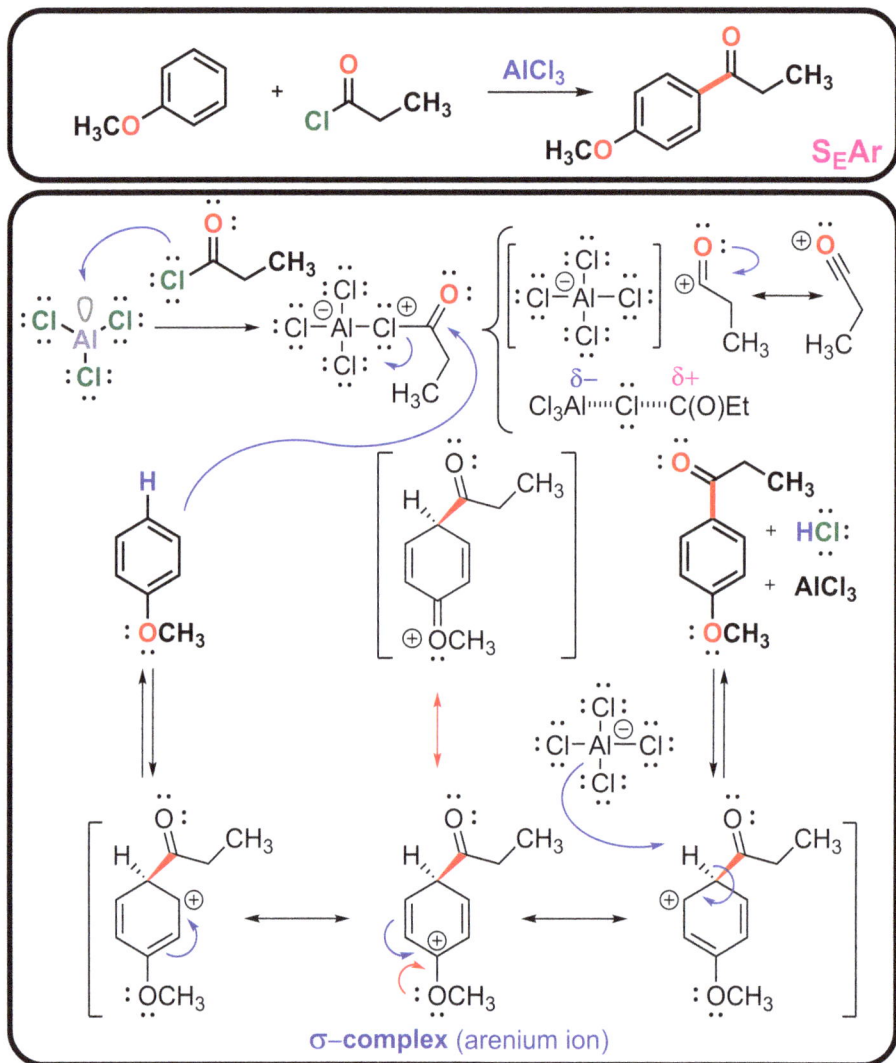

Fig. 39.5: The *Friedel–Crafts* acylation mechanism of anisole.[144]

144 An example of the *Friedel–Crafts* acylation of anisole yielding 1-(4-methoxyphenyl)propan-1-one.

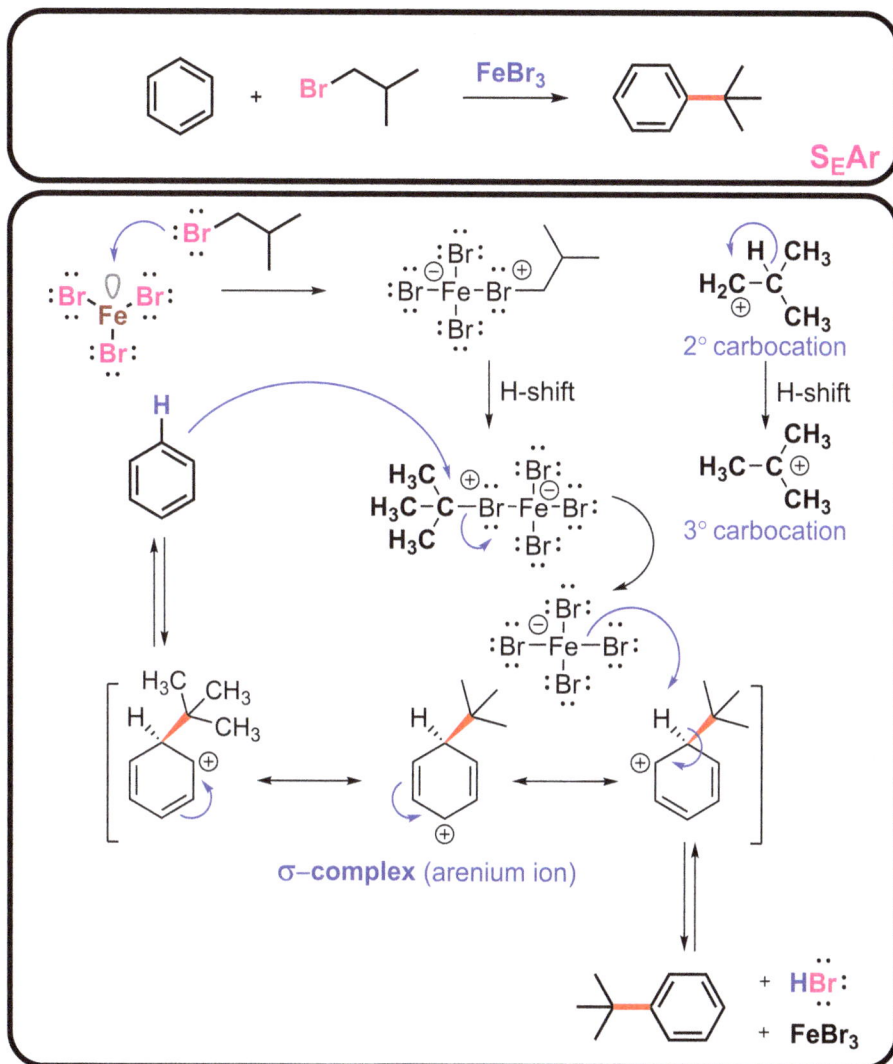

Fig. 39.6: The **Friedel–Crafts** *alkylation* mechanism of benzene.[145]

---

145 An example of the **Friedel–Crafts** *alkylation* of benzene yielding *tert*-butylbenzene.

40 Gabriel Synthesis

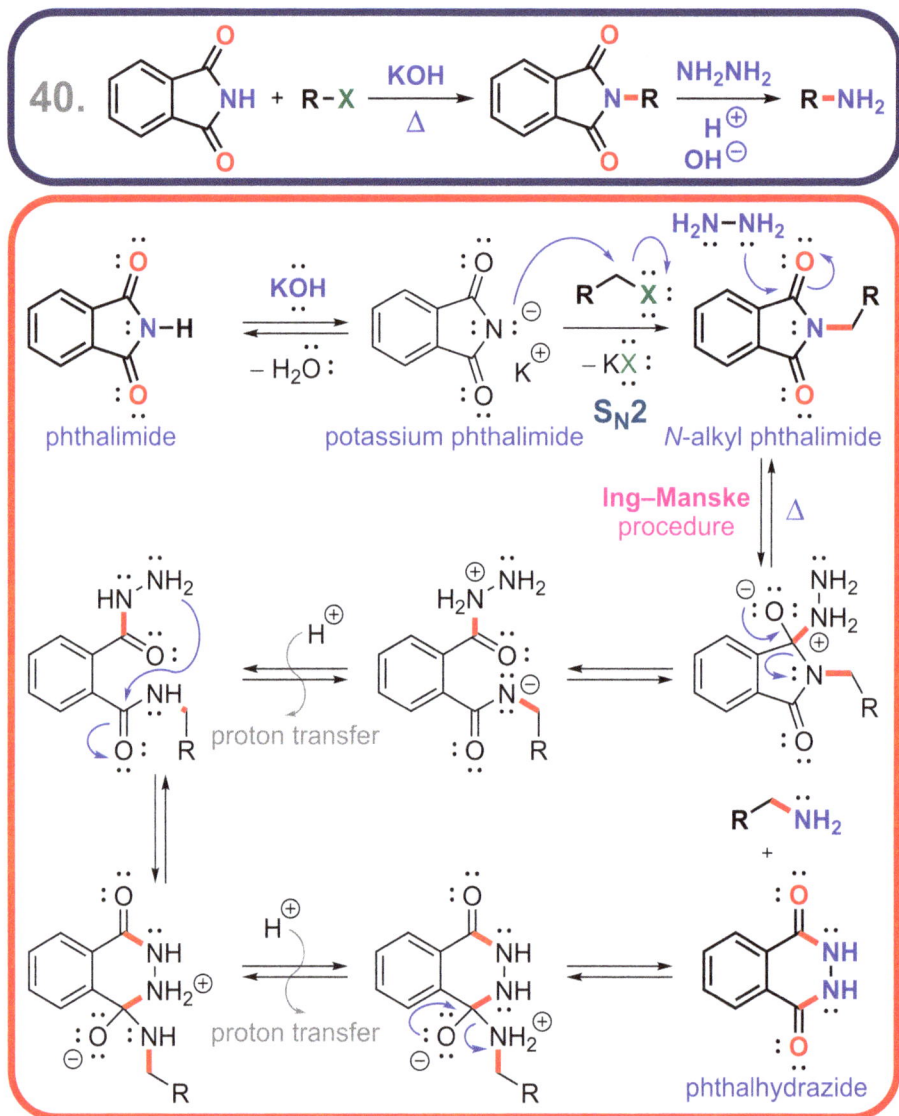

Fig. 40.1: The **Gabriel** *synthesis* mechanism.[146]

---

146 The **Gabriel** *synthesis* is a chemical reaction that converts *alkyl halides* to *primary (1°) amines* via the S_N2 reaction using *phthalimide*. The **Ing–Manske** *procedure* [40a] is a chemical reaction that converts *N-alkyl phthalimide* to *primary (1°) amine* using *hydrazine*.

https://doi.org/10.1515/9783110786835-040

Delépine Reaction

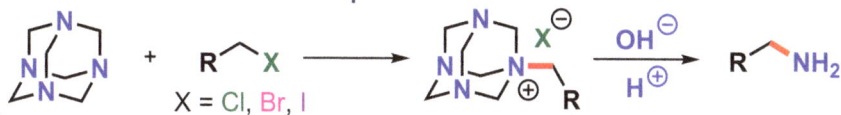

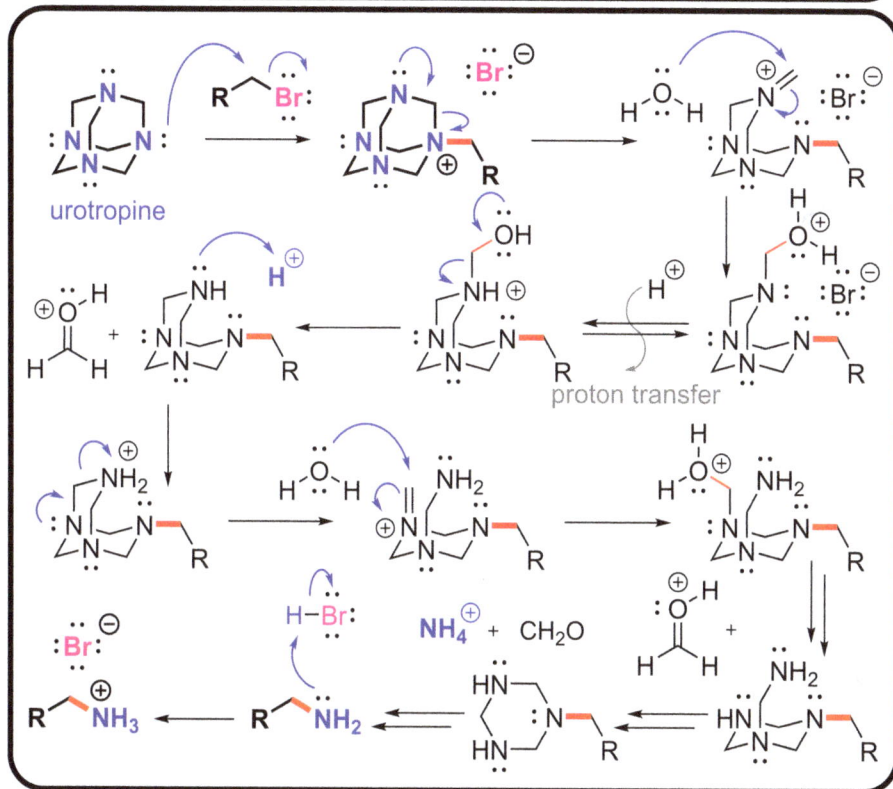

Fig. 40.2: The **Delépine** *reaction* mechanism.[147]

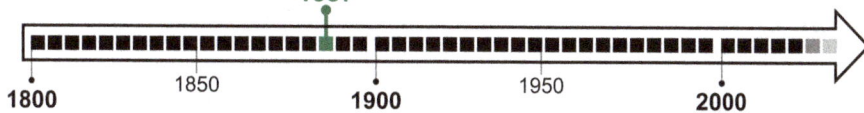

Fig. 40.3: The discovery of the **Gabriel** *synthesis*.[148]

---

147 The **Delépine** *reaction* mechanism (*urotropine* acts as the nitrogen nucleophile). There are other synthetic transformations yielding *primary amines*: the **Mitsunobu** *reaction* (Chapter 61) or other S_N2 reactions using various **N** (nitrogen) nucleophiles [40b].

148 The reaction was likely first described around 1887 [40c].

41 Gewald Reaction

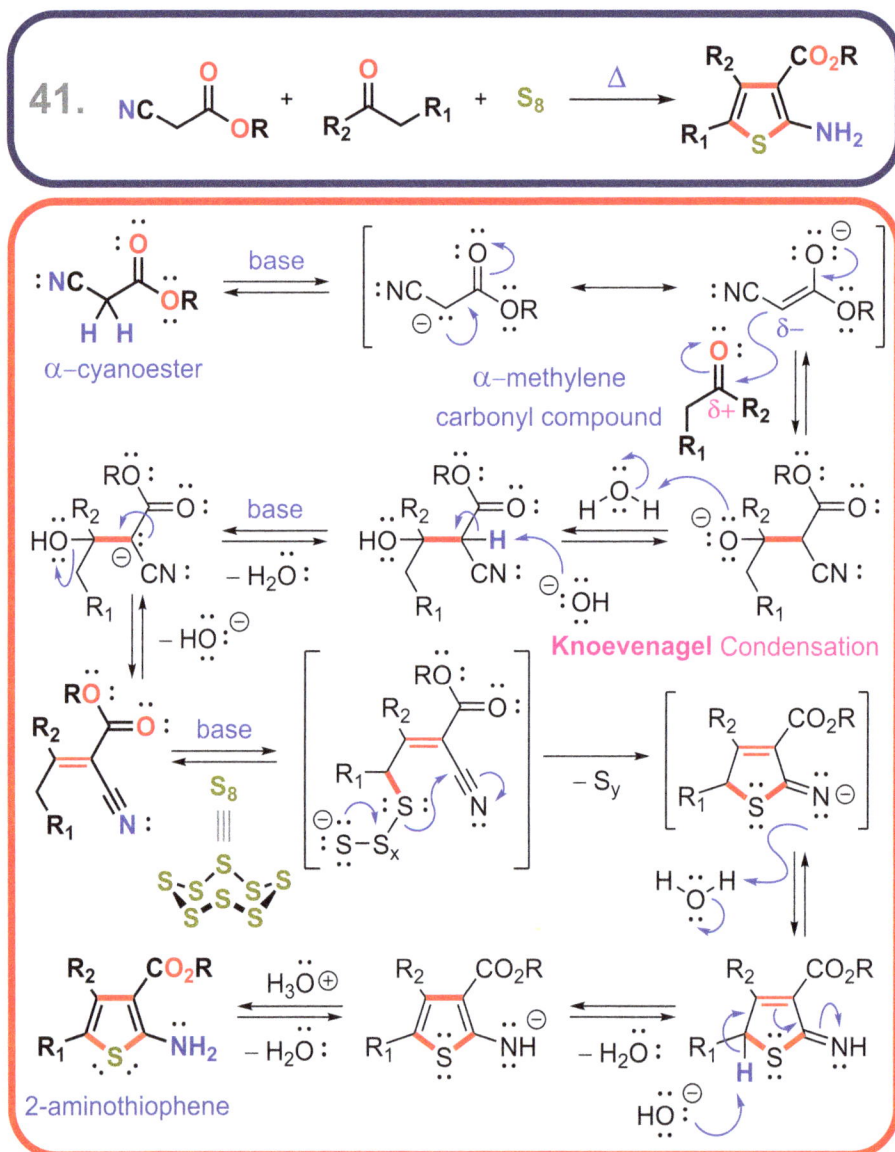

Fig. 41.1: The *Gewald* reaction mechanism.[149]

---

149 The *Gewald reaction*, also called the *Gewald condensation*, is a three-component reaction (3-CR) producing *2-aminothiophenes*. The key condensation step is the *Knoevenagel* condensation [41a].

https://doi.org/10.1515/9783110786835-041

Knoevenagel Condensation

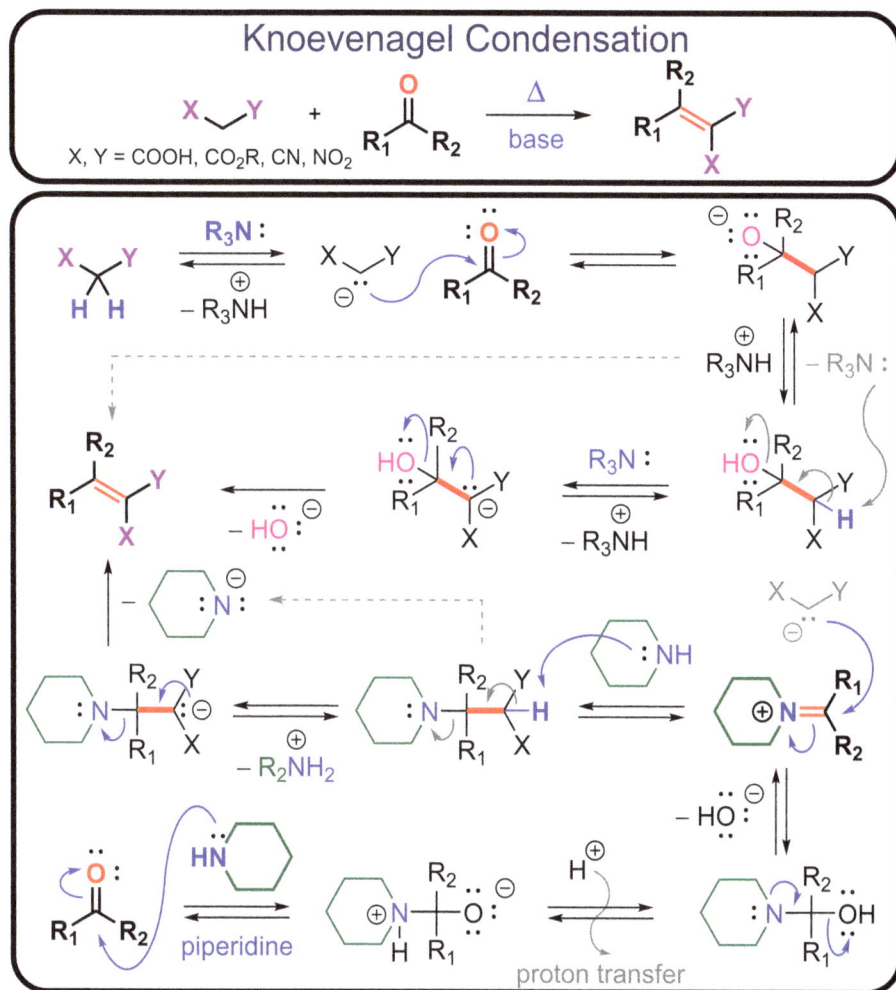

Fig. 41.2: The **Knoevenagel** condensation mechanism.[150]

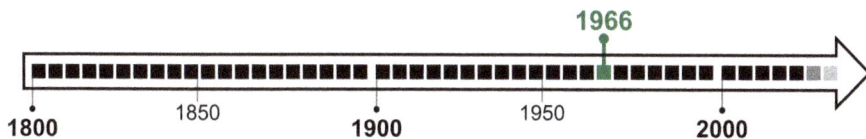

Fig. 41.3: The discovery of the **Gewald reaction**.[151]

---

150 The **Knoevenagel** condensation is a variation of the aldol condensation followed by crotonation (covered in Chapter 83). The reaction is often catalyzed by piperidine.
151 The reaction was likely first described around 1966 [41b].

42 Glaser–Eglinton–Hay Coupling

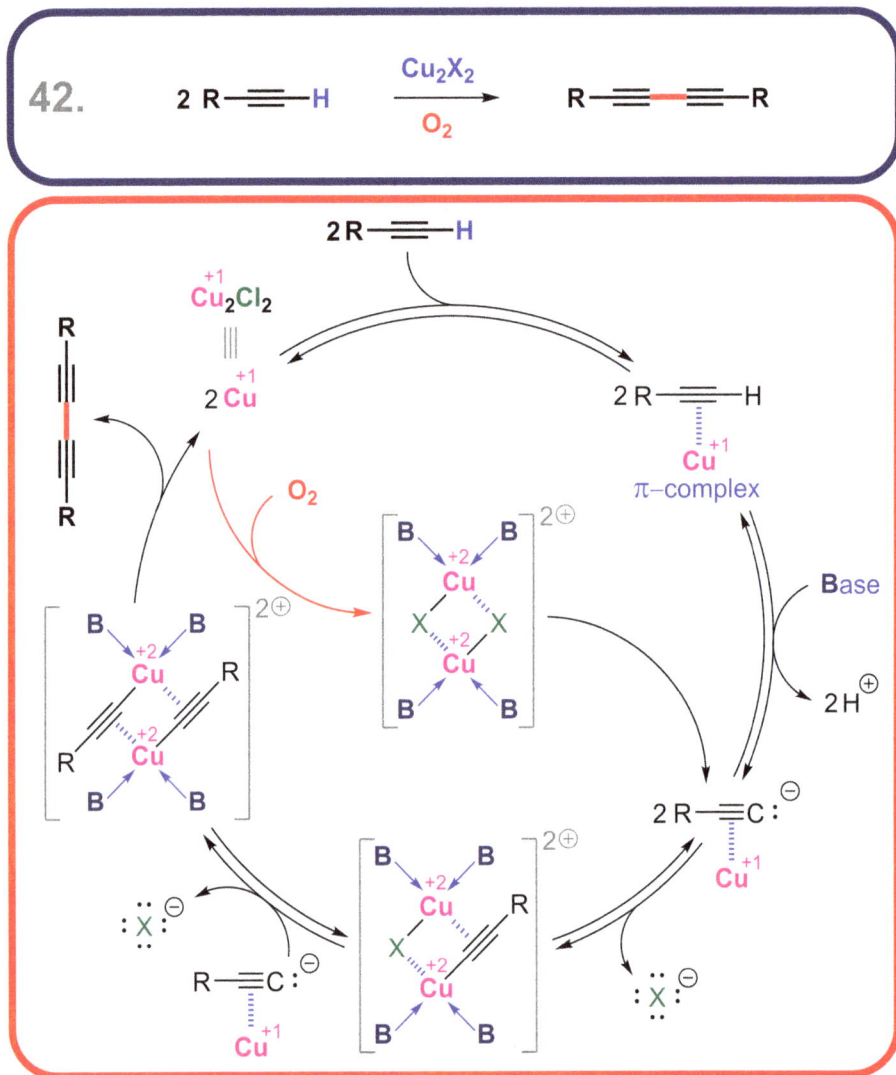

$$2\ R\!\!-\!\!\!\equiv\!\!\!-\!\!H \xrightarrow[\ O_2\ ]{\ Cu_2X_2\ } R\!\!-\!\!\!\equiv\!\!\!-\!\!\!-\!\!\!\equiv\!\!\!-\!\!R$$

Fig. 42.1: The *Glaser–Eglinton–Hay* coupling mechanism.[152]

---

152 The *Glaser–Eglinton–Hay* coupling is a general name for three named reactions: the *Glaser* coupling, the *Eglinton* coupling, and the *Hay* coupling. It is one of many examples of *Cu*-mediated dimerization of *terminal alkynes*. In all three cases, the formed products are symmetrical.

https://doi.org/10.1515/9783110786835-042

Eglinton Reaction (Coupling)

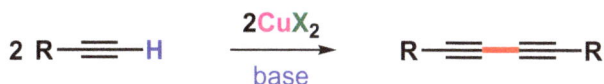

$$2\ R\text{---}\equiv\text{---}H \quad \xrightarrow[\text{base}]{\textbf{2CuX}_2} \quad R\text{---}\equiv\text{---}\equiv\text{---}R$$

Hay Coupling

$$2\ R\text{---}\equiv\text{---}H \quad \xrightarrow[\textbf{O}_2]{\substack{\textbf{CuX}\\ \text{TMEDA}}} \quad R\text{---}\equiv\text{---}\equiv\text{---}R$$

Cadiot–Chodkiewicz Cross Coupling

$$\begin{array}{c} R_1\text{---}\equiv\text{---}H \\ + \\ R_2\text{---}\equiv\text{---}Br \end{array} \quad \xrightarrow[\text{base}]{\textbf{CuX}} \quad R_1\text{---}\equiv\text{---}\equiv\text{---}R_2$$

Fig. 42.2: Reactions related to the **Glaser–Eglinton–Hay** coupling.[153]

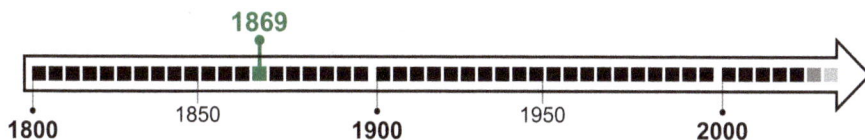

Fig. 42.3: The discovery of the **Glaser–Eglinton–Hay** coupling.[154]

---

153 More specifically, in the **Eglinton** coupling, the product is (a) symmetrical, (b) **Cu** is used as a stoichiometric reagent [42a, 42b]; in the **Glaser** coupling, the product is (a) symmetrical, (b) **CuX** is used as a catalyst with NH_3 or NH_4OH [42c]; in the **Hay** coupling, the product is (a) symmetrical, (b) **CuX·TMDA** complex is used as a catalyst [42d, 42e]; in the **Cadiot–Chodkiewicz** coupling, the product is (a) **asymmetrical**, (b) **Cu** is used as a catalyst [42f], and other examples [1, 4].
154 The reaction was likely first described around 1869 [42c].

43 Grignard Reaction

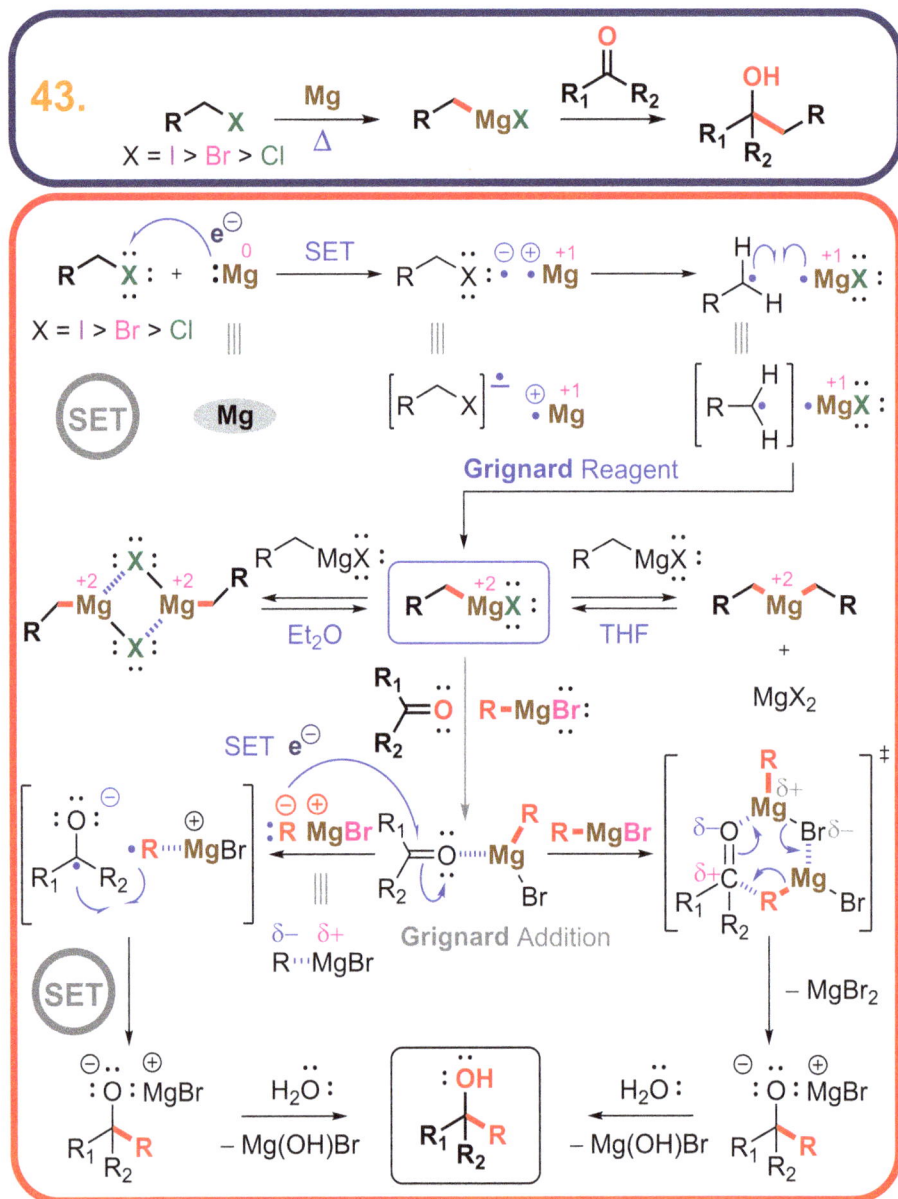

Fig. 43.1: The *Grignard reaction* mechanism.[155]

https://doi.org/10.1515/9783110786835-043

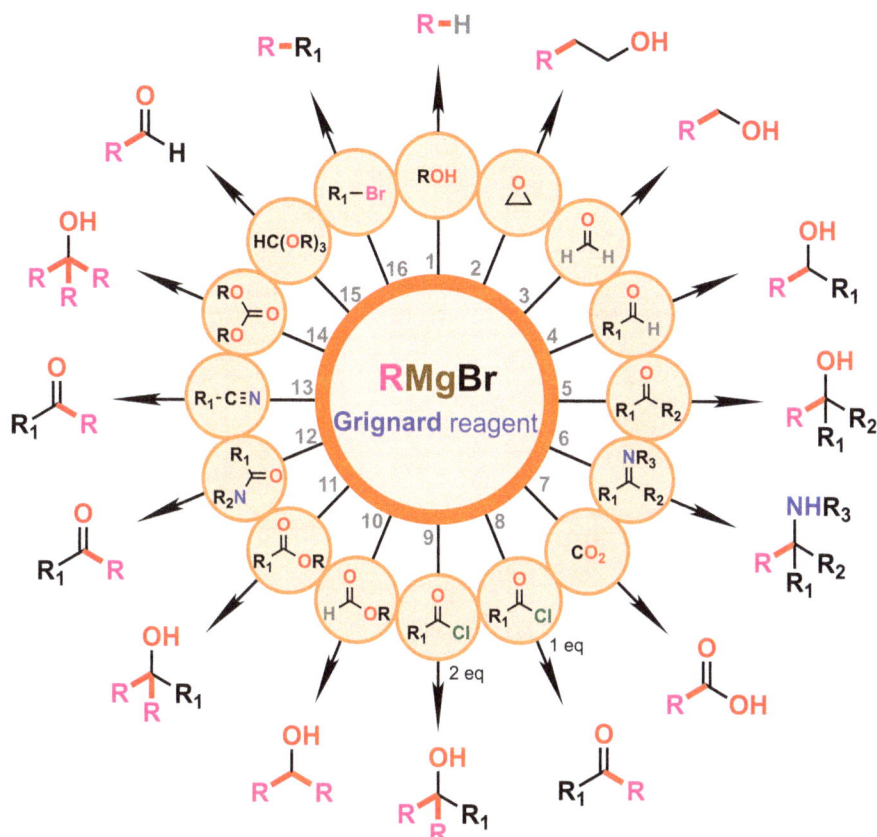

Fig. 43.2: Synthetic versatility of the **Grignard** reagent.[156]

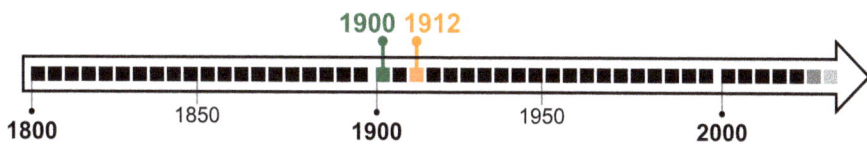

Fig. 43.3: The discovery of the **Grignard** reaction.[157]

---

155 The **Grignard** reaction is based on the use of a named reagent: the **Grignard** reagent (**RMgX**). The mechanism is not well-understood and most likely involves a **single electron transfer** (**SET**) (Chapter 5).

156 The **Grignard** reagent has wide synthetic applications. It can react with a variety of electrophiles (electrophilic centers): 1. alcohols, deuterated water; 2. epoxides; 3. formaldehyde; 4. aldehydes; 5. ketones; 6. imines; 7. carbon dioxide (disulfide); 8. acyl chlorides (1 eq); 9. acyl chlorides (excess); 10. formates; 11. esters; 12. amides; 13. nitriles; 14. carbonates; 15. orthoesters; 16. alkyl halides; and others [1].

157 The reaction was likely first described around 1900 [43a]. In **1912**, Victor Grignard (jointly with Paul Sabatier) received the Nobel Prize in Chemistry for the discovery of the **Grignard** reagent (and other achievements in chemistry) [43b].

44 Grob Fragmentation

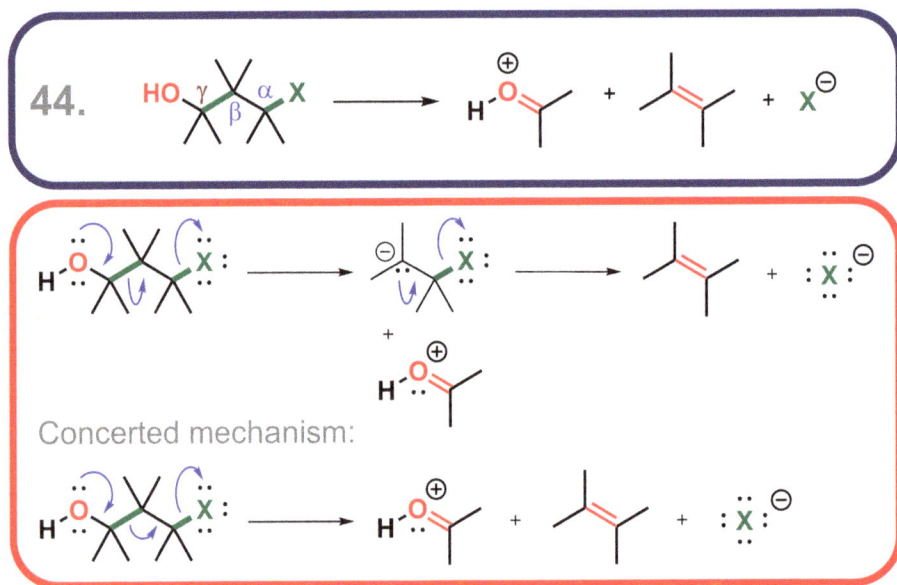

Fig. 44.1: The **Grob** *fragmentation* mechanism.[158]

Concerted mechanism:

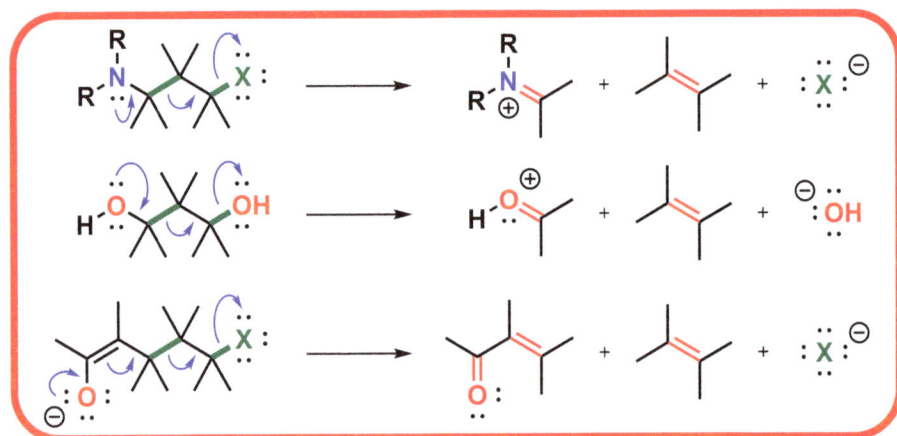

Fig. 44.2: Variations of the **Grob** *fragmentation*.[159]

---

158 The **Grob** *fragmentation* mechanism is most likely related to the **β-elimination** mechanisms (in this case, 1,4-elimination) covered in Chapter 6. The common feature of this fragmentation is the formation of three species: positively charged *(electrofuge)*, neutral unsaturated fragment, and negatively charged *(nucleofuge)*. A stepwise or concerted mechanism can take place.

https://doi.org/10.1515/9783110786835-044

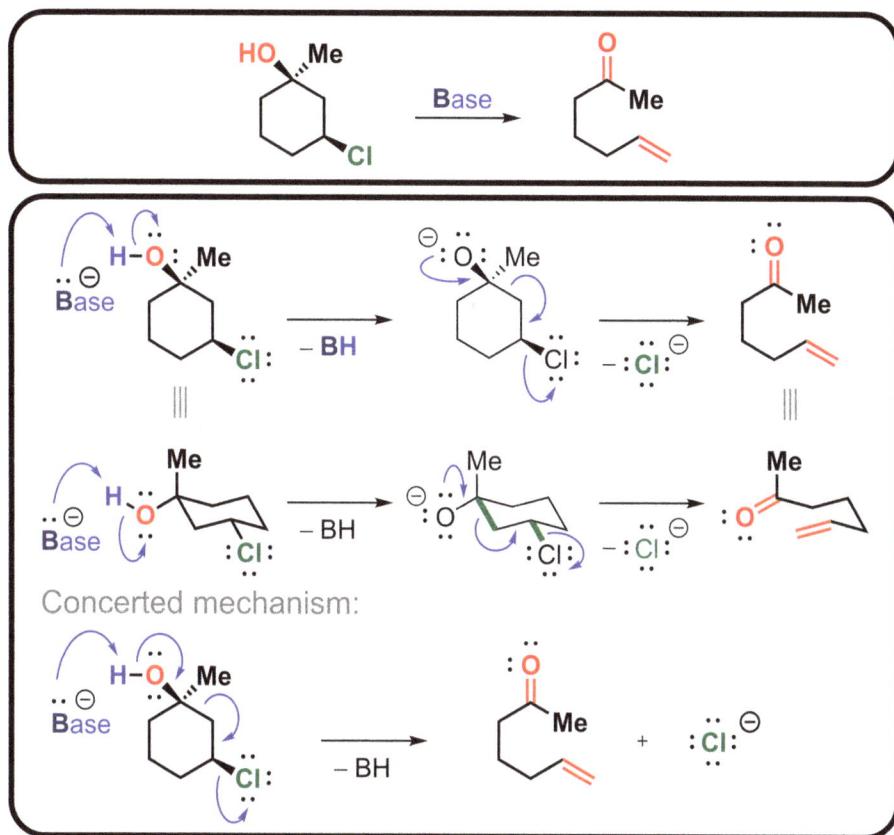

Fig. 44.3: The **Grob** *fragmentation* mechanism of (1R,3S)-3-chloro-1-methylcyclohexan-1-ol.[160]

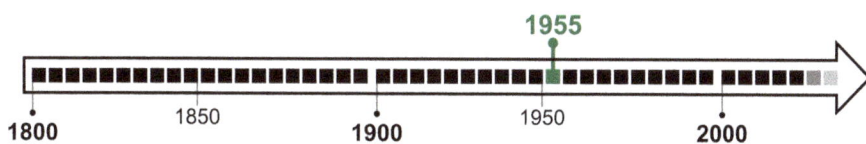

Fig. 44.4: The discovery of the **Grob** *fragmentation*.[161]

---

159 There are many variations of the **Grob** *fragmentation* involving γ-hydroxy halides (shown here), γ-amino halides, 1,3-diols, and others [44a].

160 An example of the **Grob** *fragmentation* of (1R,3S)-3-chloro-1-methylcyclohexan-1-ol yielding hept-6-en-2-one.

161 The reaction was likely first described around 1955 [44b, 44c].

45 Haloform Reaction

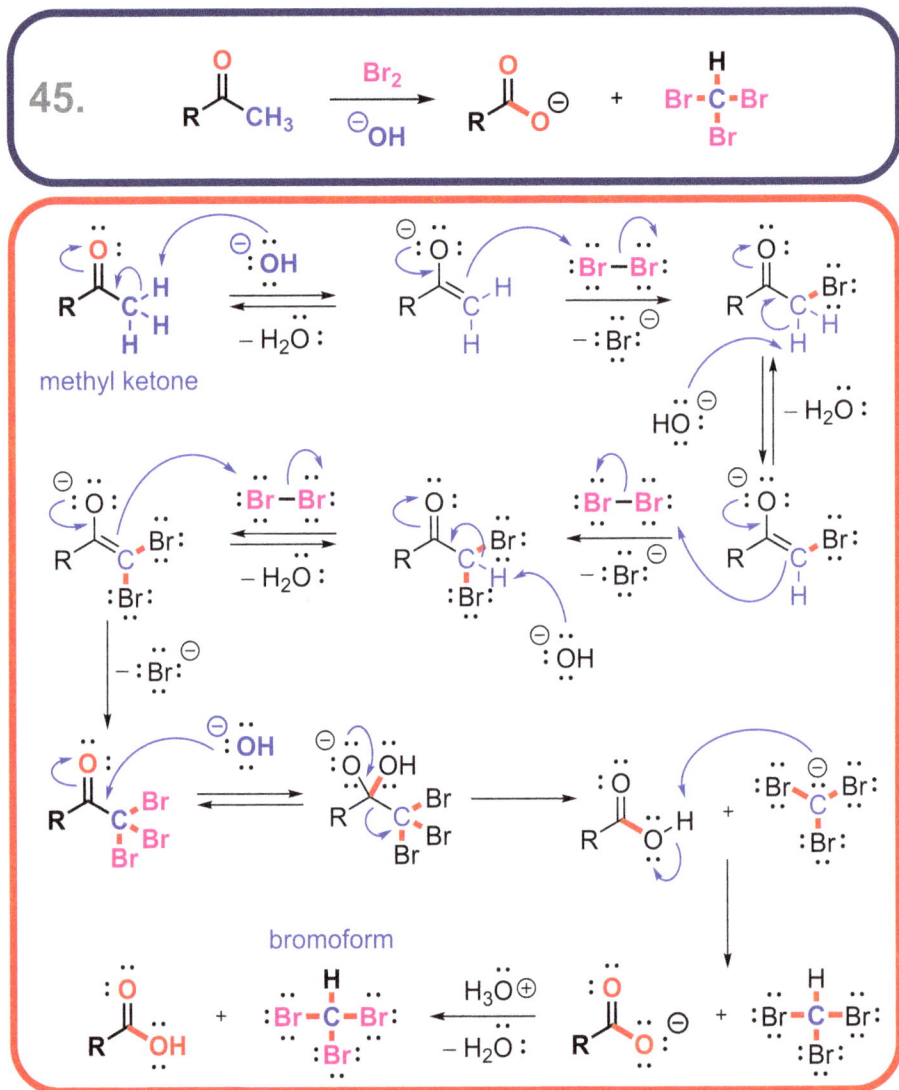

Fig. 45.1: The *haloform reaction* mechanism.[162]

---

162 The *haloform reaction* is one of the oldest reactions in organic chemistry. It is an example of **aliphatic electrophilic substitution**, which is not covered in this book (Chapter 3).

https://doi.org/10.1515/9783110786835-045

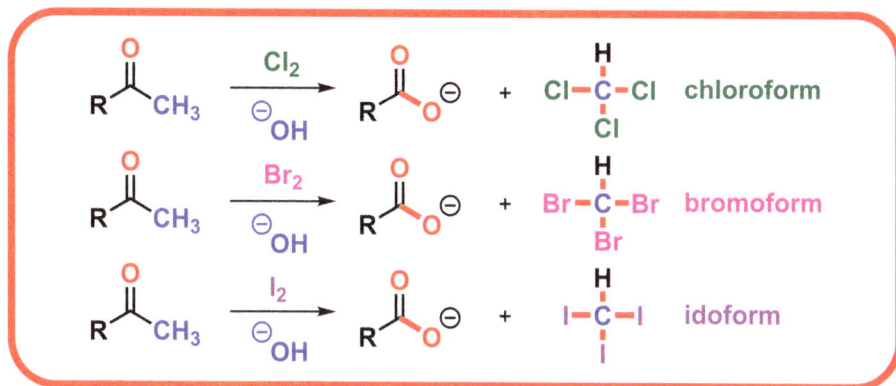

Fig. 45.2: Variations of the *haloform reaction*.[163]

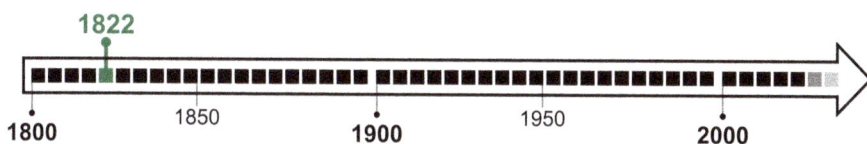

Fig. 45.3: The discovery of the *haloform reaction*.[164]

---

163 The *haloform reaction* can be carried out with most halogens: (Cl) the *chloroform reaction*, (Br) the *bromoform reaction*, and (I) the *iodoform reaction*, also known as the *iodoform test* or the **Lieben** *iodoform test* (it is used as an indication of the presence of methyl ketones) [45].
164 The reaction was likely first described between 1822 and 1870 [45].

46 Heck Cross-Coupling

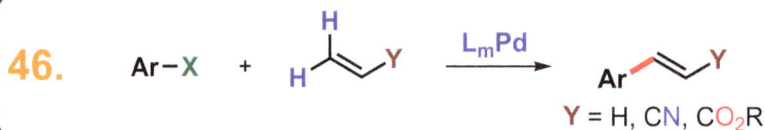

$$\textbf{46.} \quad Ar-X \; + \; \underset{H}{\overset{H}{\rangle}}=\langle Y \; \xrightarrow{\; L_mPd \;} \; Ar \diagdown = \diagup Y$$

$$Y = H, \; CN, \; CO_2R$$

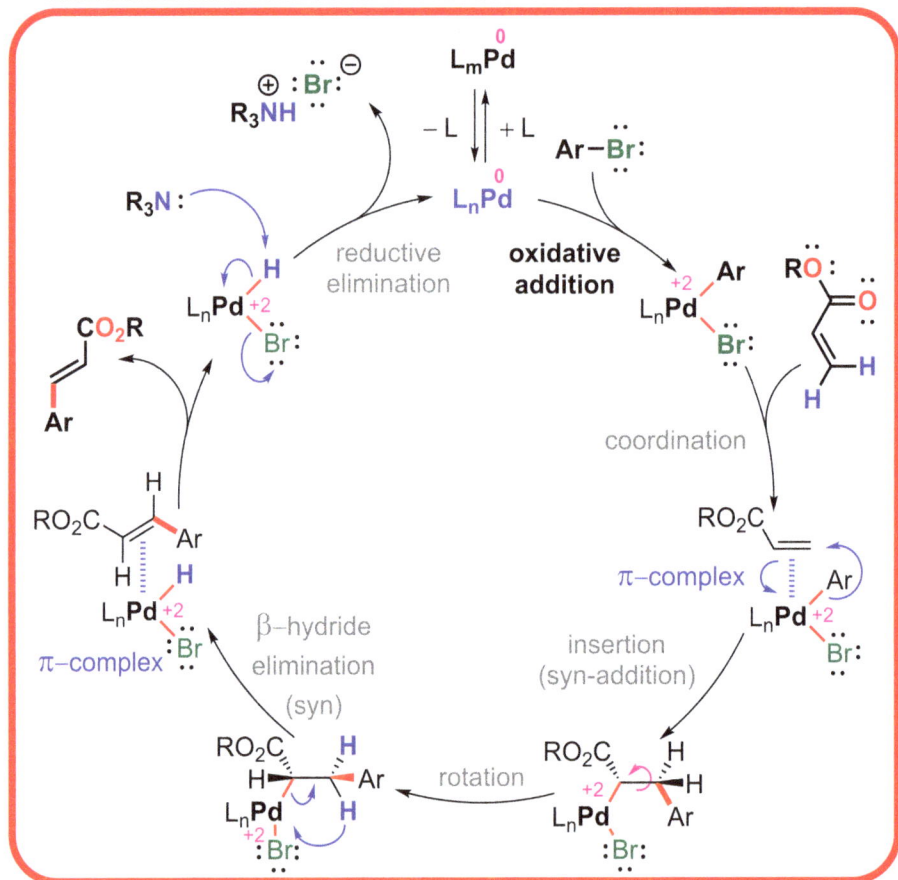

Fig. 46.1: The **Heck** *cross-coupling* mechanism.[165]

---

165 The **Heck** *cross-coupling* or the **Heck** *reaction* is also called the **Mizoroki–Heck** *reaction*. It is one of the most important types of **Pd**-*catalyzed cross-coupling* reactions (C–C bond formation using *aryl halides* and *alkenes*). For teaching purposes, a simplified and general mechanism is shown.

https://doi.org/10.1515/9783110786835-046

Oxidative Addition:

1. Not Hindered Monodentate Ligand:

18e⁻ tetrahedral → 14e⁻ linear

2. Hindered Monodentate Ligand:

12e⁻

3. Hindered Chelating Ligand:

$$L_mPd$$

$$L_nPd$$

$$L_nPd$$

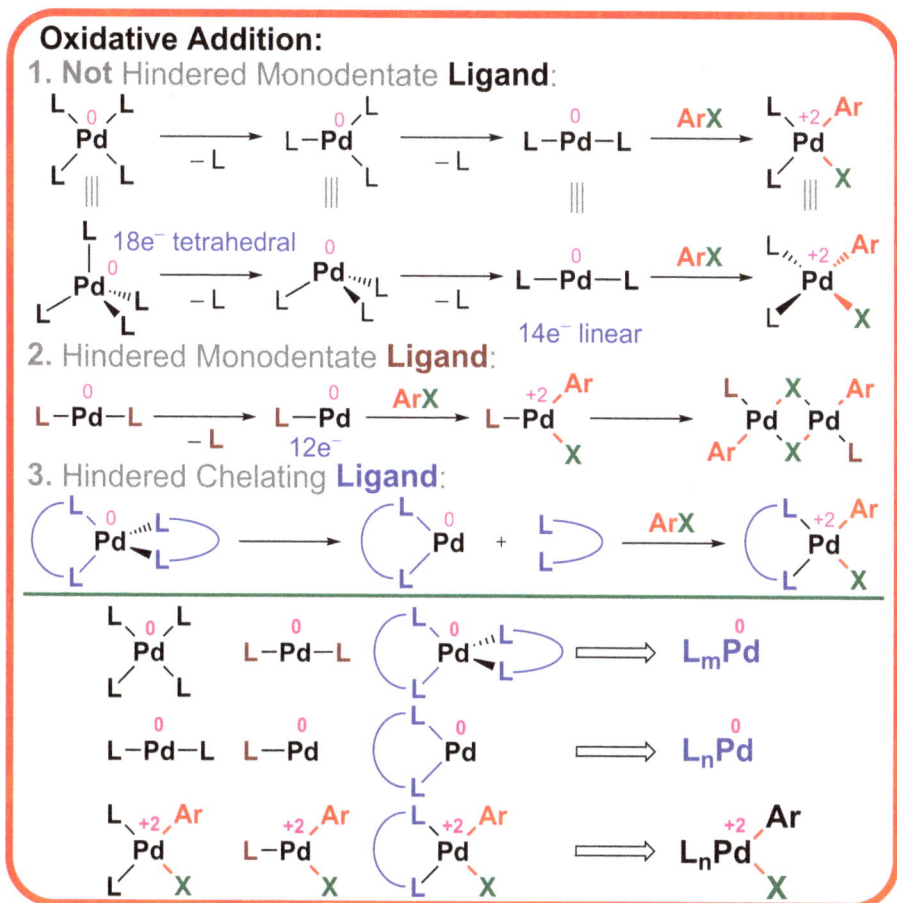

Fig. 46.2: General illustration of the *oxidative addition* step.[166]

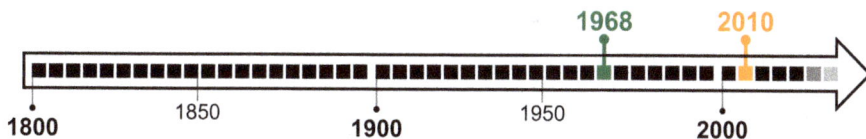

Fig. 46.3: The discovery of the **Heck** cross-coupling.[167]

---

166 The *oxidative addition* step can be represented in several ways, including a catalyst with: 1. a **not** *(less) hindered monodentate ligand*; 2. a *large hindered monodentate ligand*; and 3. a *hindered chelating (bidentate) ligand*. For simplicity, L_mPd or L_nPd representation will be used henceforth [2a].

167 The reaction was likely first described around 1968 [46a, 46b]. In **2010**, Richard F. Heck (jointly with Ei-ichi Negishi and Akira Suzuki) received the Nobel Prize in Chemistry for the development of **Pd**-catalyzed cross-coupling reactions [46c].

47 Hell–Volhard–Zelinsky Reaction

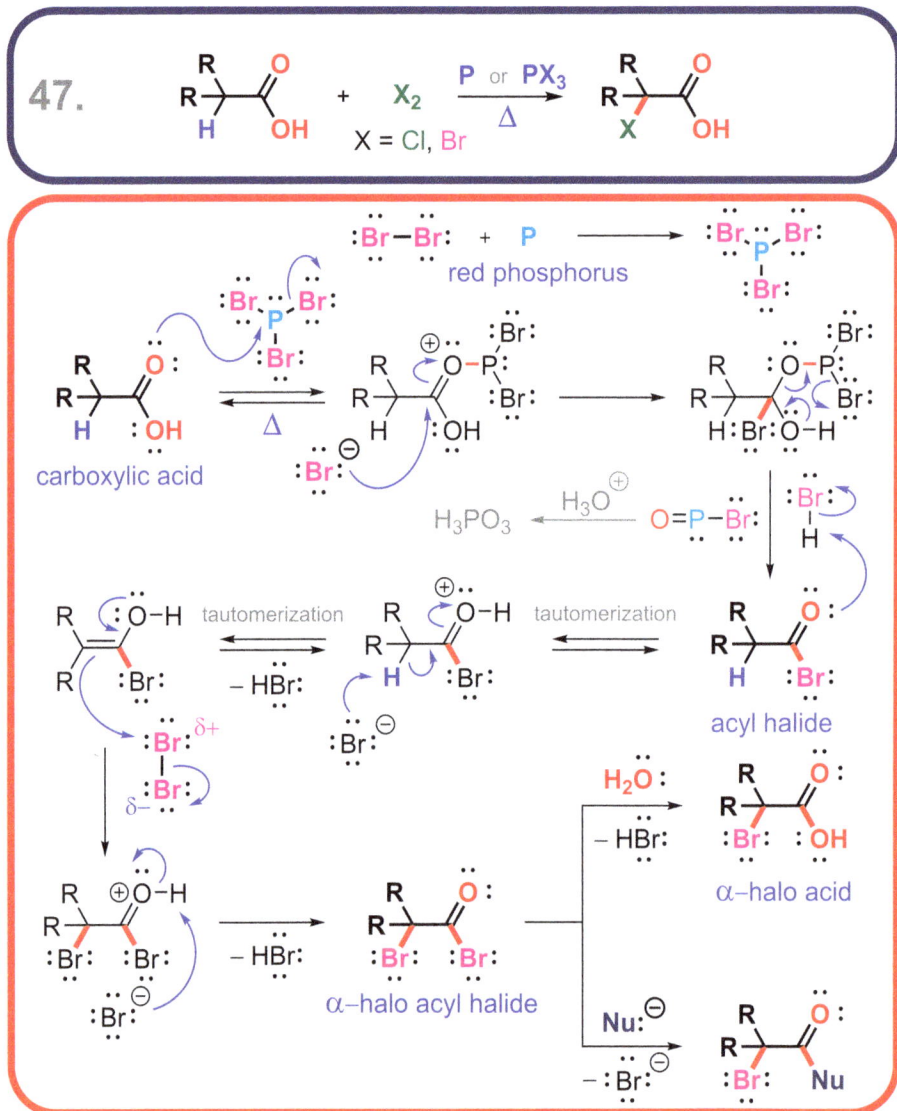

Fig. 47.1: The *Hell–Volhard–Zelinsky reaction* mechanism.[168]

---

168 The *Hell–Volhard–Zelinsky reaction* is also known as the *Hell–Volhard–Zelinsky (HVZ) halogenation*. It is a type of *aliphatic electrophilic substitution* (briefly mentioned in Chapter 3). Mechanistically, it is also related to the *haloform reaction* (see Chapter 45).

https://doi.org/10.1515/9783110786835-047

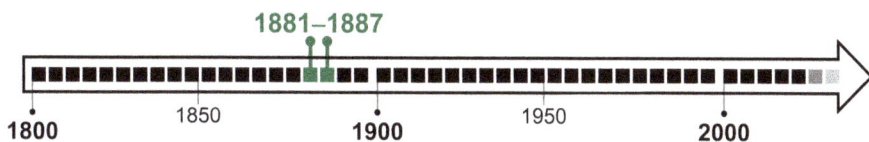

Fig. 47.2: The discovery of the **Hell–Volhard–Zelinsky** reaction.[169]

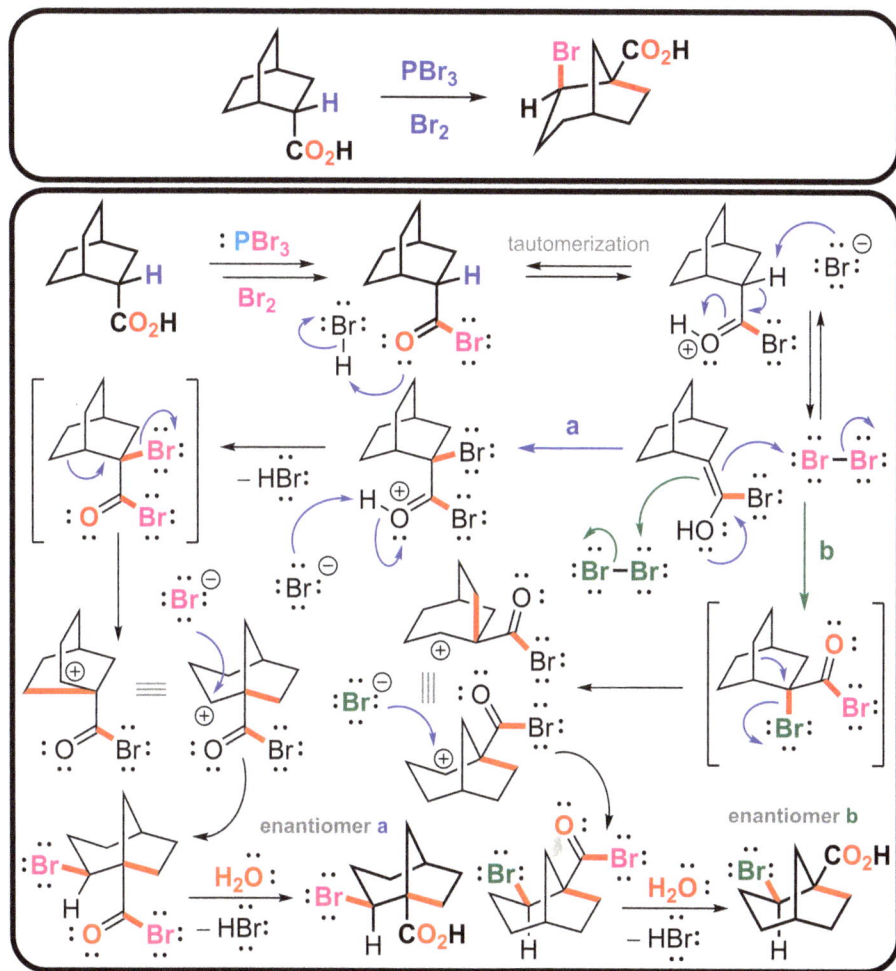

Fig. 47.3: The rearrangement mechanism of bicyclo[2.2.2]octane system.[170]

---

169 The reaction was likely first described around 1881 by Hell [47a] and around 1887 by both Volhard and Zelinsky [47b, 47c].

170 The rearrangement mechanism of bicyclo[2.2.2]octane system to bicyclo[3.2.1]octane system [47d].

48 Hiyama Cross-Coupling

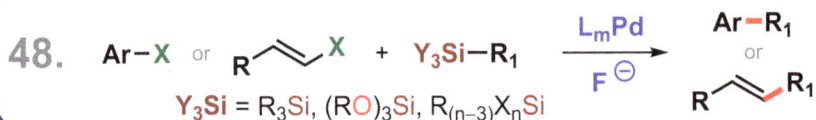

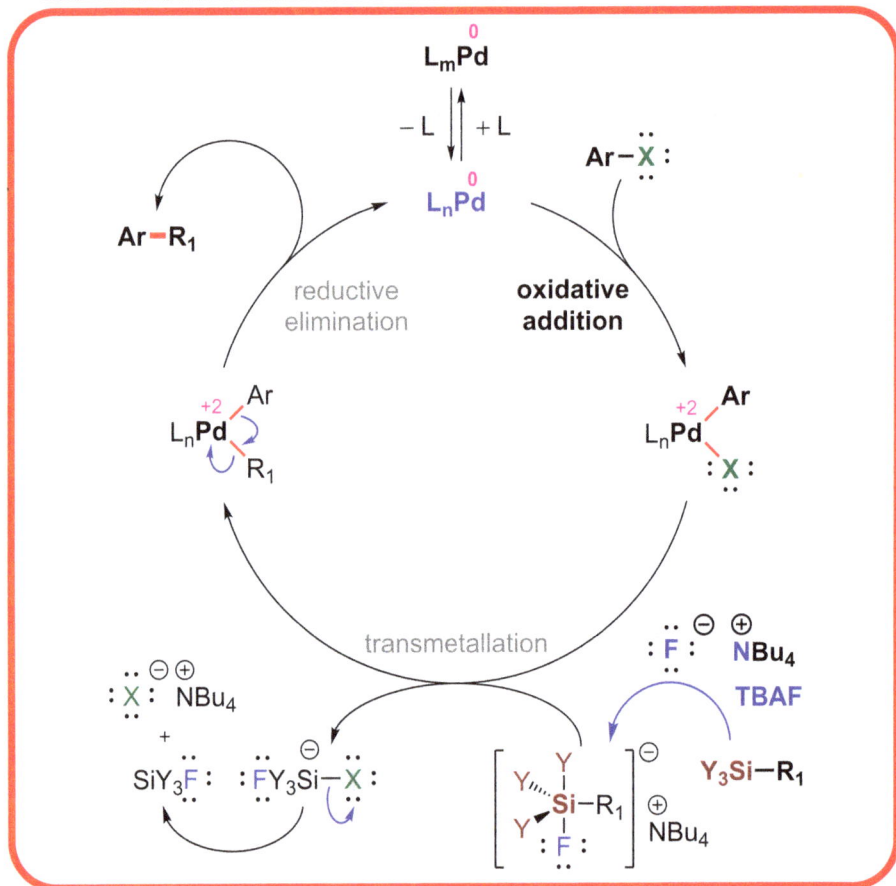

$$\textbf{48.} \quad \text{Ar}-\text{X} \;\text{or}\; \underset{R}{\diagup}\diagdown\text{X} \;+\; Y_3\text{Si}-R_1 \xrightarrow[\text{F}^{\ominus}]{L_m\text{Pd}} \begin{array}{c}\text{Ar}-R_1 \\ \text{or} \\ \underset{R}{\diagup}\diagdown R_1\end{array}$$

$$Y_3\text{Si} = R_3\text{Si},\ (\text{R}O)_3\text{Si},\ R_{(n-3)}X_n\text{Si}$$

Fig. 48.1: The *Hiyama cross-coupling* mechanism.[171]

---

171 The *Hiyama cross-coupling* is a type of *Pd-catalyzed cross-coupling* reaction (C–C bond formation using *aryl halides* and *organosilanes*). For teaching purposes, a simplified and general mechanism is shown.

https://doi.org/10.1515/9783110786835-048

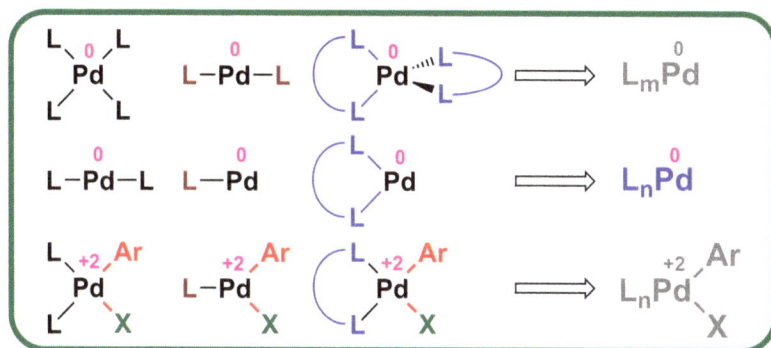

Fig. 48.2: The *oxidative addition* step representation.[172]

Hiyama–Denmark Cross Coupling

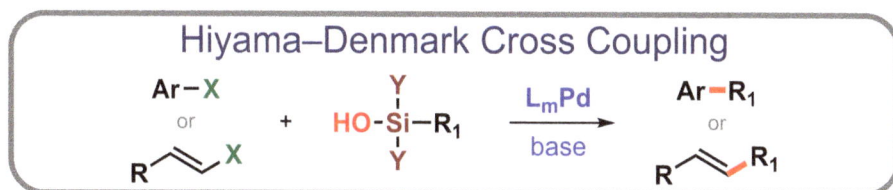

Fig. 48.3: Variations of the **Hiyama** cross-coupling.[173]

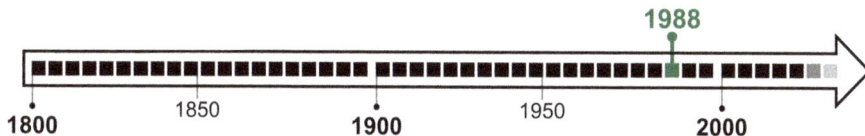

Fig. 48.4: The discovery of the **Hiyama** cross-coupling.[174]

---

172 As it was explained in Chapter 46, the representation of the *oxidative addition* step can vary. For simplicity and consistency, a general depiction of a *catalyst–ligand* complex is used: L_mPd or L_nPd [2a].
173 A modification of the **Hiyama** cross-coupling is called the **Hiyama–Denmark** cross-coupling reaction [48a]. It is also a type of **Pd**-catalyzed cross-coupling reaction (C–C bond formation using *aryl halides* and *organosilanols*).
174 The reaction was likely first described around 1988 [48b].

49 Hofmann Elimination

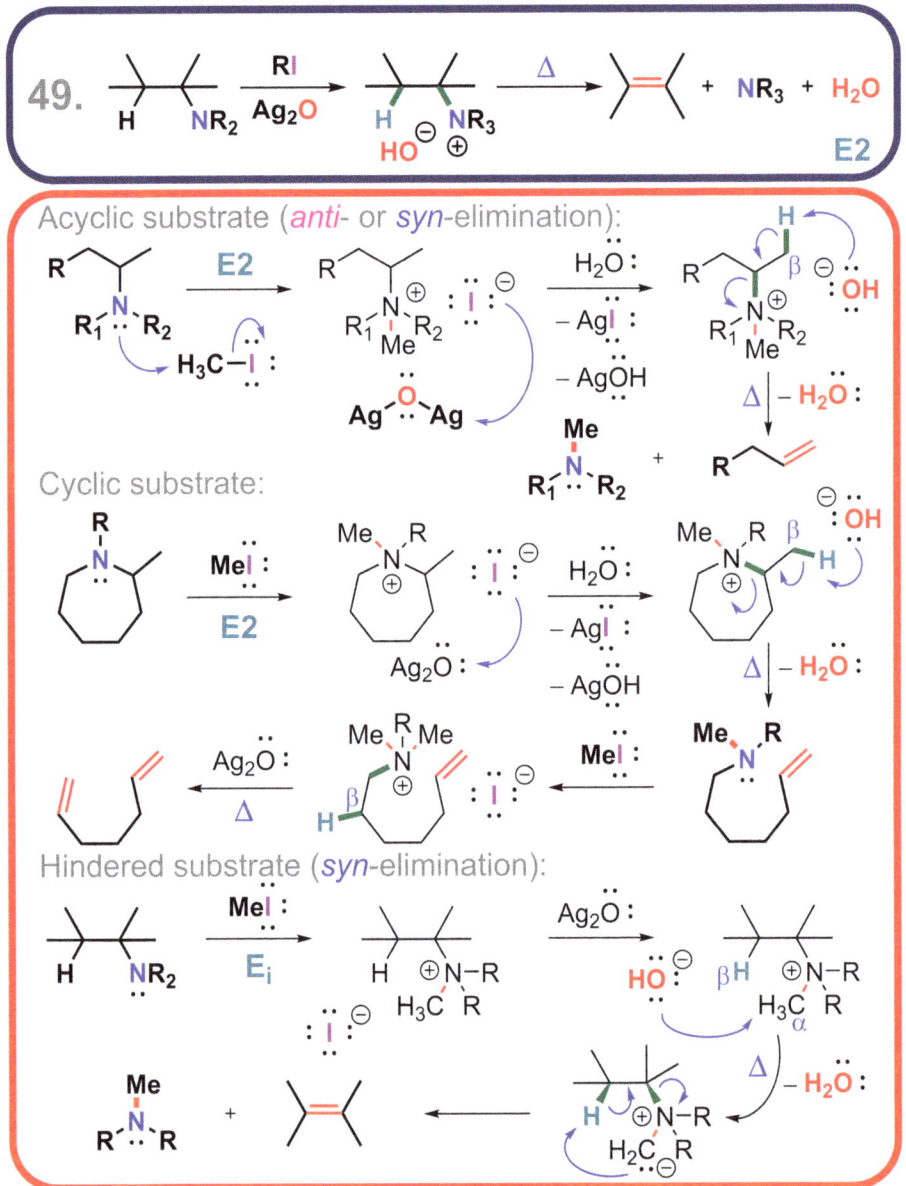

Fig. 49.1: The *Hofmann elimination* mechanism.[175]

https://doi.org/10.1515/9783110786835-049

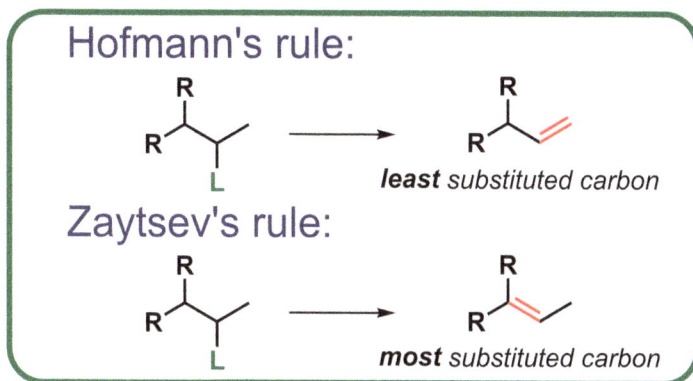

Fig. 49.2: *Hofmann's* rule and *Zaytsev's* rule.[176]

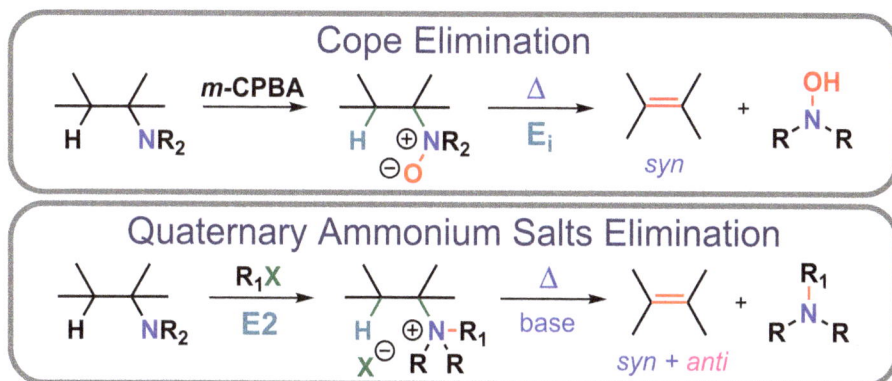

Fig. 49.3: Reactions related to the *Hofmann* elimination.[177]

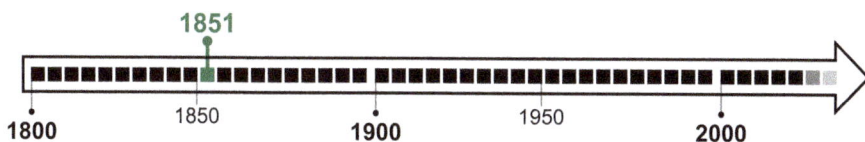

Fig. 49.4: The discovery of the *Hofmann* elimination.[178]

---

175 The *Hofmann elimination* is also known as the *Hofmann degradation*. This should not be confused with the *Hofmann rearrangement* (Chapter 31). It is an example of the β-**elimination** (Chapter 6).

176 The products of the *Hofmann elimination* obey *Hofmann's rule*: the double bond is at the *least substituted carbon*. If the double bond is at the *most substituted carbon*, then it conforms with *Zaytsev's rule* (also spelled Saytzeff, and in Russian Александр Михайлович Зайцев or А. М. Зайцев) [49a].

177 Several reactions are related to the *Hofmann elimination*: the *Cope elimination* (E_i mechanism, Chapter 27), the fragmentation of quaternary ammonium salts (**E2** mechanism), and others [1, 49b].

178 The reaction was likely first described around 1851 [49c, 49d].

50 Horner–Wadsworth–Emmons Olefination

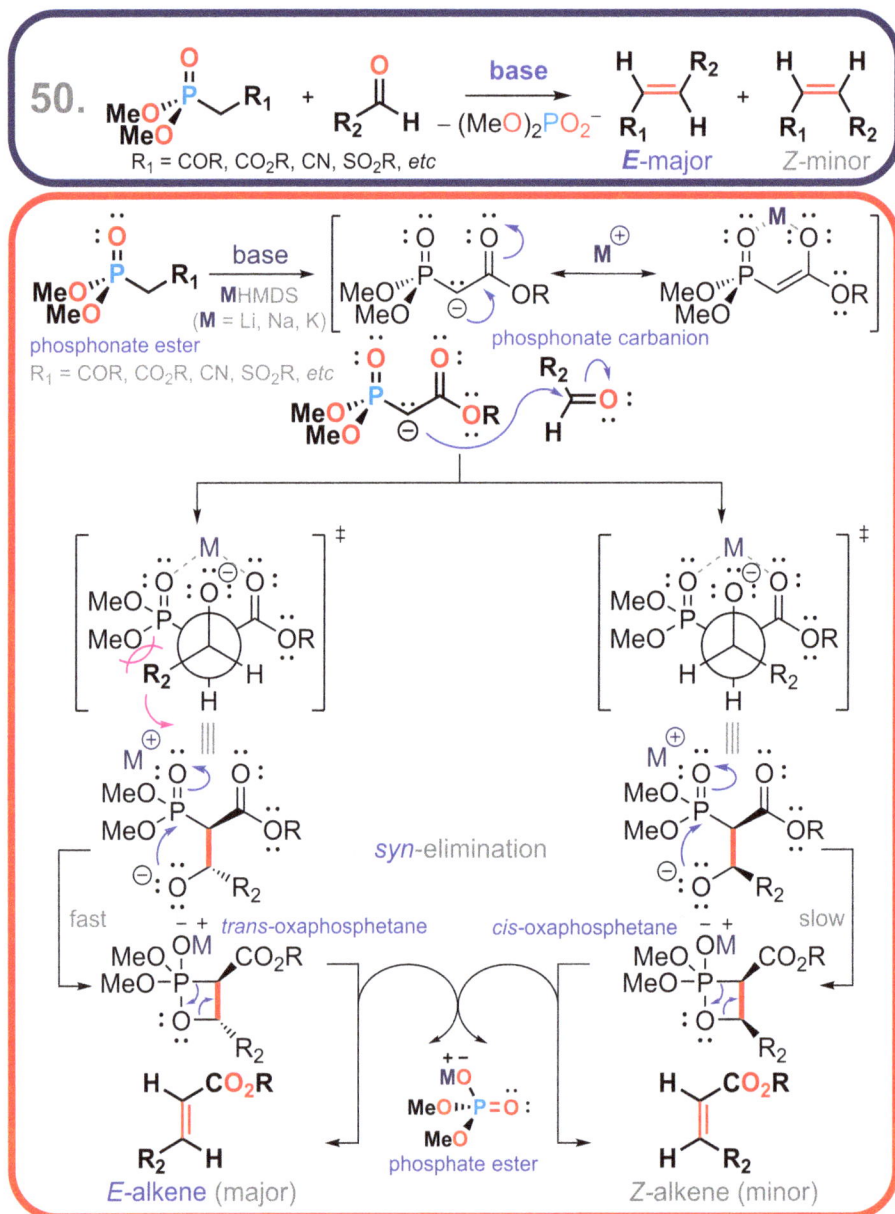

Fig. 50.1: The *Horner–Wadsworth–Emmons* olefination mechanism.[179]

https://doi.org/10.1515/9783110786835-050

Wittig Reaction

$Ph_3\overset{\oplus}{P}\overset{\ominus}{\frown}R$ + $R_1\overset{O}{\underset{R_2}{\bigwedge}}$ ⟶ $\overset{H}{\underset{R}{\diagdown}}=\overset{R_2}{\underset{R_1}{\diagup}}$ + $Ph_3P{=}O$

phosphorus ylide   phosphine oxide

Horner–Wittig Reaction

$R\overset{O}{\underset{R}{\overset{\|}{P}}}R$ + $R_1\overset{O}{\underset{R_2}{\bigwedge}}$ $\xrightarrow{\text{BuLi}}$ $\overset{H}{\underset{R}{\diagdown}}=\overset{R_2}{\underset{R_1}{\diagup}}$ + $R\overset{O}{\underset{R}{\overset{\|}{P}}}\overset{-\ +}{OLi}$

phosphine oxide   phosphinate

Peterson Olefination

$Me_3Si\overset{\ominus}{\frown}R$ + $R_1\overset{O}{\underset{R_2}{\bigwedge}}$ $\xrightarrow[\overset{\oplus}{H}]{\overset{\ominus}{OH}}$ $\overset{H}{\underset{R}{\diagdown}}=\overset{R_2}{\underset{R_1}{\diagup}}$ + $Me_3Si{-}\overset{\ominus}{O}$

α–silylcarbanion

Corey–Chaykovsky Reaction

$Me_2\overset{\oplus}{S}\overset{\ominus}{\frown}R$ + $R_1\overset{O}{\underset{R_2}{\bigwedge}}$ ⟶ $\overset{O}{\underset{R}{\triangle}}\overset{R_2}{\underset{R_1}{}}$ + Me_2S

sulfur ylide

$Me\overset{O}{\underset{Me}{\overset{\|}{\underset{\ominus}{\overset{\oplus}{S}}}}}R$ + $R_1\overset{O}{\underset{R_2}{\bigwedge}}$ ⟶ $\overset{O}{\underset{R}{\triangle}}\overset{R_2}{\underset{R_1}{}}$ + $Me_2\overset{\oplus\ \ominus}{S{-}O}$

sulfur ylide

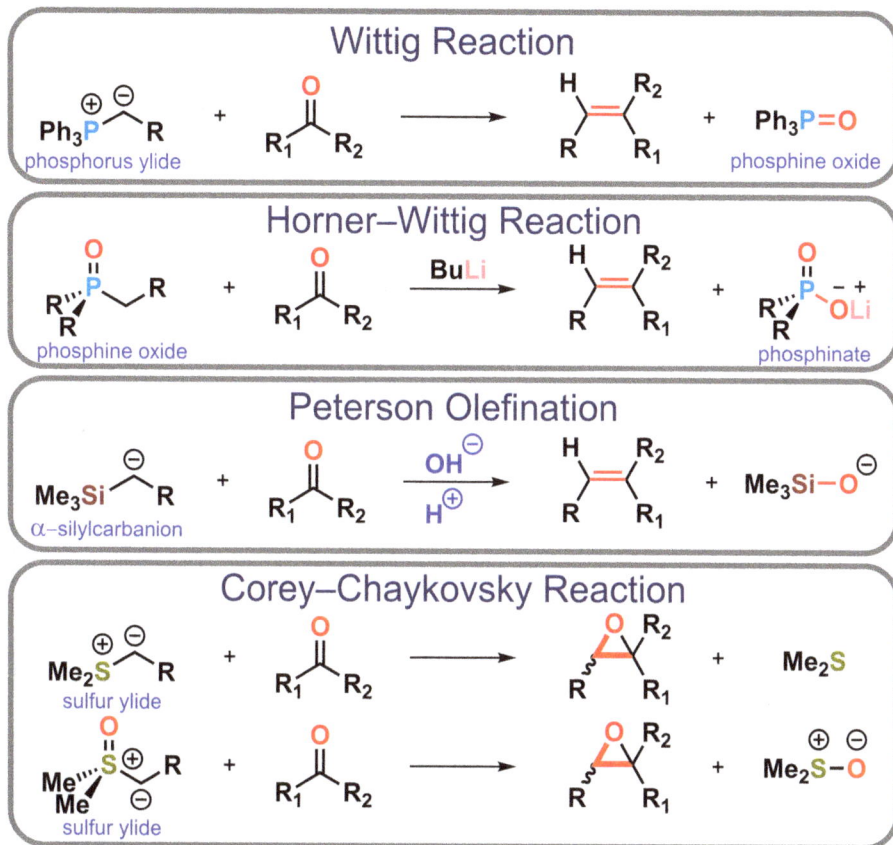

Fig. 50.2: Reactions related to the *Horner–Wadsworth–Emmons olefination.*[180]

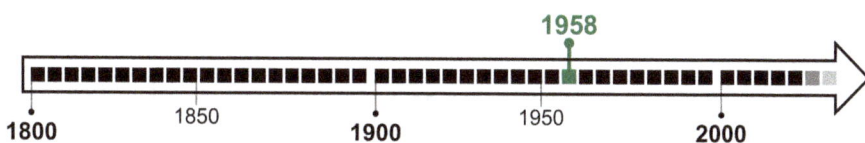

1958

```
1800        1850        1900        1950        2000
```

Fig. 50.3: The discovery of the *Horner–Wadsworth–Emmons olefination.*[181]

---

179 The *Horner–Wadsworth–Emmons (HWE) olefination* is also called the **HWE** *reaction*. The reaction relies on the use of *phosphonates* prepared via the **Arbuzov** *reaction* (Chapter 9).

180 Several reactions are related to the **HWE** *olefination*: the **Wittig** *reaction* (relies on the *phosphorus ylides* formed from the *phosphonium salts*, Chapter 98), the **Horner–Wittig** *reaction* (relies on the *ylides* formed from the *phosphine oxides*) [1, 50a], the **Peterson** *olefination* (relies on the *organosilanes*) [50b], and the **Corey–Chaykovsky** *reaction* (relies on the *sulfur ylides*, Chapter 32).

181 The reaction was likely first described around 1958 [50c, 50d, 50e].

Peterson Olefination

$Me_3Si^\ominus$—CH_3 + Ph—CHO $\xrightarrow[H^\oplus]{OH^\ominus}$ Ph/H C=C CH$_3$/H + Me_3Si–$O^\ominus$

α–silylcarbanion

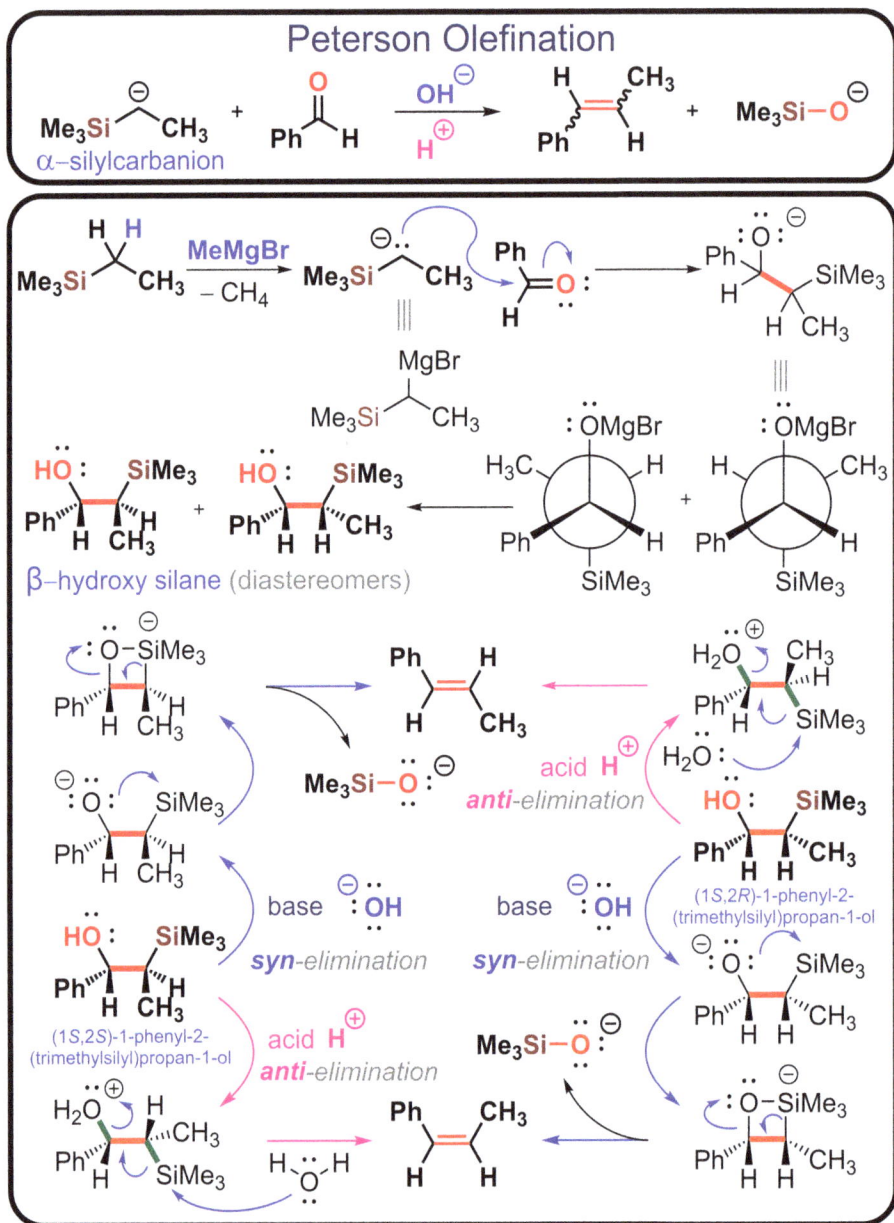

Fig. 50.4: The **Peterson** *olefination* mechanism.[182]

---

182 The **Peterson** *olefination* reaction relies on the use of *organosilanes*. The outcome of the **Peterson** *elimination* step differs depending on the basic or acidic conditions [50b].

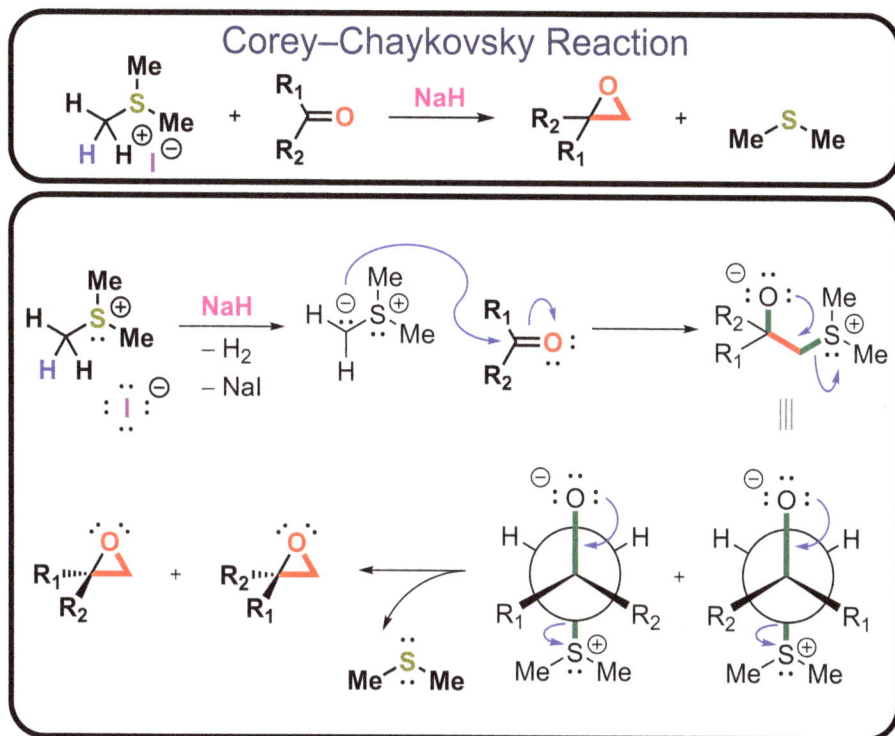

183 The *Corey–Chaykovsky reaction* (also known as the *Johnson–Corey–Chaykovsky reaction*) [32a, 32b] is related to both the *Darzens condensation* (Chapter 32) and the *Wittig reaction* (Chapter 98).

Fig. 50.5: The *Corey–Chaykovsky reaction* mechanism.[183]

51 Jones Oxidation

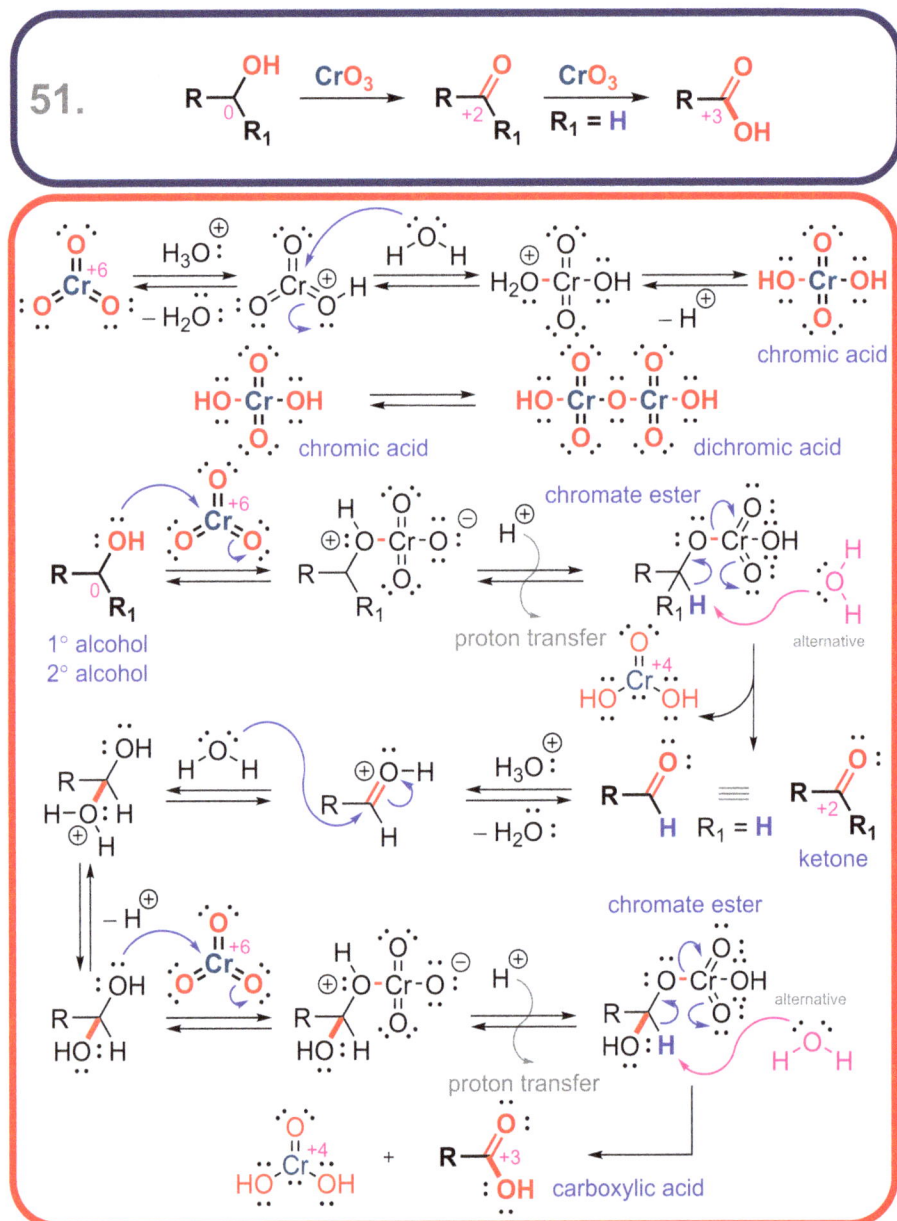

Fig. 51.1: The *Jones* oxidation mechanism.[184]

https://doi.org/10.1515/9783110786835-051

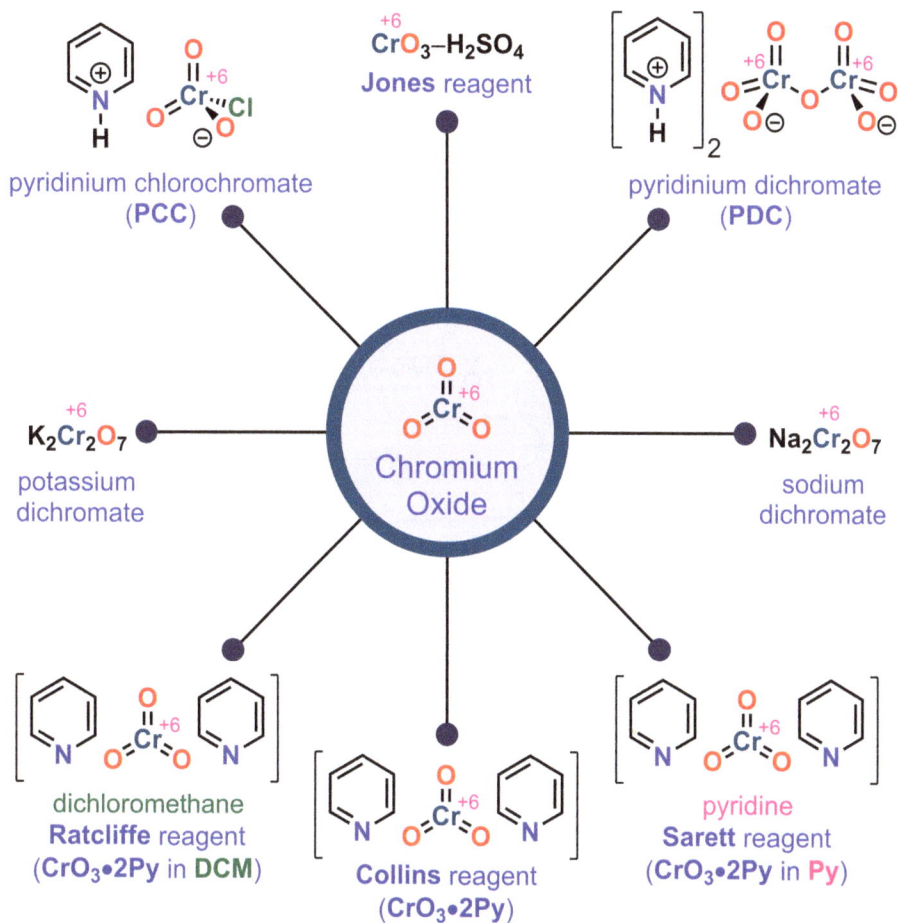

pyridinium chlorochromate
(**PCC**)

CrO_3–H_2SO_4
Jones reagent

pyridinium dichromate
(**PDC**)

$K_2Cr_2O_7$
potassium
dichromate

Chromium
Oxide

$Na_2Cr_2O_7$
sodium
dichromate

dichloromethane
Ratcliffe reagent
($CrO_3 \cdot 2Py$ in **DCM**)

Collins reagent
($CrO_3 \cdot 2Py$)

pyridine
Sarett reagent
($CrO_3 \cdot 2Py$ in **Py**)

Fig. 51.2: Various oxidizing reagents formed from chromium oxide(VI).[185]

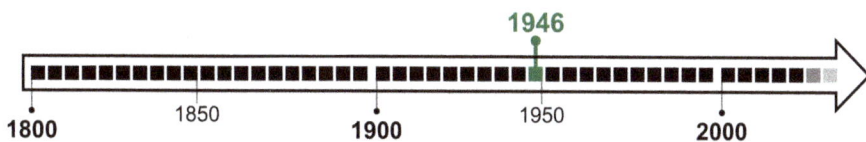

Fig. 51.3: The discovery of the **Jones** oxidation.[186]

---

184 The **Jones** oxidation is based on the use of the same named reagent: the **Jones** reagent [51a].
185 There are numerous examples of chromium oxidizing reagents, derived from chromium oxide (VI): pyridinium chlorochromate (PCC) [51b, 51c] is one of the most important.
186 The reaction was likely first described around 1946 [51d].

52 Kucherov Reaction

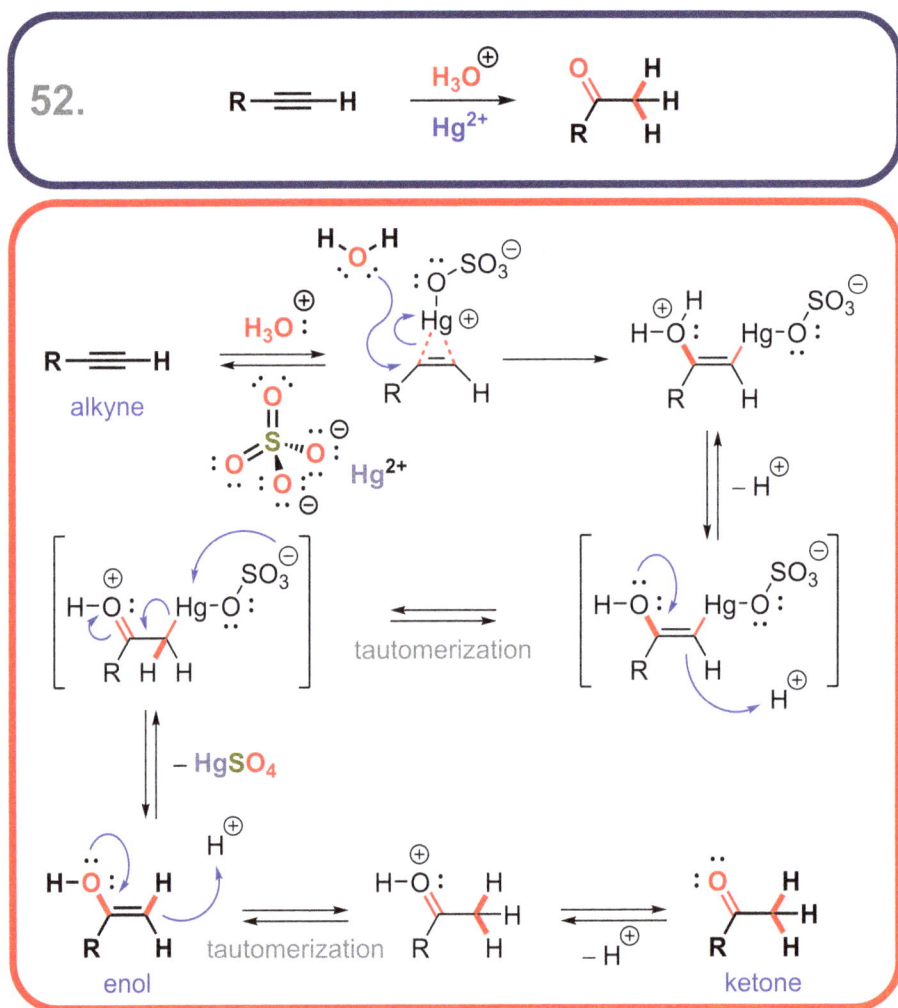

Fig. 52.1: The **Kucherov** *reaction* mechanism.[187]

---

187 The **Kucherov** *reaction* (in Russian Кучеров) is rare and very seldom called by its name. Mechanistically, it is an example of the **electrophilic addition** (to an alkyne) more broadly covered in Chapter 1. The reaction follows *Markovnikov's rule* (in Russian Владимир Васильевич Марковников or В. В. Марковников): hydrogen (H⁺, or any other electrophilic part of a molecule) is at the least substituted carbon (or H adds to the carbon with more H atoms) [52a].

https://doi.org/10.1515/9783110786835-052

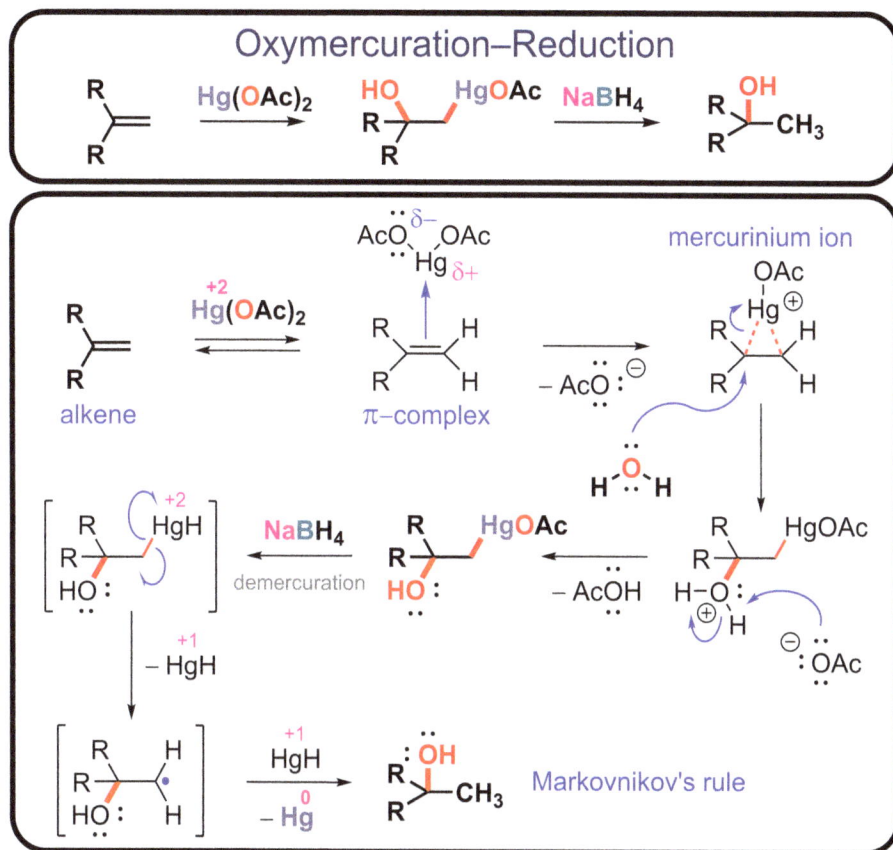

Fig. 52.2: The *oxymercuration reaction* mechanism.[188]

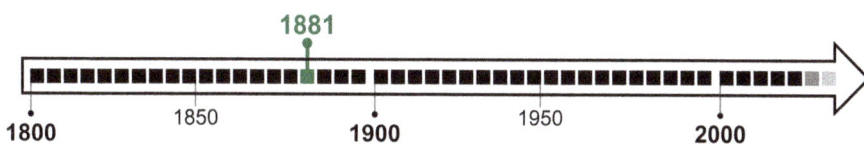

Fig. 52.3: The discovery of the **Kucherov** *reaction*.[189]

---

188 The *oxymercuration reaction* (the *oxymercuration–reduction reaction*) is related to the **Kucherov** *reaction*. It is also an **electrophilic addition** reaction predominantly forming products (alcohols) according to *Markovnikov's rule*. Please note that the *hydroboration–oxidation* (Chapter 20) yields **anti-Markovnikov's** products: hydrogen is at the most substituted carbon (or H adds to the carbon with less H atoms).

189 The reaction was likely first described around 1881 [52b].

53 Kumada Cross-Coupling

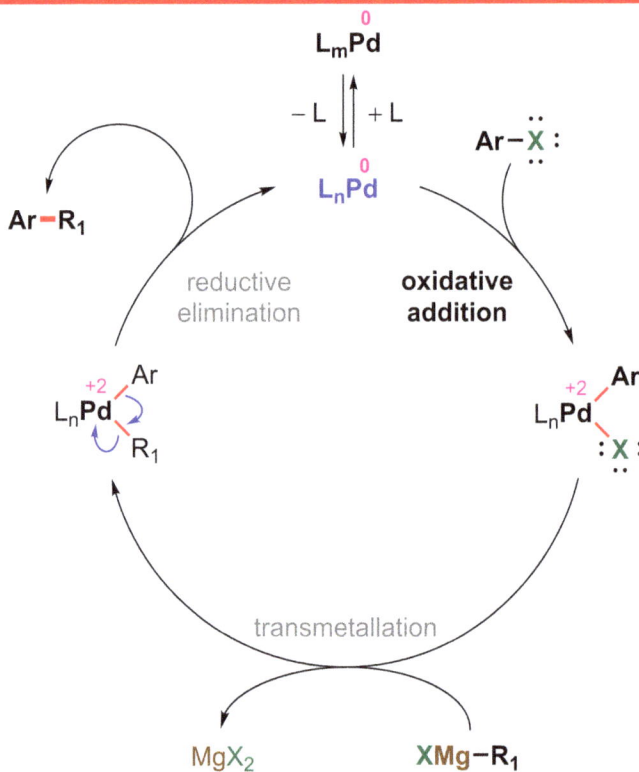

1. Concerted Mechanism:

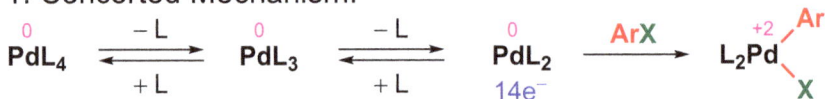

Fig. 53.1: The **Pd**-catalyzed **Kumada** cross-coupling mechanism.[190]

190 The **Kumada** cross-coupling (or the **Kumada–Corriu** cross-coupling) is a type of **Pd**-catalyzed cross-coupling reaction (C–C bond formation using aryl halides and the **Grignard** reagent = organomagnesium compound). For teaching purposes, a simplified and general mechanism is shown. Note that the (1) concerted oxidative addition step to a low-coordinate (14e⁻) **Pd**-complex is more complicated [2a].

https://doi.org/10.1515/9783110786835-053

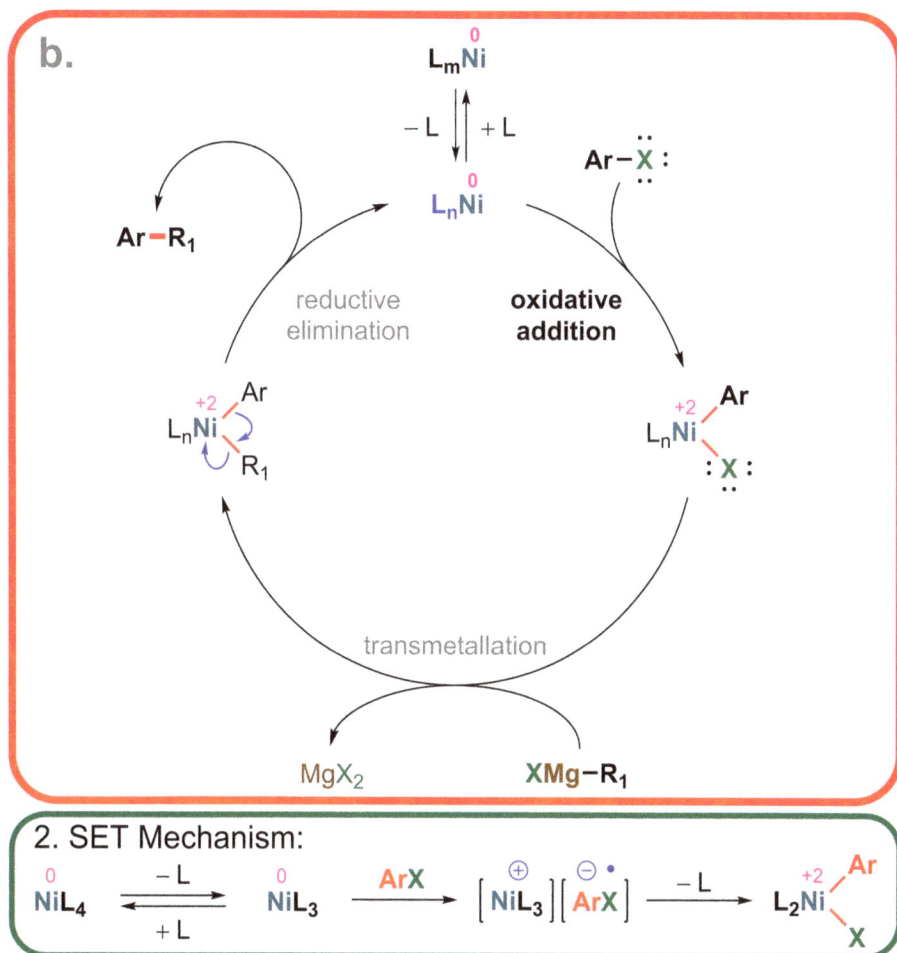

Fig. 53.2: The Ni-catalyzed *Kumada cross-coupling* mechanism.[191]

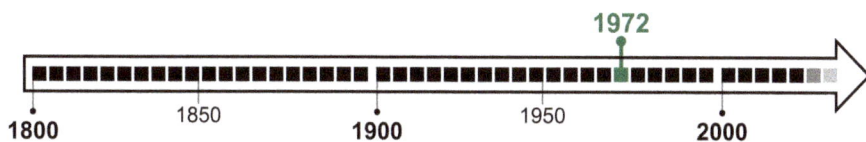

Fig. 53.3: The discovery of the *Kumada cross-coupling*.[192]

---

191 The **Kumada** cross-coupling can be **Ni-catalyzed**. Note the possible example of an (2) SET oxidative addition step to a **Ni**-complex (not necessarily at play in the example shown) [2a].
192 The reaction was likely first described around 1972 [53].

54 Ley–Griffith Oxidation

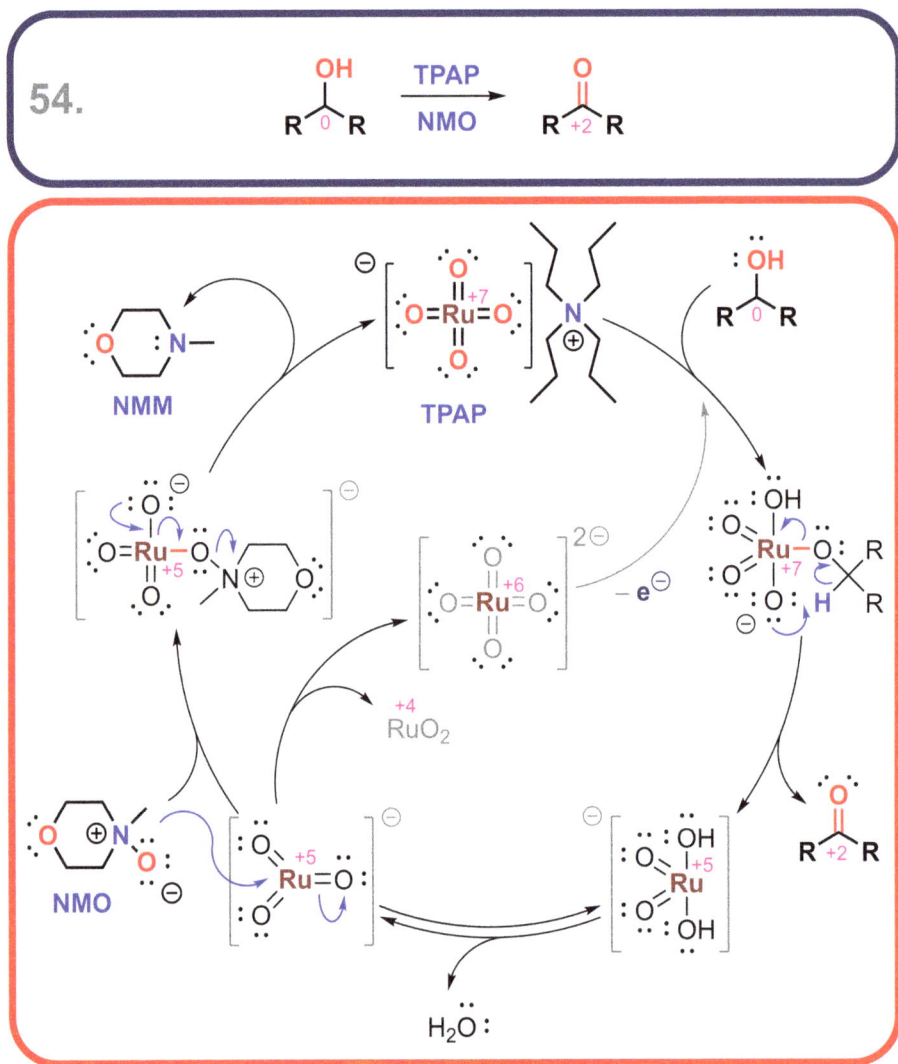

Fig. 54.1: The *Ley–Griffith* oxidation mechanism.[193]

---

193 The *Ley–Griffith* oxidation is based on the use of a named reagent: the *Ley–Griffith* reagent (**TPAP**) [54a].

https://doi.org/10.1515/9783110786835-054

Upjohn Dihydroxylation

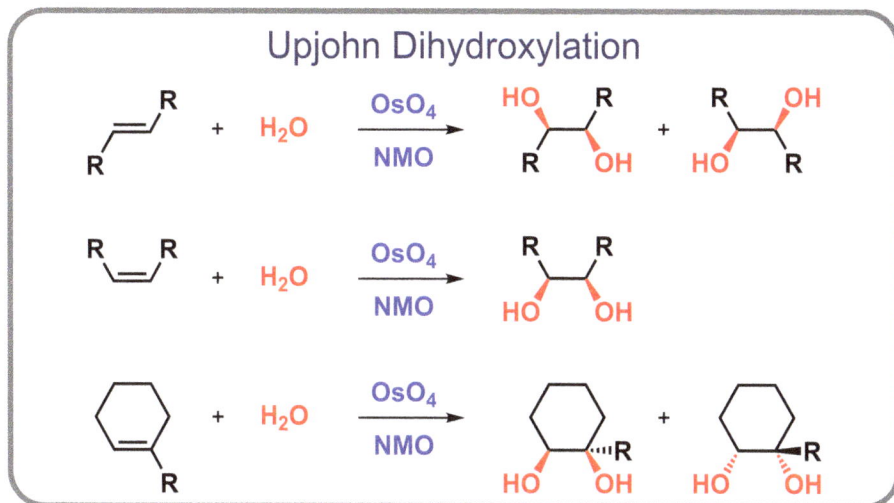

Fig. 54.2: Reactions related to the *Ley–Griffith oxidation*.[194]

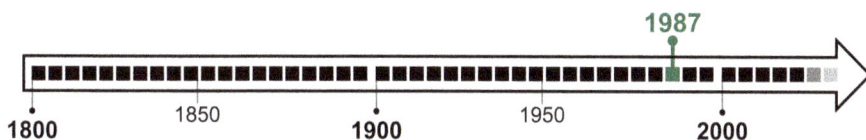

Fig. 54.3: The discovery of the *Ley–Griffith oxidation*.[195]

---

194 The *Upjohn dihydroxylation* (covered in Chapter 93) is related to the *Ley–Griffith oxidation*. Please note that the reaction outcome is different depending on the stereochemistry of the starting alkene.

195 The reaction was likely first described around 1987 [54b].

55 Liebeskind–Srogl Cross-Coupling

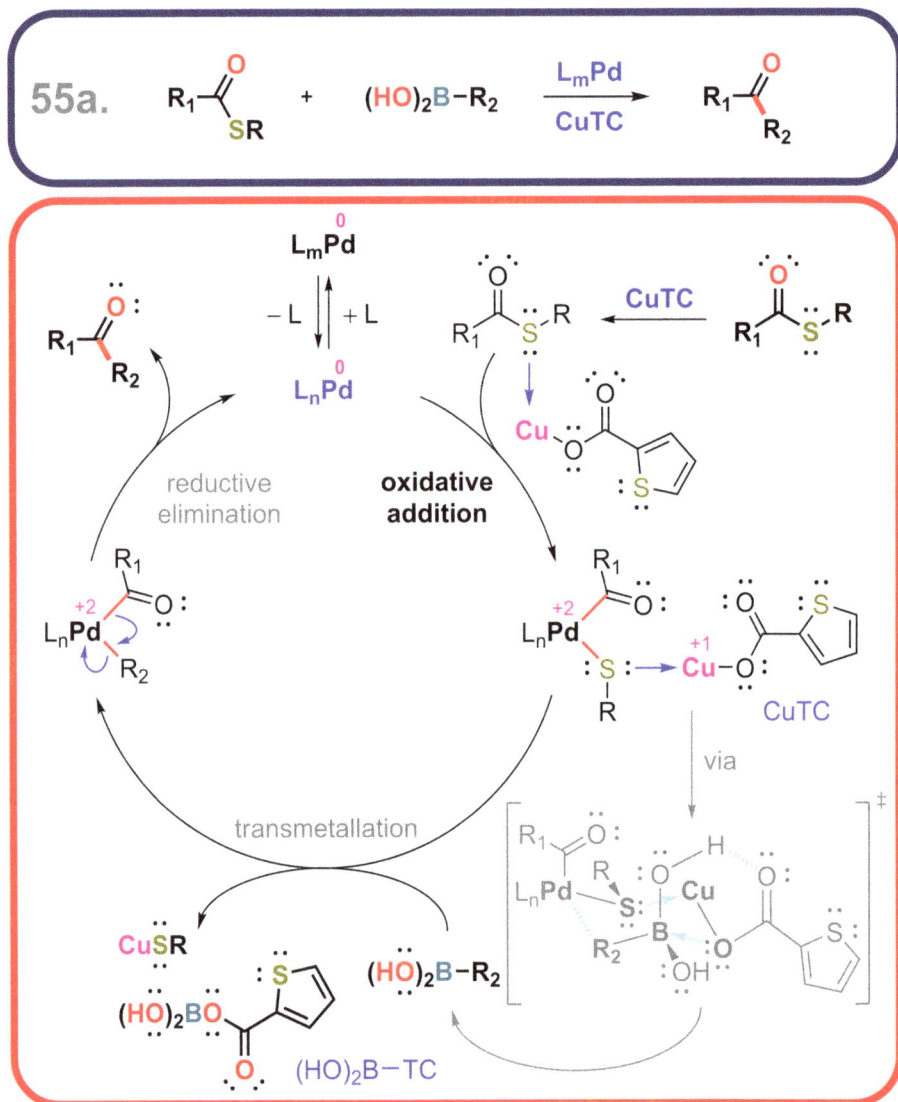

55a. $R_1\text{C(O)}SR$ + $(HO)_2B-R_2$ $\xrightarrow[\text{CuTC}]{L_mPd}$ $R_1\text{C(O)}R_2$

Fig. 55.1: The *Liebeskind–Srogl* cross-coupling (thioesters) mechanism.[196]

---

196 The *Liebeskind–Srogl* cross-coupling of *thioesters* is a type of *Pd-catalyzed cross-coupling* reaction (C–C bond formation using *thioesters* and *boronic acids*). For teaching purposes, only a simplified general mechanism is shown.

https://doi.org/10.1515/9783110786835-055

55b. $^{HET}Ar-SR$ + $\begin{matrix}(HO)_2B-R_2 \\ or \\ Bu_3Sn-R_2\end{matrix}$ $\xrightarrow[\text{CuTC}]{L_mPd}$ $^{HET}Ar=R_2$

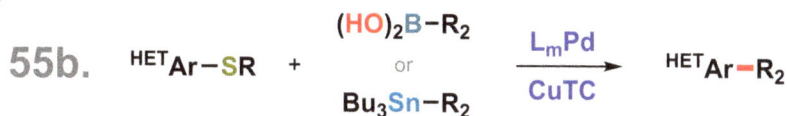

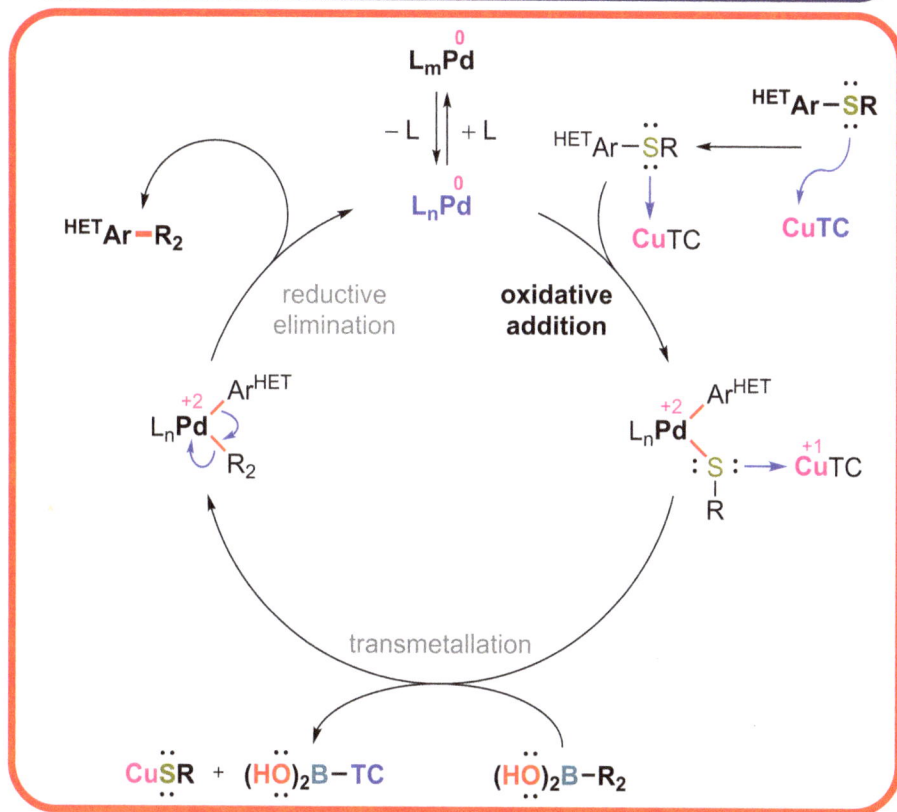

Fig. 55.2: The *Liebeskind–Srogl* cross-coupling (thioethers) mechanism.[197]

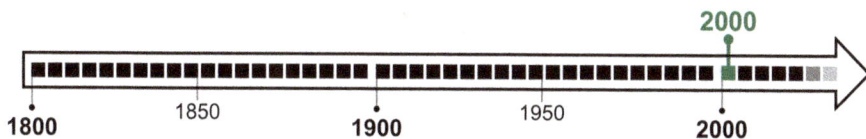

Fig. 55.3: The discovery of the *Liebeskind–Srogl* cross-coupling.[198]

---

197 The *Liebeskind–Srogl* cross-coupling of *thioethers* is a variation (C–C bond formation using *thioethers (ArSR)* and *boronic acids* or *organotin reagents = organostannanes*). For teaching purposes, only a simplified general mechanism is shown.
198 The reaction was likely first described around 2000 [55].

56 Mannich Reaction

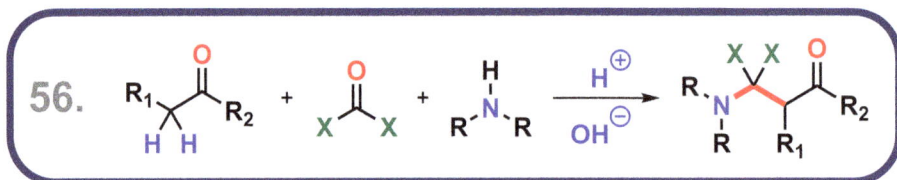

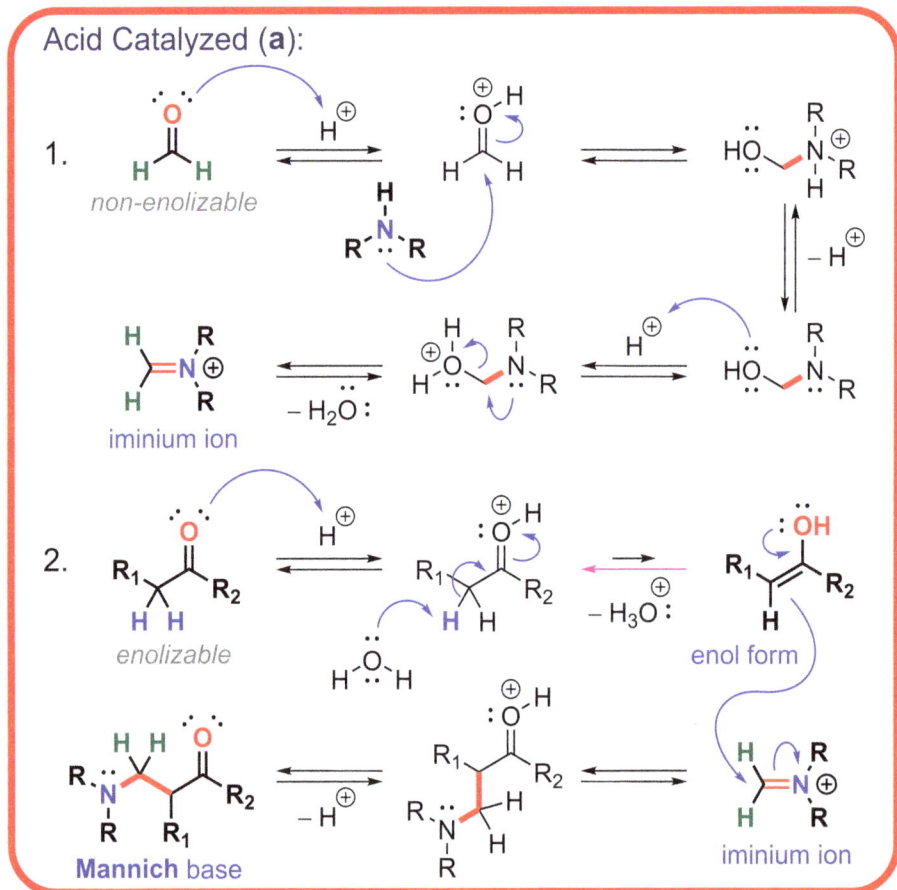

56. $R_1 \overset{O}{\underset{H\ H}{\wedge}} R_2$ + $X \overset{O}{\underset{X}{\wedge}}$ + $R \overset{H}{\underset{N}{\wedge}} R$ $\xrightarrow[OH^{\ominus}]{H^{\oplus}}$ $R \underset{R}{\overset{X\ X}{\underset{N}{\wedge}}} \overset{O}{\underset{R_1}{\wedge}} R_2$

Acid Catalyzed (a):

Fig. 56.1: The **Mannich** reaction mechanism (acid catalyzed).[199]

199 The **Mannich** reaction is also known as *the **Mannich** condensation*. This three-component reaction (3-CR) can be catalyzed in (a) _acidic_ media (via an *iminium ion* intermediate). The final product (β-amino carbonyl) is also called a **Mannich** base.

https://doi.org/10.1515/9783110786835-056

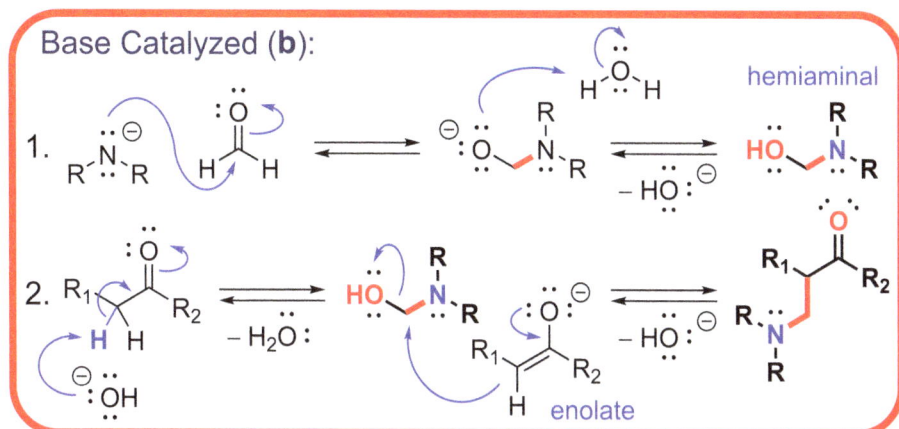

Fig. 56.2: The **Mannich** *reaction* mechanism (base catalyzed).[200]

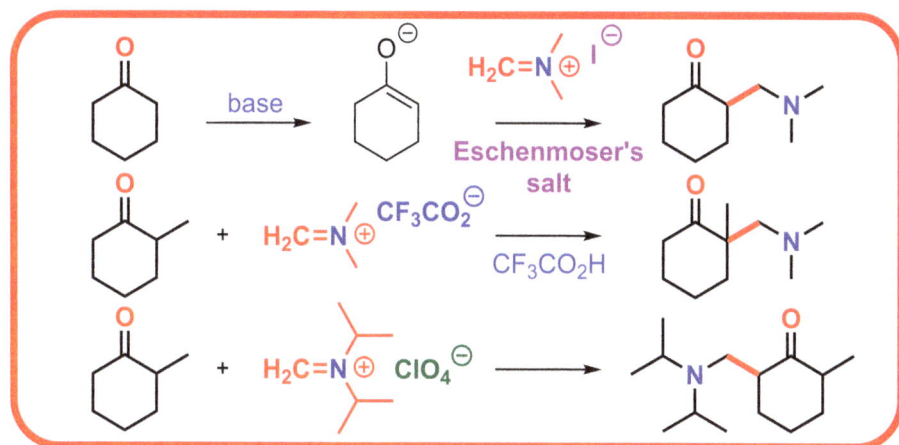

Fig. 56.3: Variations of the **Mannich** *reaction*.[201]

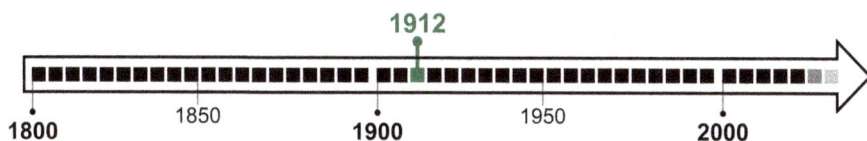

Fig. 56.4: The discovery of the **Mannich** *reaction*.[202]

---

200 The **Mannich** *reaction* can be also catalyzed in (b) _basic_ media (via a *hemiaminal* intermediate).
201 There are several iterations of the **Mannich** *reaction* based on availability of the preformed iminium ions: **Eschenmoser's** *salts* or **Böhme's** *salts* (not shown here) [56a].
202 The reaction was likely first described around 1912 [56b].

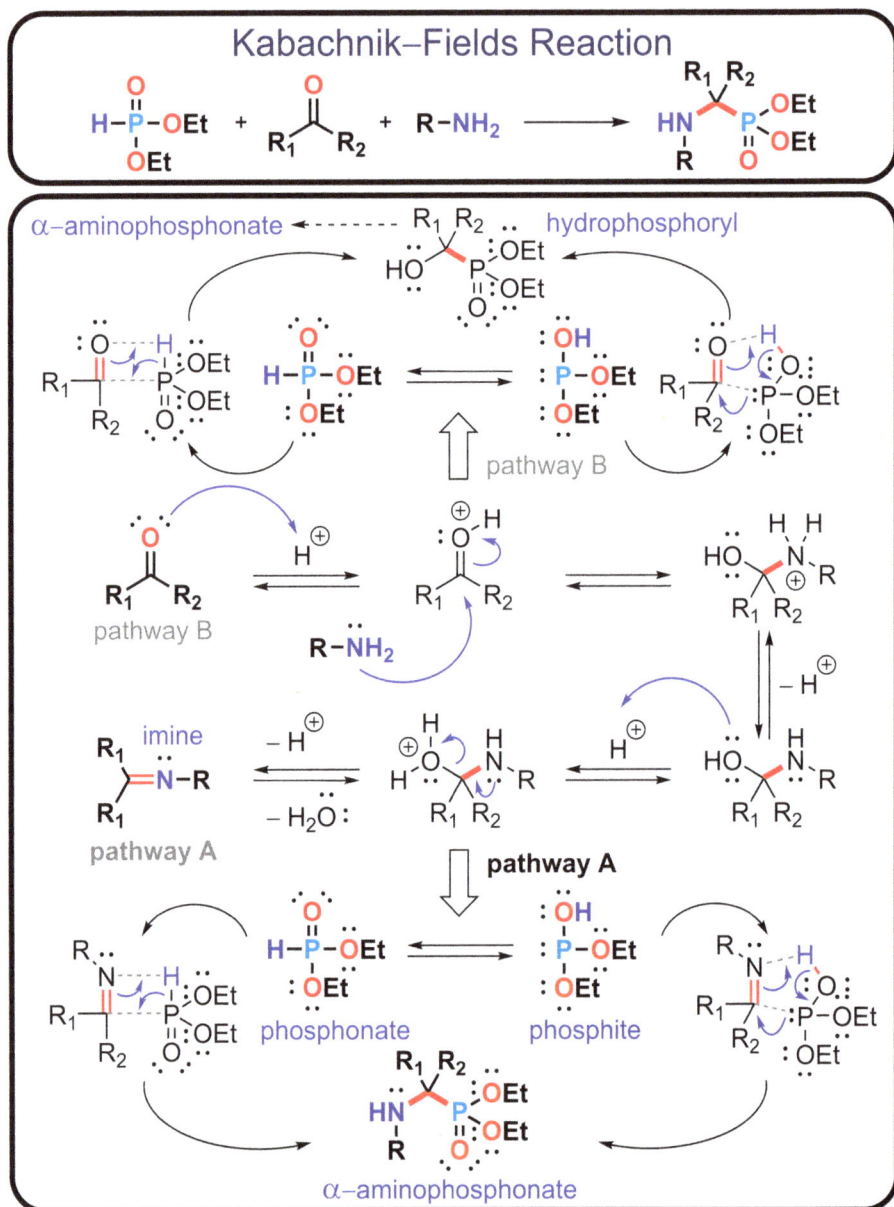

Fig. 56.5: The *Kabachnik–Fields* reaction mechanism.[203]

---

203 The *Kabachnik–Fields reaction* is a 3-CR yielding peptidomimetic compounds [56c, 56d].

Petasis Reaction

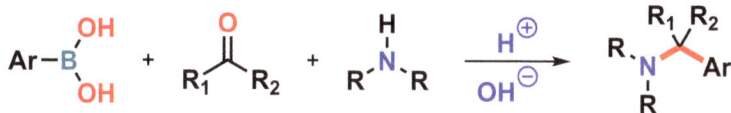

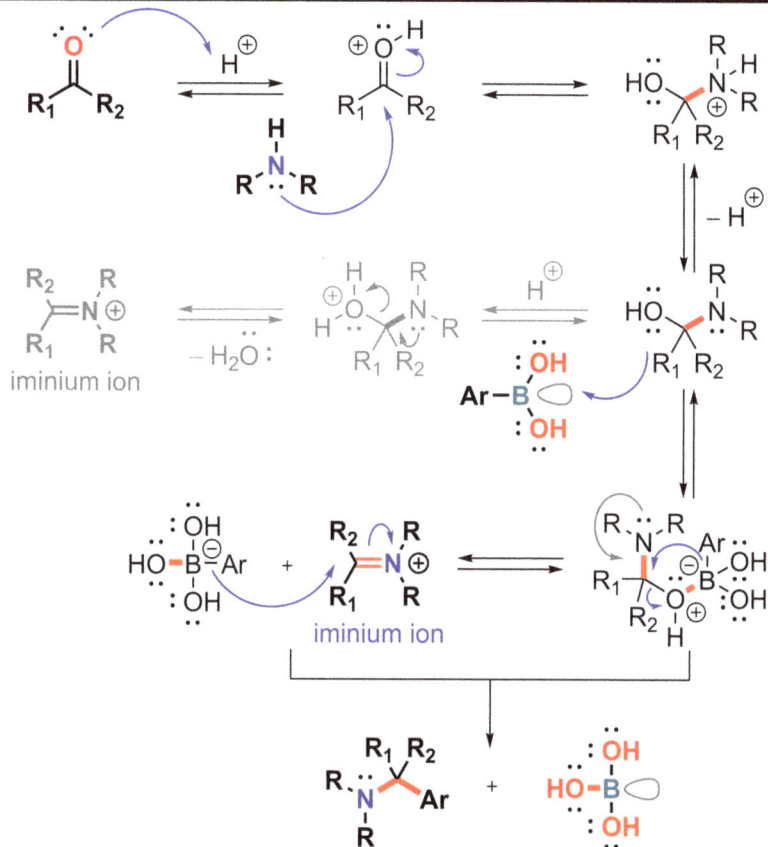

---

204 The **Petasis** *reaction* (also known as the **Petasis** *boronic acid–***Mannich** *reaction*) is a 3-CR reaction. Its mechanism is not fully understood [62b].

57 McMurry Coupling

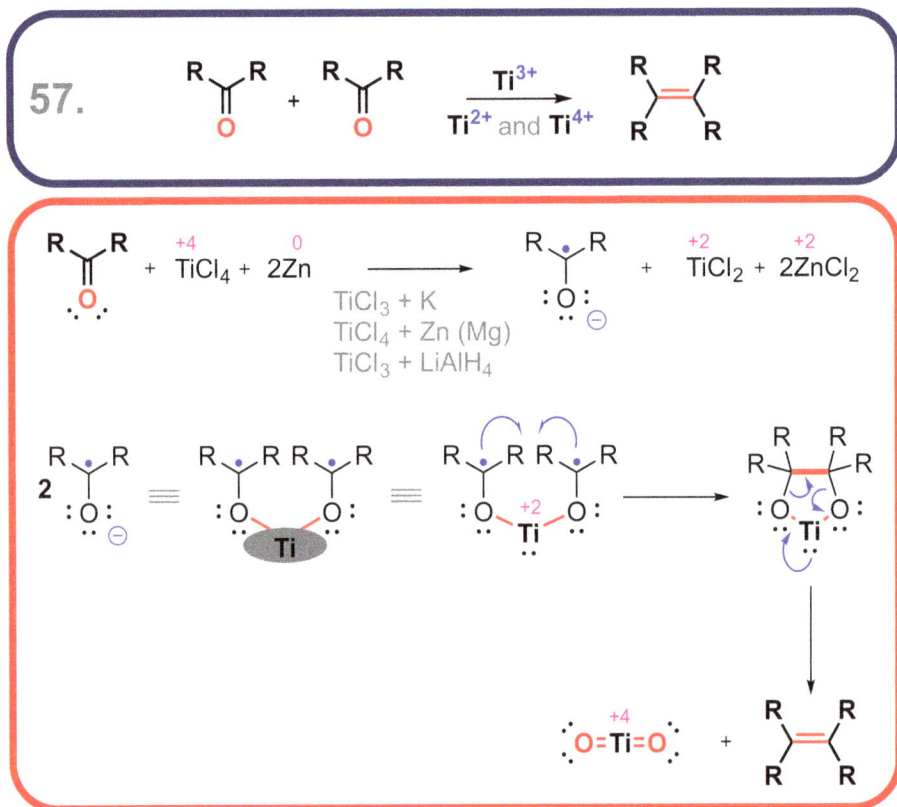

57. $R \overset{R}{\underset{O}{\bigvee}} + R \overset{R}{\underset{O}{\bigvee}} \xrightarrow[\text{Ti}^{2+} \text{ and Ti}^{4+}]{\text{Ti}^{3+}} \overset{R}{\underset{R}{\bigvee}} = \overset{R}{\underset{R}{\bigvee}}$

Fig. 57.1: The *McMurry* coupling mechanism.[205]

205 The *McMurry* coupling or the *McMurry* reaction mechanism is not fully understood. It is believed that the *low-valent titanium* species play a major role: Ti(0) + Ti(II) + Ti(III).

https://doi.org/10.1515/9783110786835-057

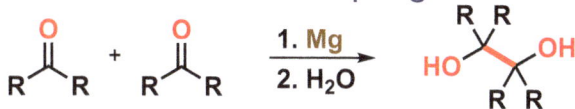

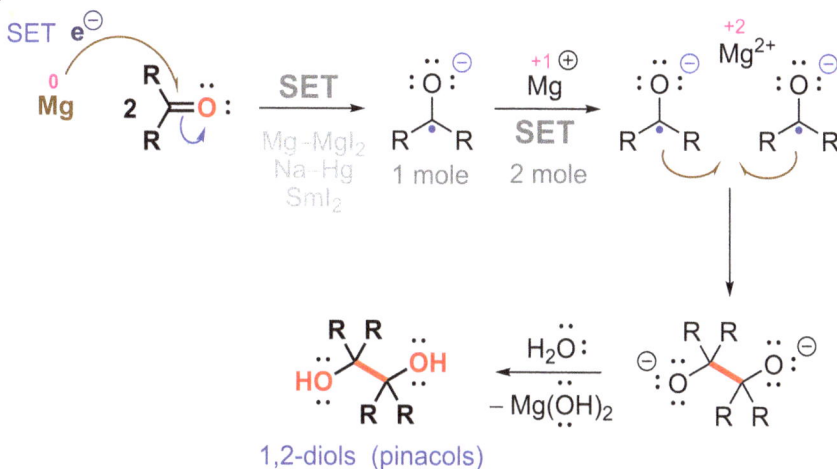

Fig. 57.2: The *pinacol coupling* mechanism.[206]

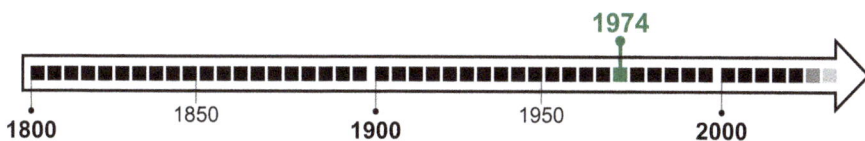

Fig. 57.3: The discovery of the **McMurry** coupling.[207]

---

206 The *pinacol coupling* undergoes a **single electron transfer** (SET) mechanism [57a, 57b]. This reaction is related to the **McMurry** *coupling* and the *acyloin condensation* (covered in Chapter 7). Please do not confuse the *pinacol coupling* with the *pinacol–pinacolone rearrangement* covered in Chapter 76.
207 The reaction was likely first described around 1974 [57c].

58 Meerwein–Ponndorf–Verley Reduction

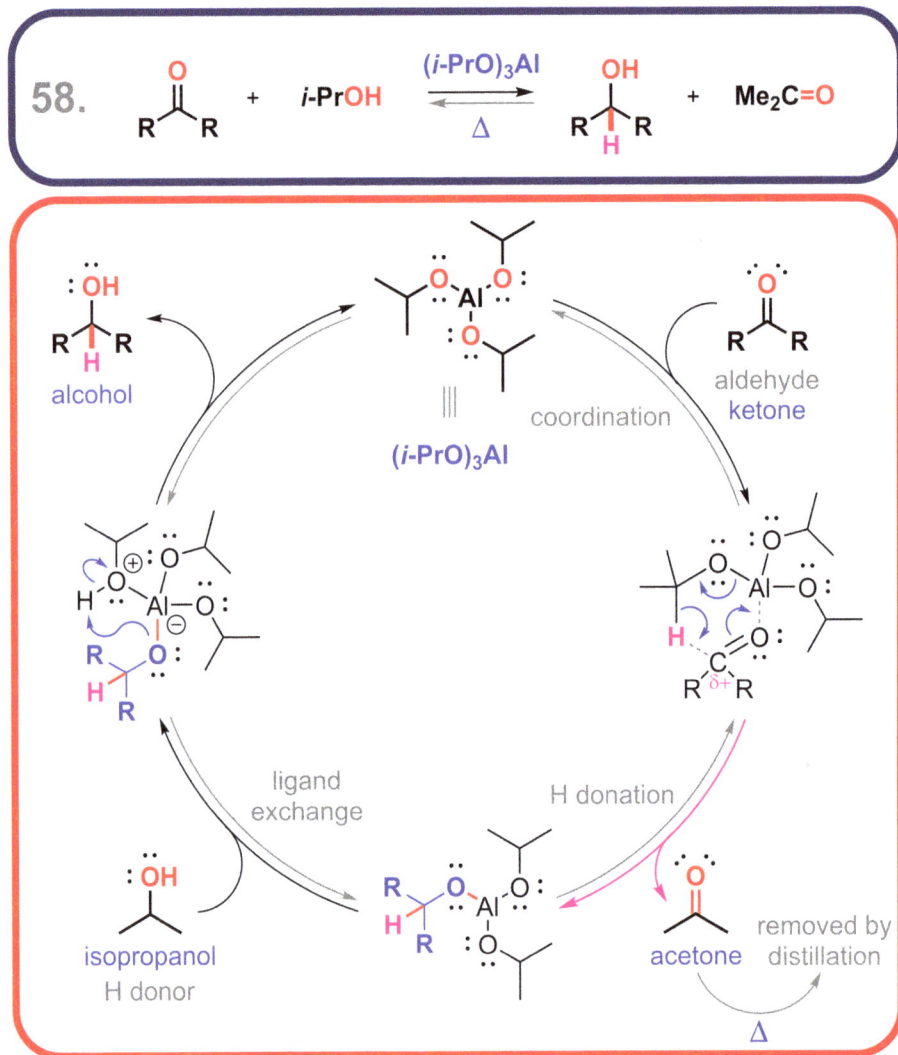

58. R–C(=O)–R + *i*-PrOH $\underset{\Delta}{\overset{(\textit{i-PrO})_3\text{Al}}{\rightleftharpoons}}$ R–CH(OH)–R + Me₂C=O

alcohol

(*i*-PrO)₃Al

coordination

aldehyde
ketone

ligand
exchange

H donation

isopropanol
H donor

acetone   removed by
distillation

Fig. 58.1: The *Meerwein–Ponndorf–Verley reaction* mechanism.[208]

---

208 The *Meerwein–Ponndorf–Verley (MPV) reduction* is reversible. The reversed oxidation is called the *Oppenauer* oxidation. The equilibrium can be shifted toward *reduction* by removing the formed *acetone* from the reaction mixture (via distillation).

https://doi.org/10.1515/9783110786835-058

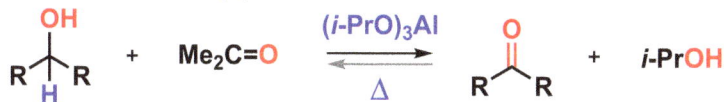

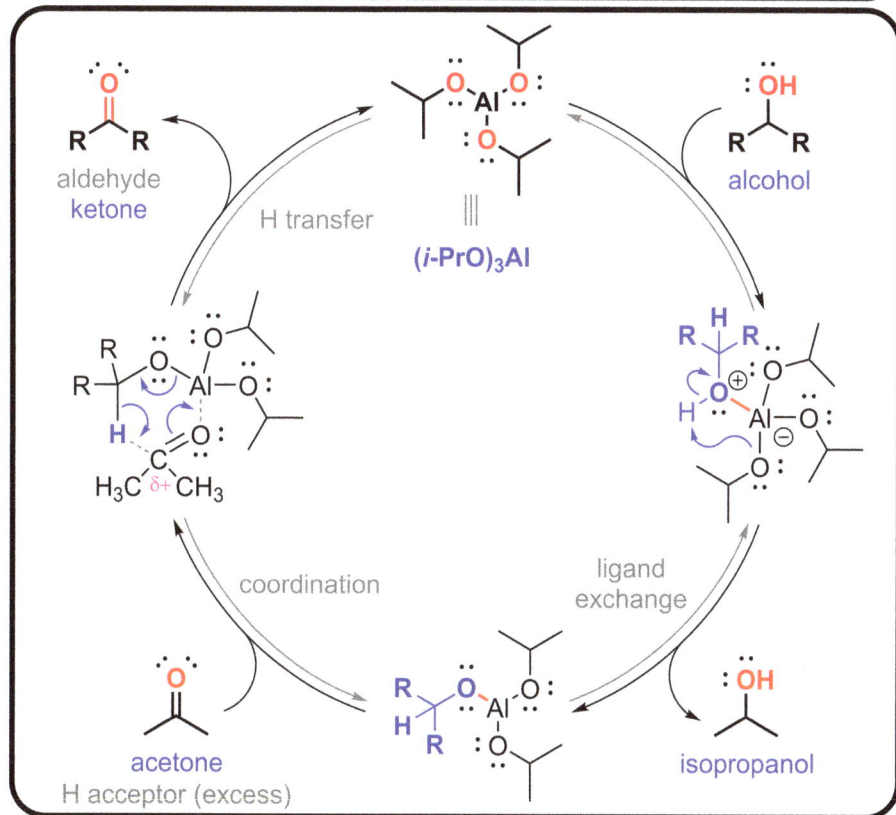

Fig. 58.2: The *Oppenauer* oxidation mechanism.[209]

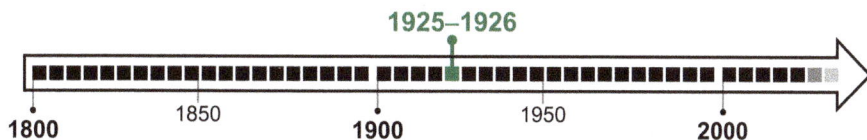

Fig. 58.3: The discovery of the *Meerwein–Ponndorf–Verley reaction*.[210]

---

209 The *Oppenauer oxidation* is a reversed process of the *MPV reduction* (see Chapter 69).
210 The reaction was likely first described around 1925 by Meerwein and Verley [58a, 58b], and then in 1926 by Ponndorf [58c].

59 Michael Addition

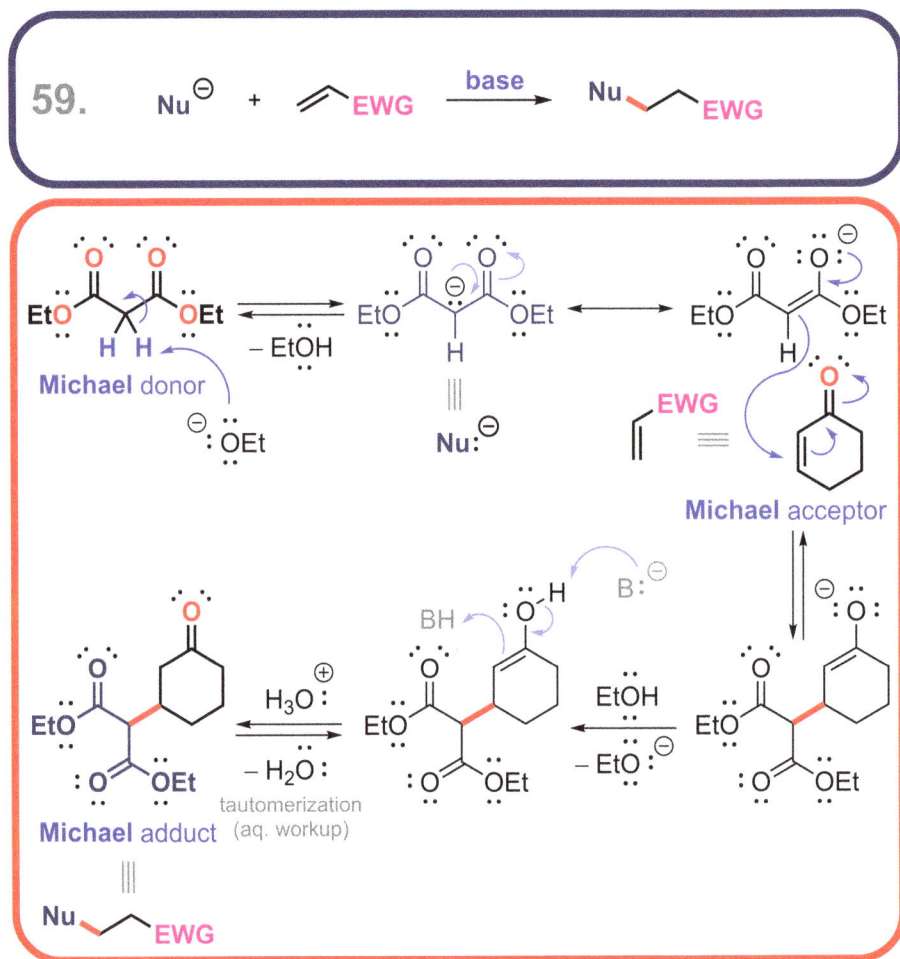

Fig. 59.1: The *Michael* addition mechanism.[211]

211 The *Michael* addition or the *Michael* conjugate addition is also simply called *the Michael reaction*. The products are known as *Michael adducts*. It is one of the most important reactions in organic chemistry.

https://doi.org/10.1515/9783110786835-059

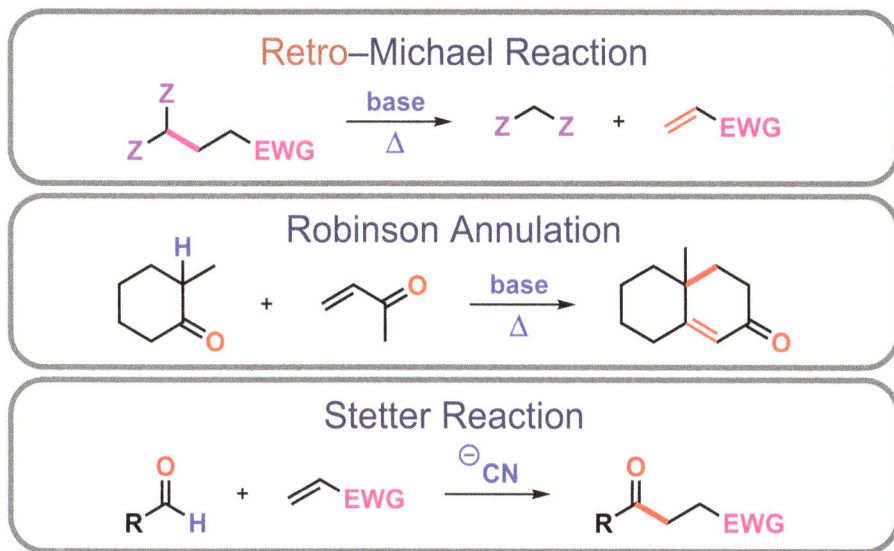

Fig. 59.2: Reactions related to the **Michael** addition.[212]

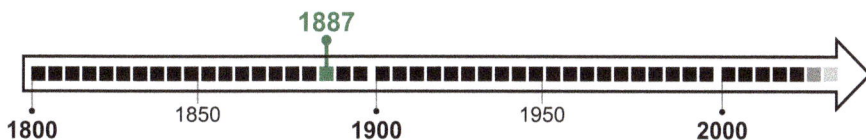

Fig. 59.3: The discovery of the **Michael** addition.[213]

---

212 There are variations of this reaction; for example, the retro-**Michael** addition and the **Robinson** annulation (covered in Chapter 83).

213 The reaction was likely first described around 1887 [59b].

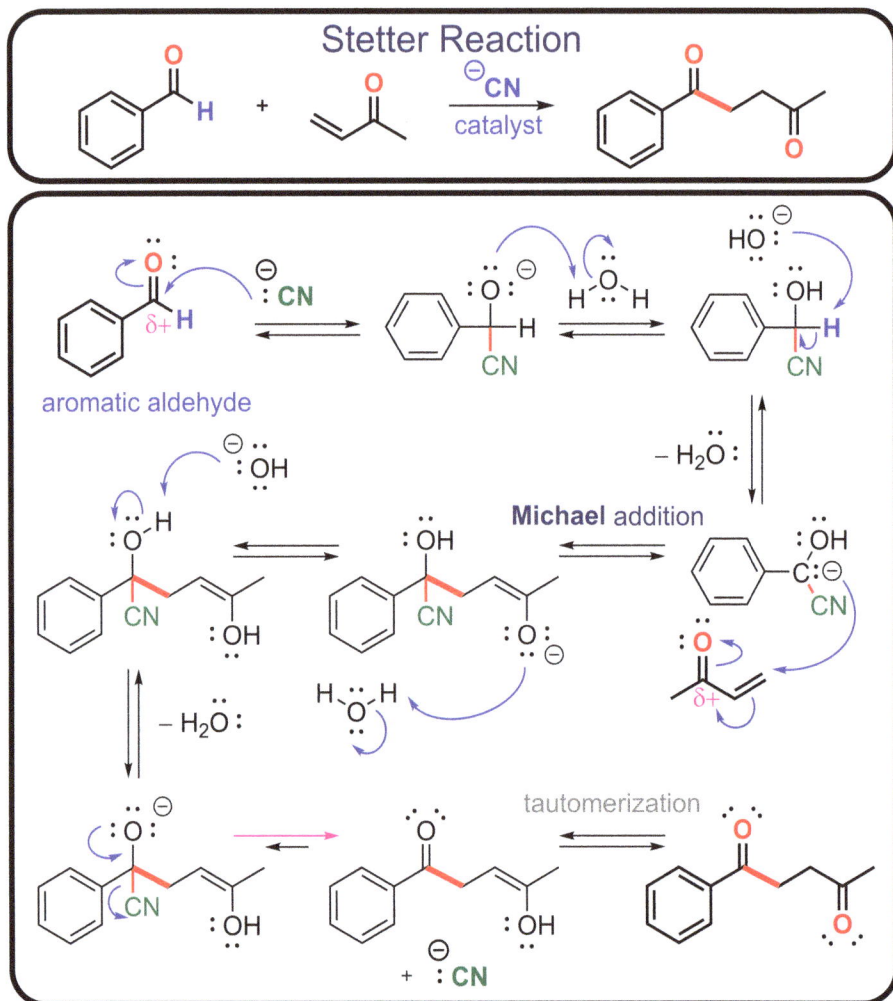

Fig. 59.4: The *Stetter reaction* mechanism (aromatic aldehyde).[214]

---

214 The mechanism of the *Stetter reaction* [59a] is related to both the *Michael addition* and the *benzoin condensation* (Chapter 15). In case of *aromatic* aldehydes, it is catalyzed by **cyanide ions**.

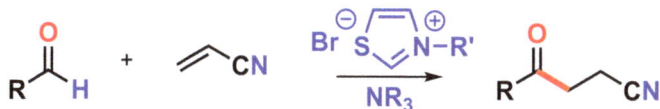

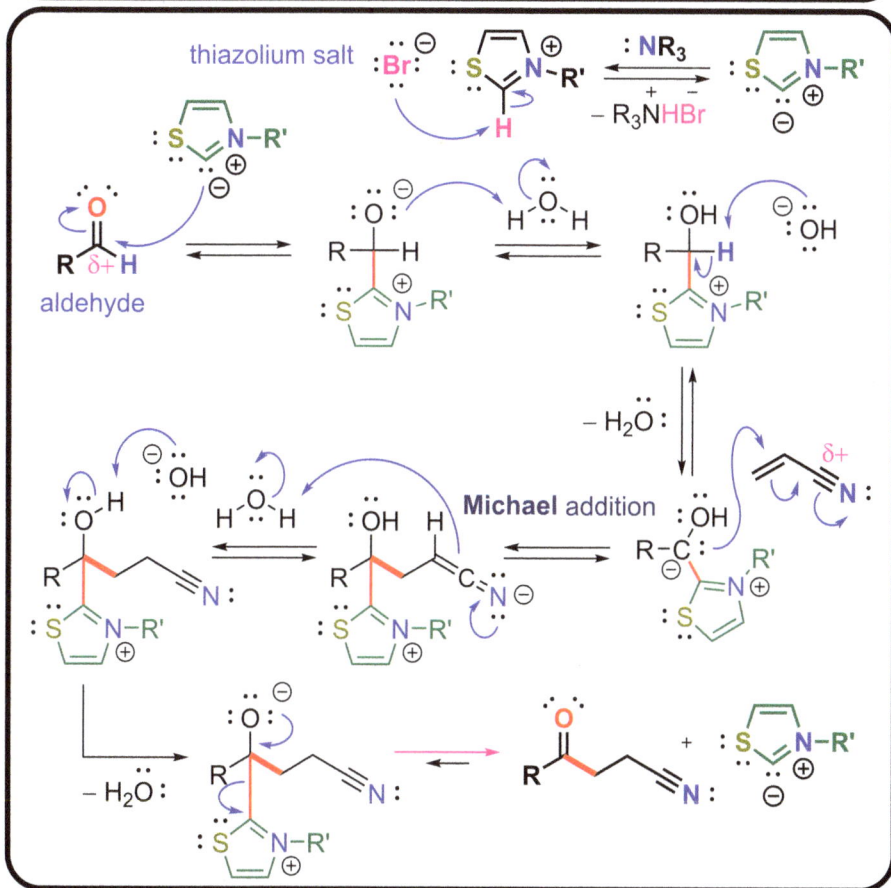

Fig. 59.5: The **Stetter** *reaction* mechanism (aliphatic aldehyde).[215]

---

215 The mechanism of the **Stetter** *reaction* [59a] is related to both the **Michael** *addition* and the *ben-zoin condensation* (Chapter 15). In case of *aliphatic* aldehydes, it is catalyzed by **thiazolium salts**.

60 Minisci Reaction

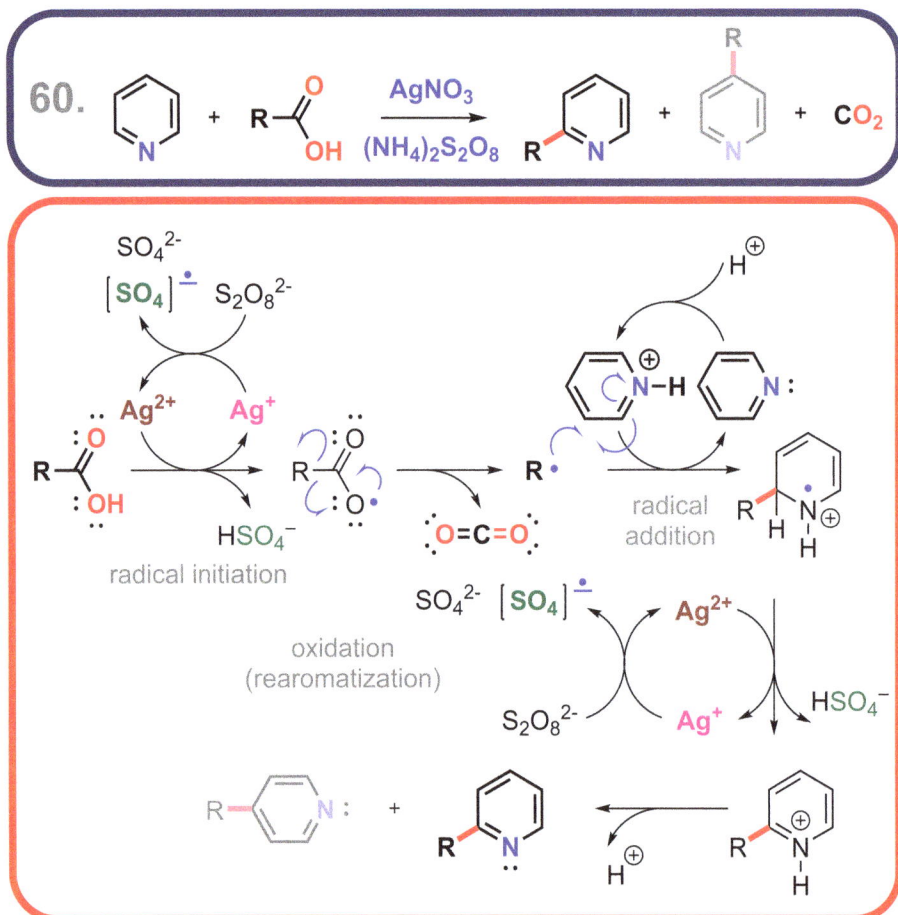

Fig. 60.1: The **Minisci** *reaction* mechanism.[216]

---

216 The **Minisci** *reaction* is a type of **free radical substitution** (not covered in this book). The closely related mechanistic examples are the $S_{RN}1$ mechanism (covered in Chapter 5), the **Barton** *decarboxylation* (covered in Chapter 12), and the **Wohl–Ziegler** *reaction* (covered in Chapter 99).

https://doi.org/10.1515/9783110786835-060

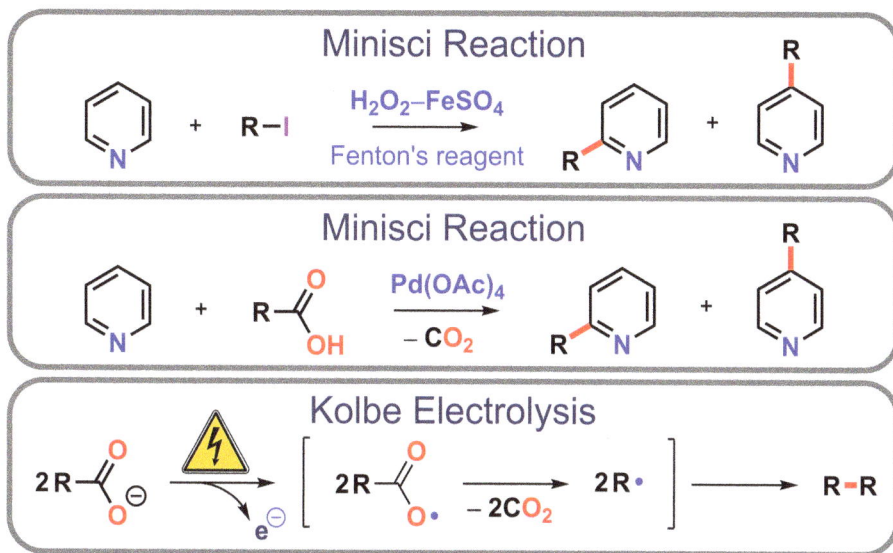

Fig. 60.2: Variations of the *Minisci reaction.*[217]

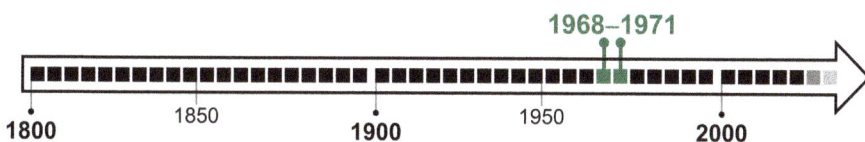

Fig. 60.3: The discovery of the *Minisci reaction.*[218]

---

217 There are several variations of the *Minisci reaction* depending on the free radical sources: *Fenton's reagent* [60a] and alkyl iodides; lead(IV) acetate [60b] and carboxylic acids. The *Kolbe electrolysis* or the *Kolbe reaction* is also related [60c].

218 The reaction was likely first described between 1968 and 1971 [60d, 60e].

61 Mitsunobu Reaction

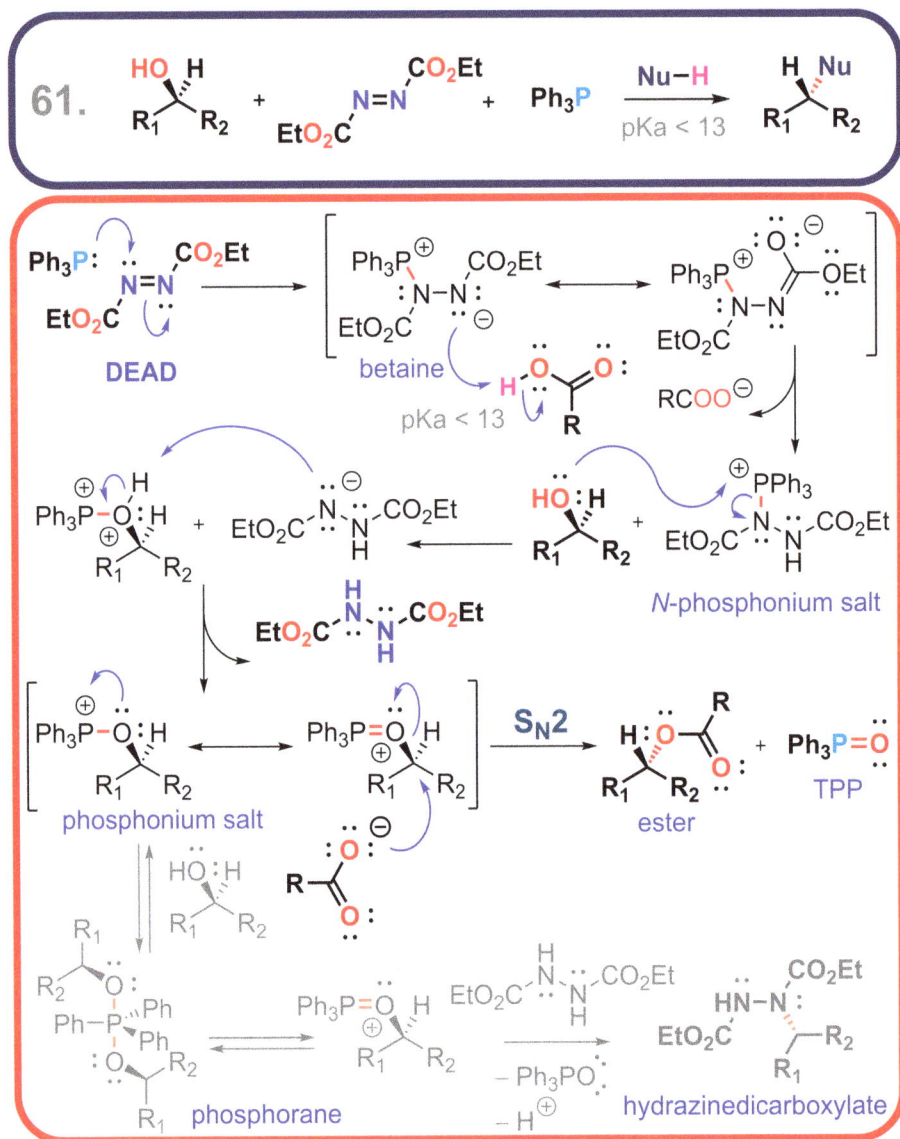

Fig. 61.1: The *Mitsunobu reaction* mechanism.[219]

---

219 The *Mitsunobu reaction* mechanism is complicated but related to the (aliphatic) **nucleophilic substitution** (S_N2) covered in Chapter 2. Note that the pK_a of the NuH acid should be generally < 13 [61a].

https://doi.org/10.1515/9783110786835-061

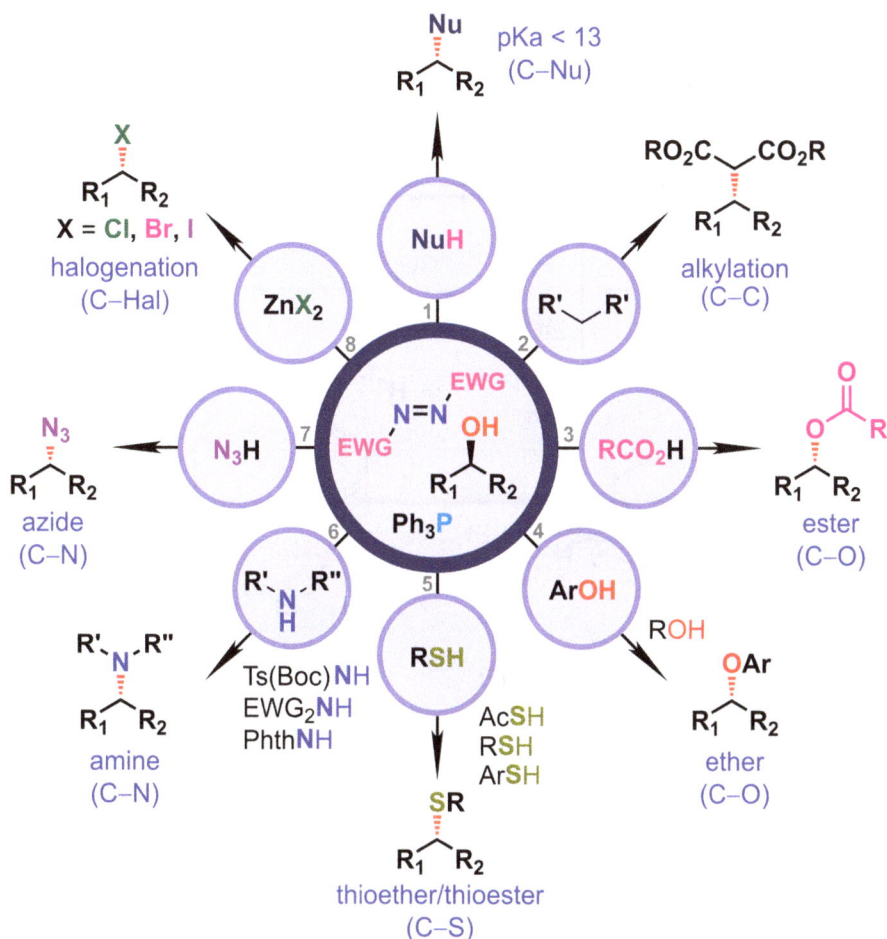

Fig. 61.2: Synthetic versatility of the *Mitsunobu reaction*.[220]

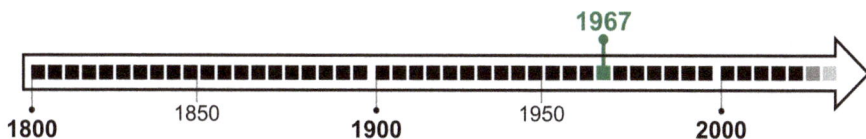

Fig. 61.3: The discovery of the *Mitsunobu reaction*.[221]

---

220 The *Mitsunobu reaction* has wide synthetic application and can convert alcohols into various products using different nucleophiles (Nu): 1. R–Nu, $pK_a < 13$; 2. alkylated products C–C; 3. esters C–O; 4. ethers C–O; 5. thioethers or thioesters C–S; 6. amines C–N; 7. azides C–N; 8. alkyl halides C–X; and so on [61b, 61c].

221 The reaction was likely first described around 1967 [61d, 61e].

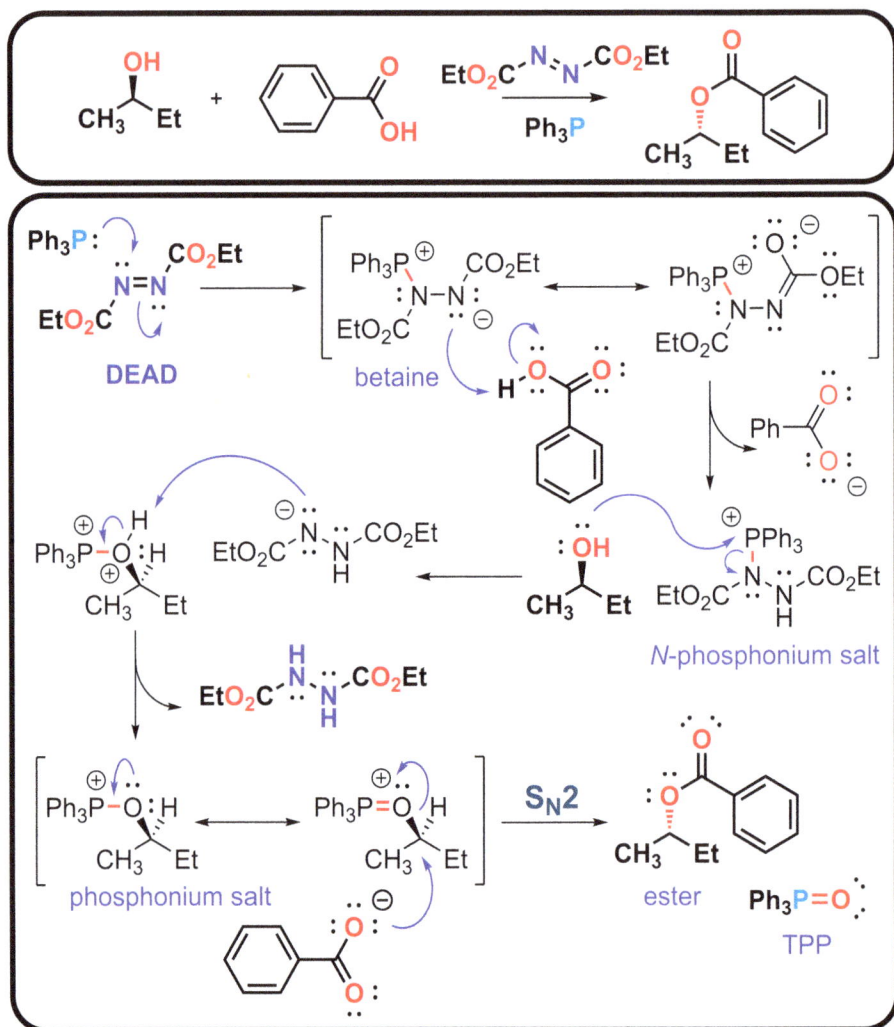

Fig. 61.4: The **Mitsunobu** reaction mechanism (ester synthesis).[222]

---

[222] In this example of ester formation via the **Mitsunobu** reaction mechanism, please notice the inversion of the stereochemistry in the final product. Compare with the **Fischer** esterification (Fig. 61.5) [61f].

Fischer Esterification

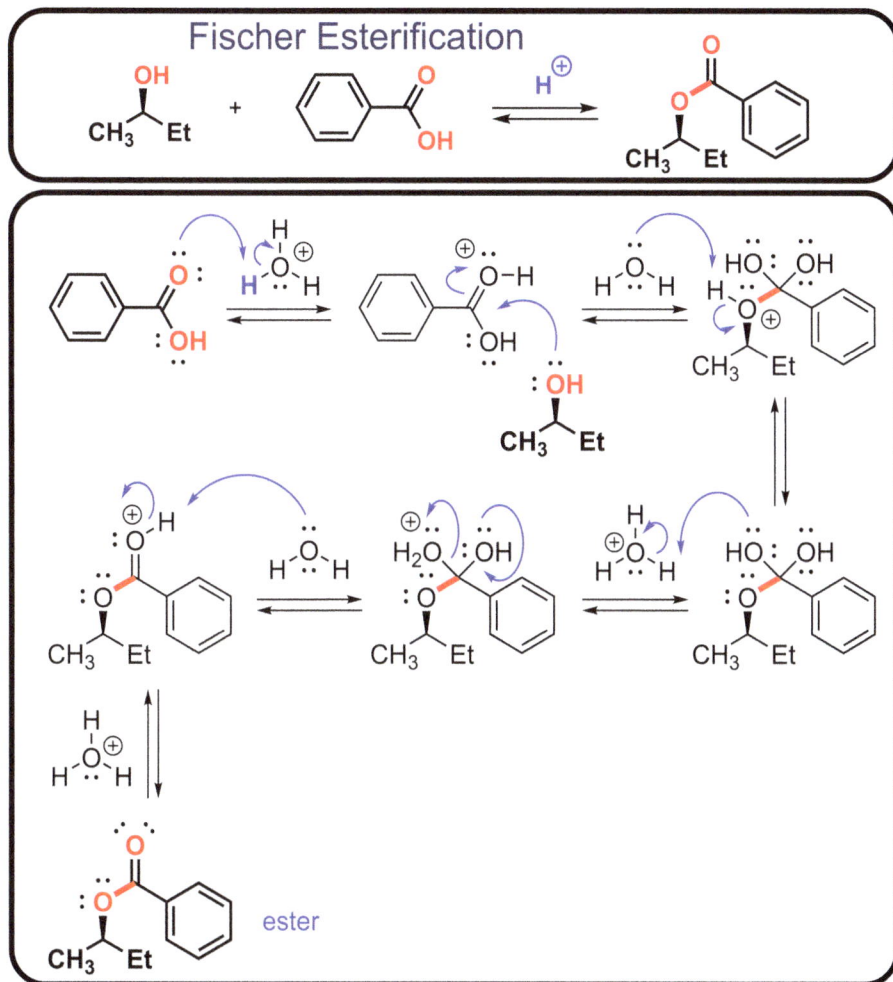

Fig. 61.5: The *Fischer esterification* mechanism.[223]

---

223 In the case of the **Fischer** *esterification* mechanism [61f], please notice the *retention* of the configuration of the chiral center in the final product. Compare with the **Mitsunobu** *reaction* (Fig. 61.4).

62 Miyaura Borylation

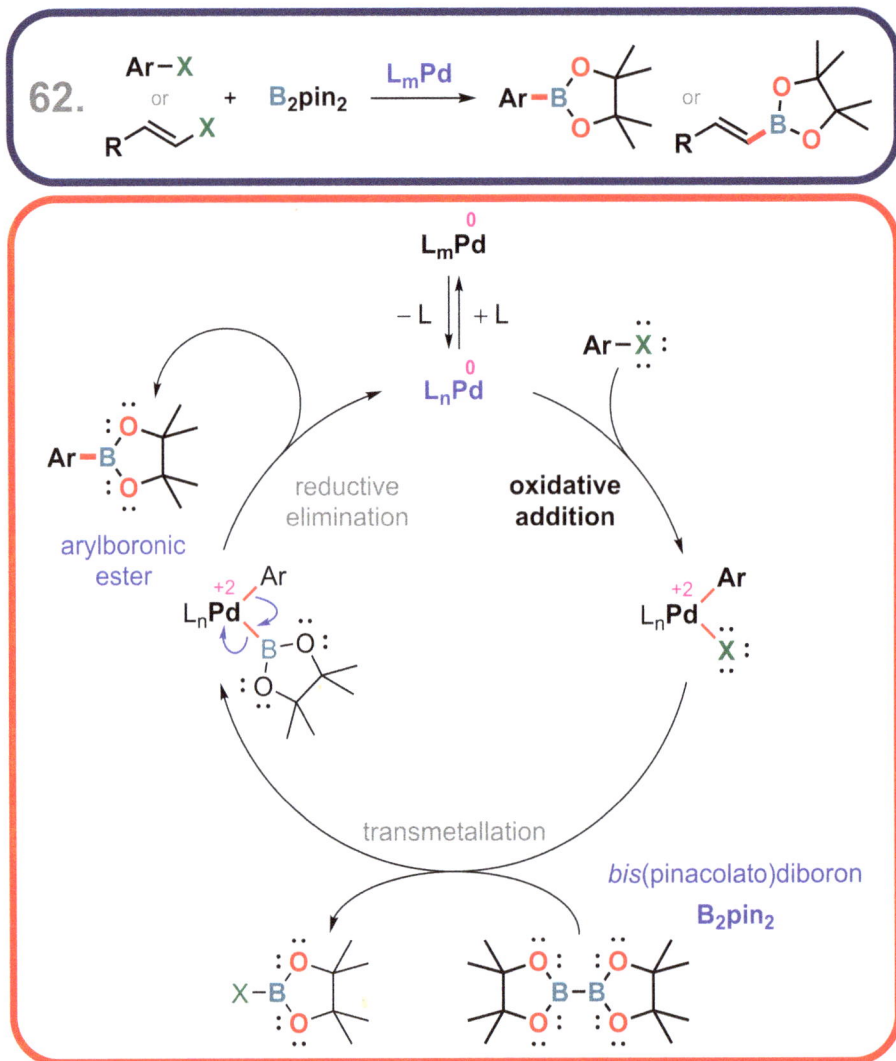

Fig. 62.1: The *Miyaura borylation* mechanism.[224]

---

224 The *Miyaura borylation* is a type of **Pd**-*catalyzed cross-coupling* reaction (C–B bond formation using *aryl halides* and *bis(pinacolato)diboron* or **B$_2$pin$_2$** [62a]). For teaching purposes, a simplified and general mechanism is shown. The synthesized *boronic esters* (and their related *boronic acids*) are one of the most important reagents in synthetic organic and medicinal chemistry.

https://doi.org/10.1515/9783110786835-062

Suzuki Cross Coupling

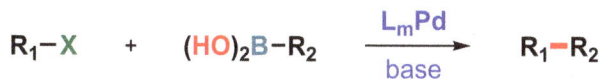

$$R_1-X \quad + \quad (HO)_2B-R_2 \quad \xrightarrow[\text{base}]{L_mPd} \quad R_1-R_2$$

Chan–Evans–Lam Cross Coupling

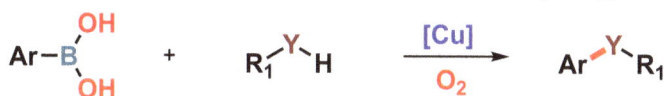

$$Ar-B\overset{OH}{\underset{OH}{|}} \quad + \quad R_1\overset{Y}{\diagdown}H \quad \xrightarrow[O_2]{[Cu]} \quad Ar\overset{Y}{\diagdown}R_1$$

Liebeskind–Srogl Cross Coupling

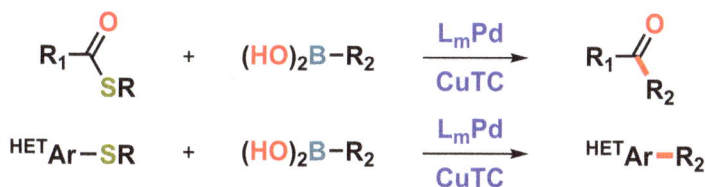

$$R_1\overset{O}{\diagup}{SR} \quad + \quad (HO)_2B-R_2 \quad \xrightarrow[CuTC]{L_mPd} \quad R_1\overset{O}{\diagup}R_2$$

$$^{HET}Ar-SR \quad + \quad (HO)_2B-R_2 \quad \xrightarrow[CuTC]{L_mPd} \quad ^{HET}Ar-R_2$$

Petasis Reaction

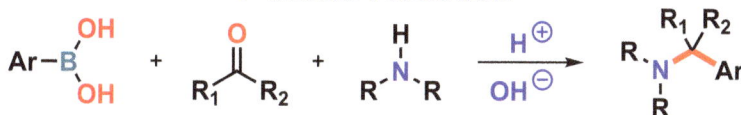

$$Ar-B\overset{OH}{\underset{OH}{|}} \quad + \quad R_1\overset{O}{\diagup}R_2 \quad + \quad R\overset{H}{\underset{N}{\diagup}}R \quad \xrightarrow[OH^{\ominus}]{H^{\oplus}} \quad R\cdot\overset{R_1 \ R_2}{\underset{N}{\diagup}}Ar \ \overset{}{\underset{R}{|}}$$

Fig. 62.2: Synthetic application of boronic esters and acids.[225]

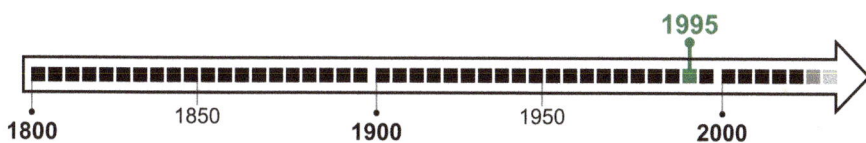

Fig. 62.3: The discovery of the *Miyaura borylation*.[226]

---

225 Many *key cross-coupling* reactions utilize *boronic esters* (and their related *boronic acids*): the *Suzuki cross-coupling* (covered in Chapter 89), the *Chan–Evans–Lam cross-coupling* (covered in Chapter 23), and *Liebeskind–Srogl cross-coupling* (covered in Chapter 55). The *Petasis reaction* is a mechanistically different three-component reaction, but it uses boronic acids as well [62b].
226 The reaction was likely first described around 1995 [62c].

63 Mukaiyama RedOx Hydration

$$63. \quad R\diagup\!\!\!\!\diagdown \quad + \quad O_2 \quad + \quad PhSiH_3 \quad \xrightarrow{\text{Co(acac)}_2} \quad R\overset{\text{OH}}{\underset{}{\diagup}}CH_3$$

Markovnikov's rule

silyl peroxide

cobalt-peroxide adduct

Mechanism by **Nojima**

$$O_2 \equiv \text{:O=O:} \equiv \text{:O-O:}$$

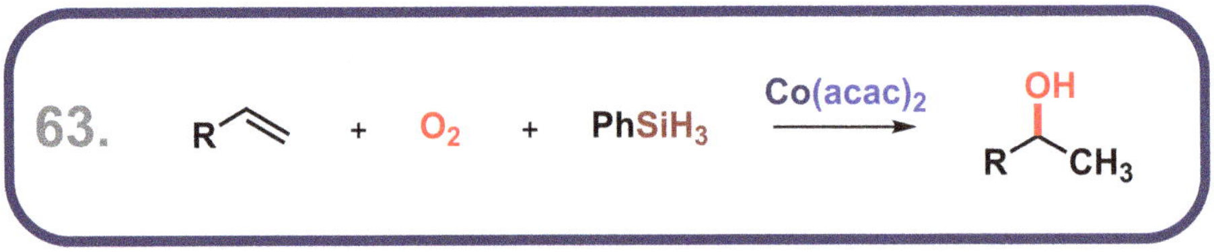

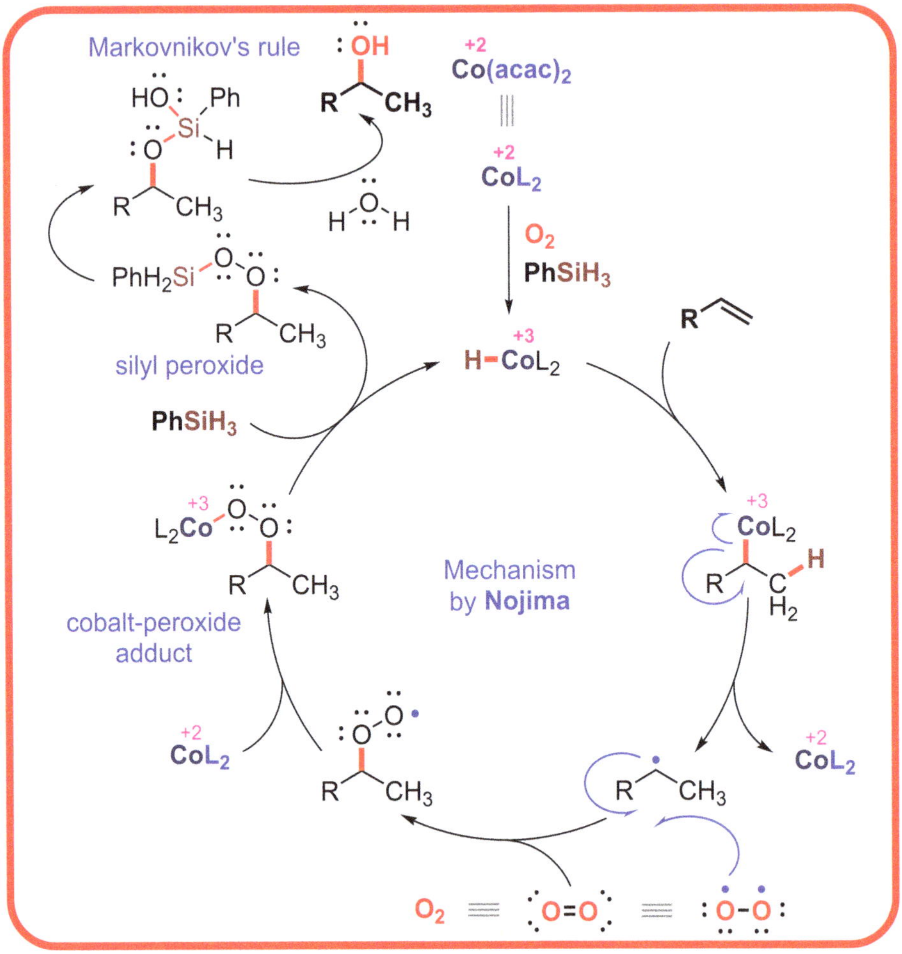

Fig. 63.1: The *Mukaiyama RedOx hydration* mechanism by **Nojima**.[227]

227 The revised *Mukaiyama RedOx hydration* mechanism was recently proposed by **Nojima** [63a].

https://doi.org/10.1515/9783110786835-063

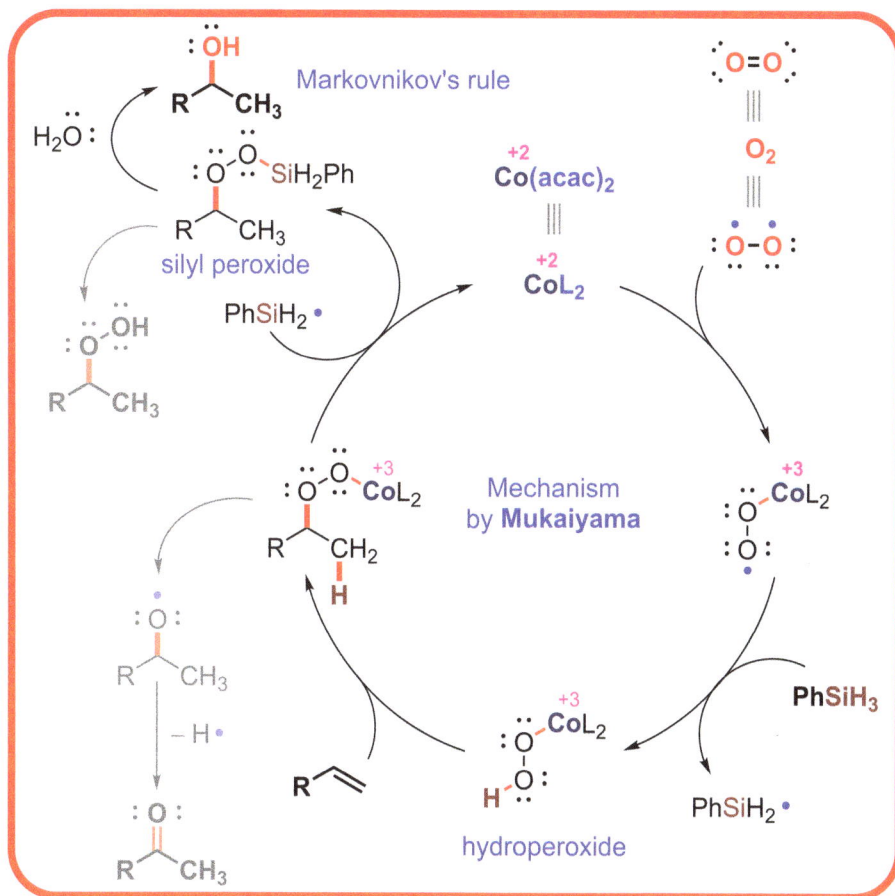

Fig. 63.2: The *Mukaiyama oxidation–reduction hydration* mechanism by **Mukaiyama**.[228]

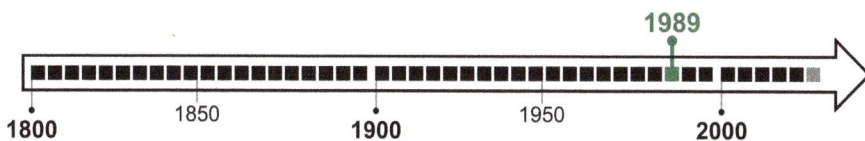

Fig. 63.3: The discovery of the *Mukaiyama oxidation–reduction hydration*.[229]

228 The original *Mukaiyama oxidation–reduction hydration* mechanism was by **Mukaiyama** [63b, 63c, 63d]. The *Mukaiyama oxidation–reduction hydration* should not be confused with the *Mukaiyama aldol addition reaction* (not shown). It follows *Markovnikov's rule*. The *Mukaiyama oxidation–reduction hydration* is a safe alternative to the *oxymercuration–reduction reaction* (Chapters 20 and 52).
229 The reaction was likely first described around 1989 [63b, 63c, 63d].

64 Nazarov Cyclization

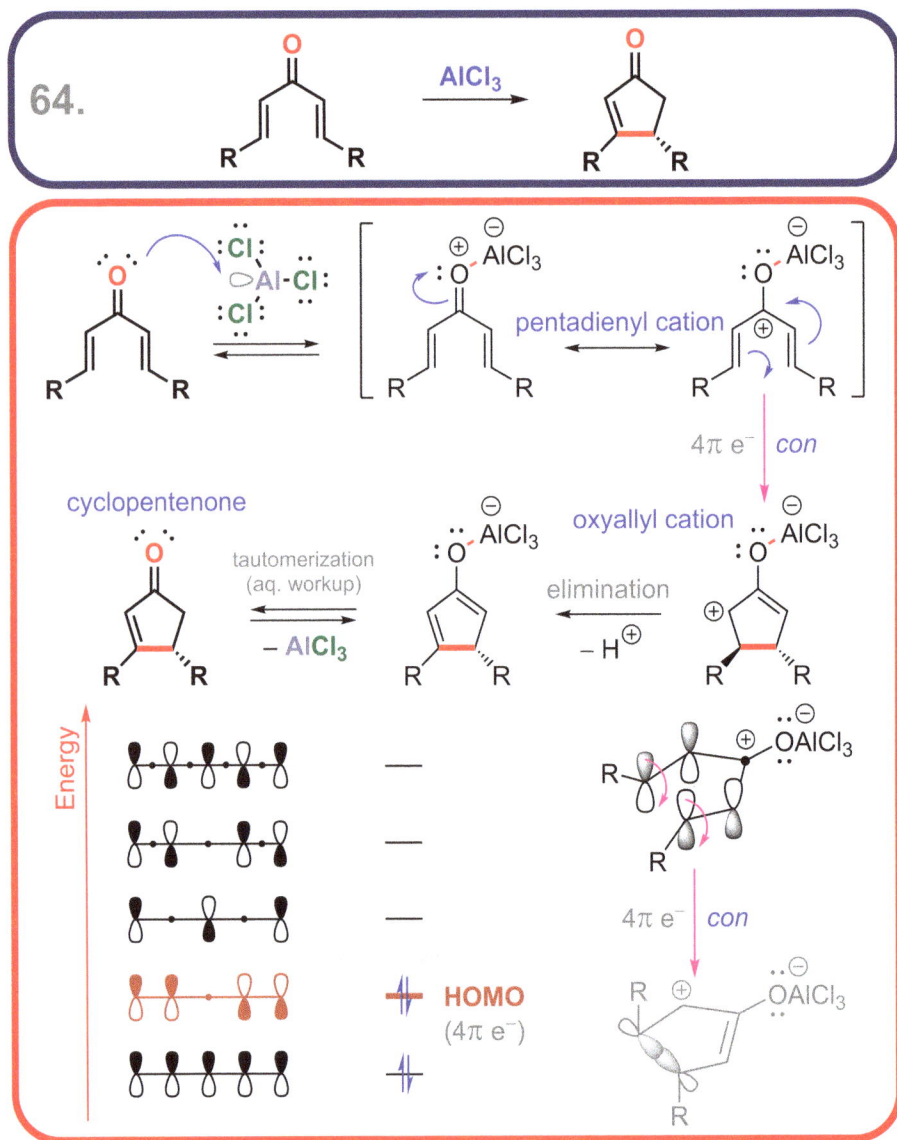

Fig. 64.1: The *Nazarov* cyclization mechanism.[230]

---

230 The *Nazarov cyclization reaction* is a pericyclic reaction with a concerted mechanism. This is an example of a [4π] **con**rotatory *electrocyclization*.

https://doi.org/10.1515/9783110786835-064

Fig. 64.2: The **Woodward–Hoffmann** *rules* (the pericyclic selection rules).[231]

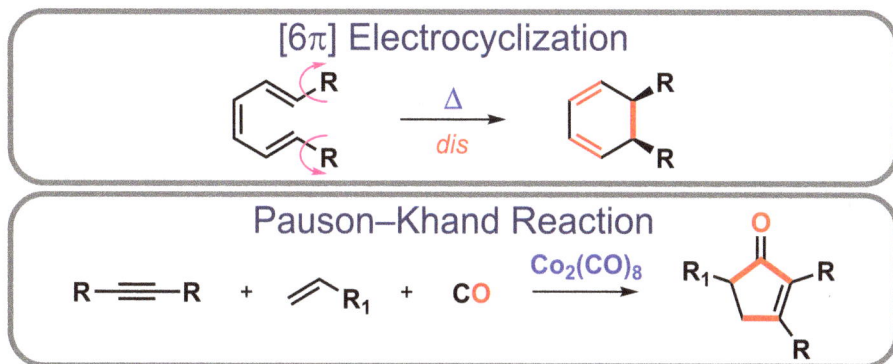

Fig. 64.3: Reactions related to the **Nazarov** *cyclization*.[232]

Fig. 64.4: The discovery of the **Nazarov** *cyclization*.[233]

---

231 The **Woodward–Hoffmann** *rules* (the pericyclic selection rules) [64a, 64b] for the *electrocyclization reactions*. Please note that the **Nazarov** *cyclization* is a **conrotatory** process (**4n** = 4π), which is allowed at the ground state = under thermal conditions or control (Δ). An example of [6π] *electrocyclization* below should be a **disrotatory** process (**4n + 2** = 6π), which is allowed at the ground state (Δ). The outcome at the excited state = under photochemical conditions or control (hν) should be reverse [64c].

232 There are numerous examples of other [4n] *electrocyclic* and [4n + 2] *electrocyclic reactions*. The **Pauson–Khand** *reaction* (see Chapter 73) undergoes a different mechanism, but it also yields *cyclopentenones*.

233 The reaction was likely first described around 1941 [64d, 64e]; see also [64f, 64g].

65 Nef Reaction

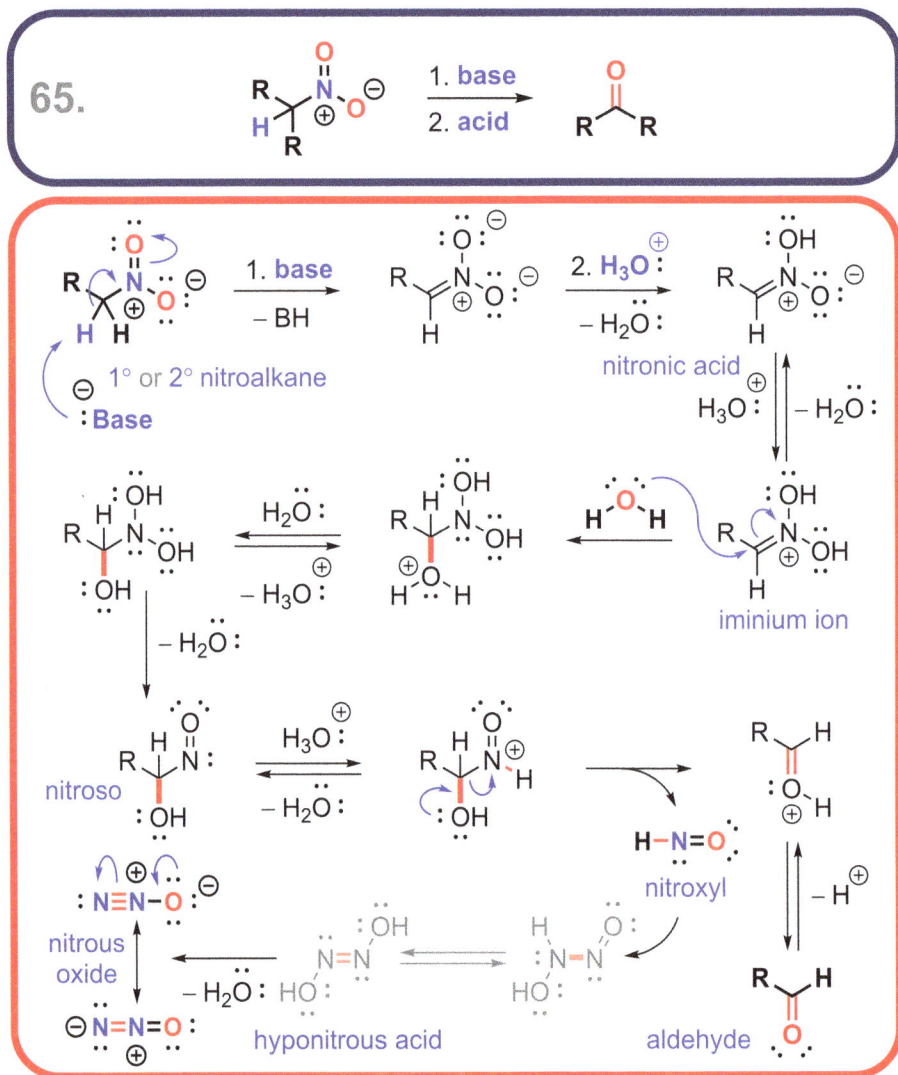

65.

Fig. 65.1: The *Nef reaction* mechanism (base–acid-catalyzed).[234]

---

234 The classic *Nef reaction* is catalyzed by an acid and yields *aldehydes* and *ketones*. A base is needed to convert a primary (1°) or secondary (2°) *nitroalkane* into its conjugate base (*nitronic acid*). The tertiary (3°) nitroalkanes do not react.

https://doi.org/10.1515/9783110786835-065

Nef Reaction (acid)

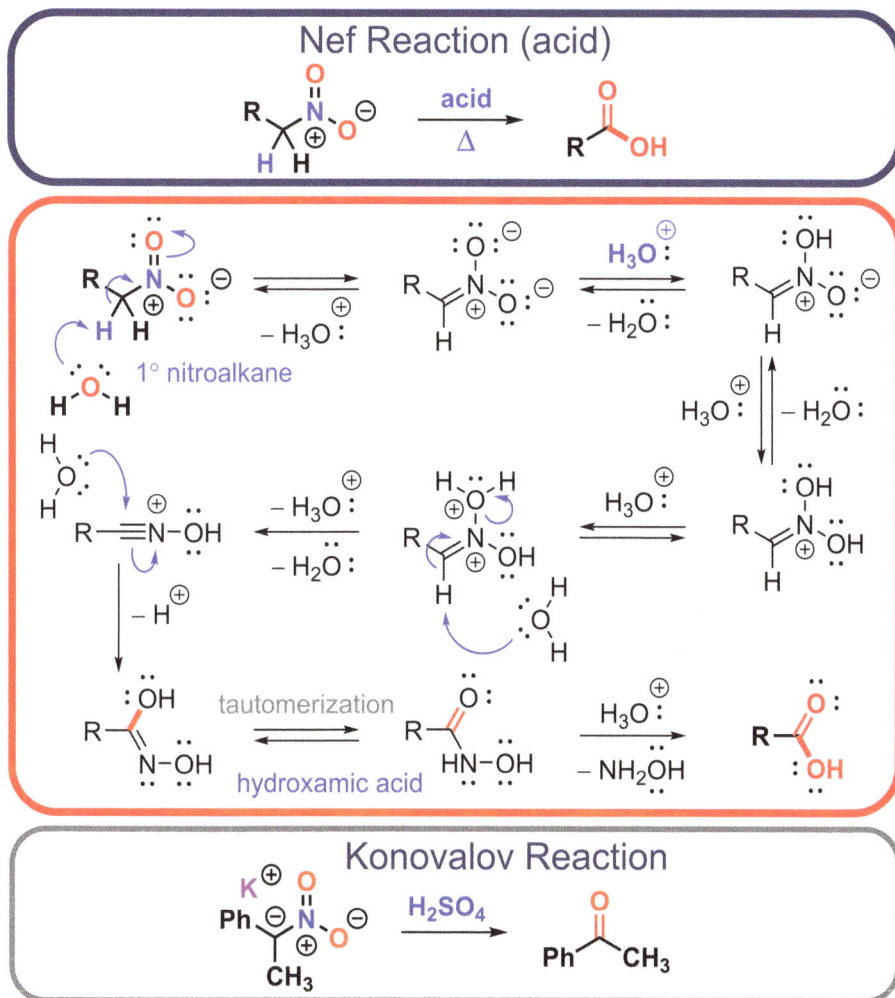

Konovalov Reaction

Fig. 65.2: The **Nef** reaction mechanism (acid-catalyzed).[235]

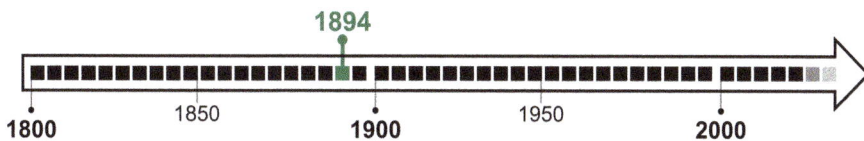

1894

1800   1850   1900   1950   2000

Fig. 65.3: The discovery of the **Nef** reaction.[236]

---

235 The mechanism of the **Nef** reaction can change and go through a *hydroxamic acid* intermediate if a strong acid (exclusively) is used with a primary (1°) *nitroalkane*. In this case, a *carboxylic acid* is formed [1, 65a]. Please note that the reaction was likely first reported by Konovalov [65b].
236 The reaction was likely first described around 1894 [65c, 65d].

66 Negishi Cross-Coupling

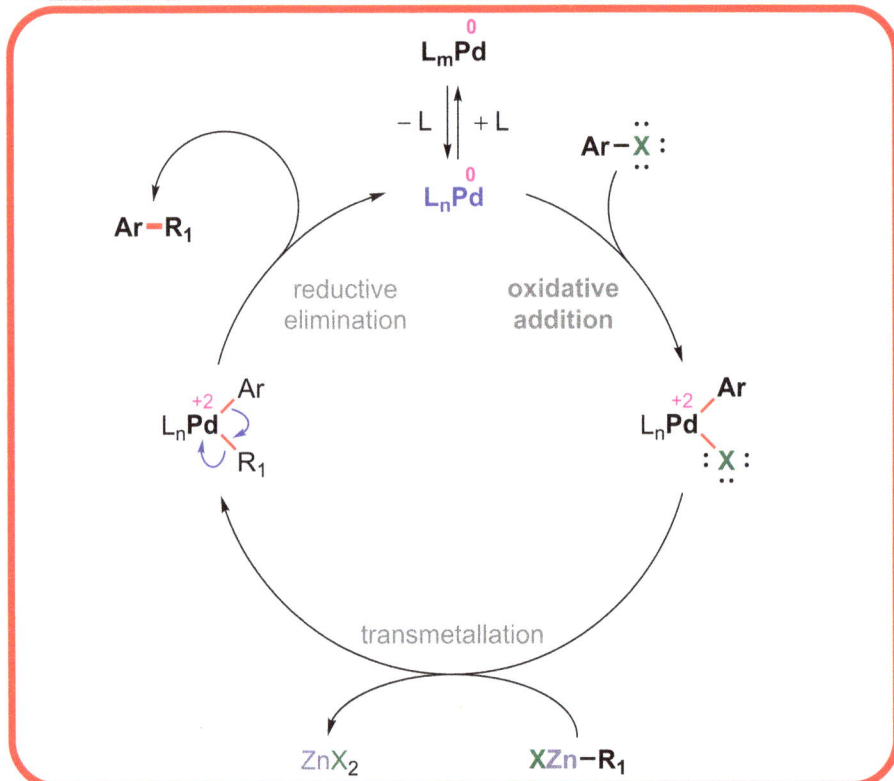

66.

Ar–X
or
R $\diagdown$ X

+ XZn–R$_1$ $\xrightarrow{\text{L}_m\text{Pd} \\ \text{L}_m\text{Ni}}$

Ar–R$_1$
or
R $\diagdown$ R$_1$

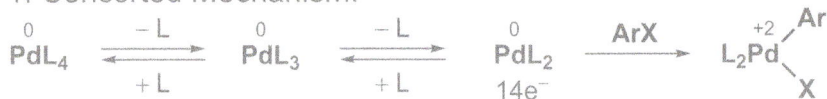

Fig. 66.1: The **Pd**-catalyzed *Negishi cross-coupling* mechanism.[237]

---

237 The *Negishi cross-coupling* is a type of *Pd-catalyzed cross-coupling* reaction (C–C bond formation using *aryl halides* and *organozinc compounds*). For teaching purposes, a simplified and general mechanism is shown. Note that the (1) *concerted oxidative addition* step to a low-coordinate (14e$^-$) **Pd**-complex is more complicated [2a].

https://doi.org/10.1515/9783110786835-066

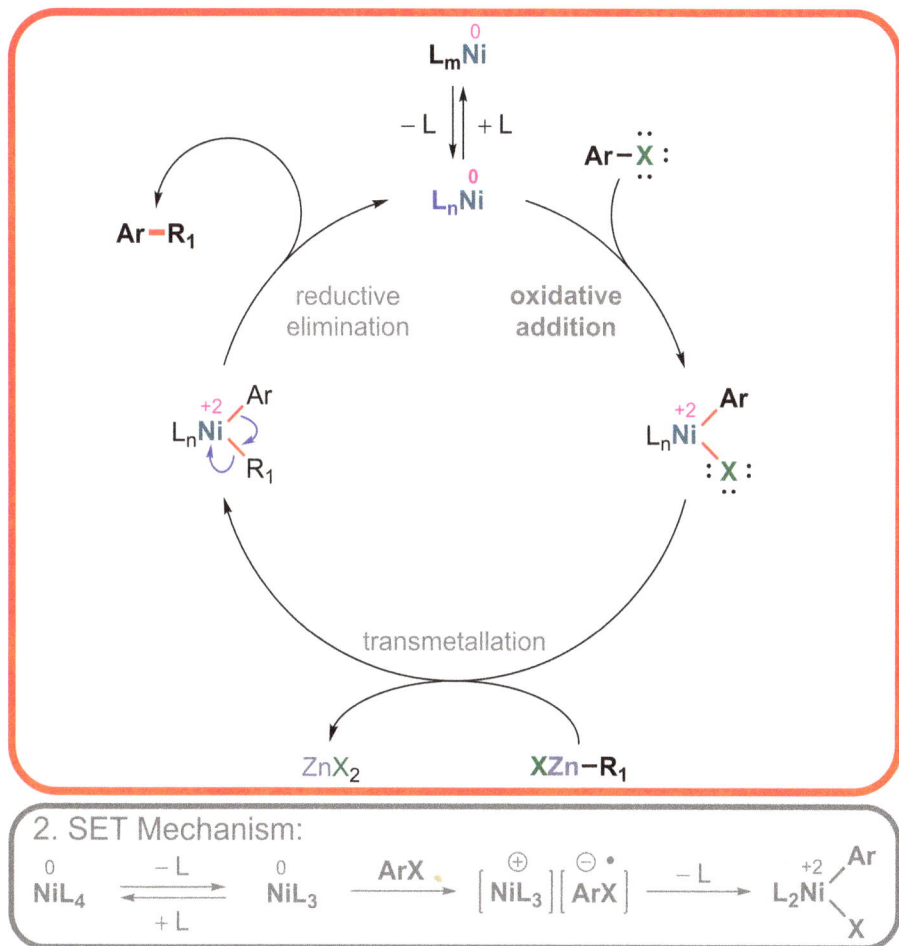

Fig. 66.2: The **Ni**-catalyzed *Negishi cross-coupling* mechanism.[238]

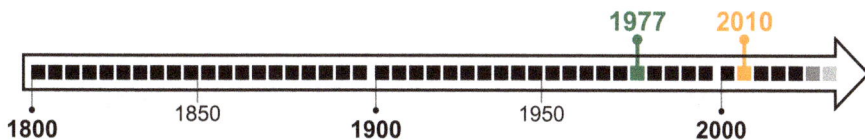

Fig. 66.3: The discovery of the *Negishi cross-coupling*.[239]

---

238 The *Negishi cross-coupling* can be *Ni-catalyzed*. Note the possible example of an (2) *SET oxidative addition* step to a **Ni**-complex (not necessarily at play in the example shown) [2a].

239 The reaction was likely first described around 1977 [66]. In **2010**, Ei-ichi Negishi (jointly with Richard F. Heck and Akira Suzuki) received the Nobel Prize in Chemistry for the development of **Pd**-catalyzed cross-coupling reactions [46c].

67 Norrish Type I and II Reactions

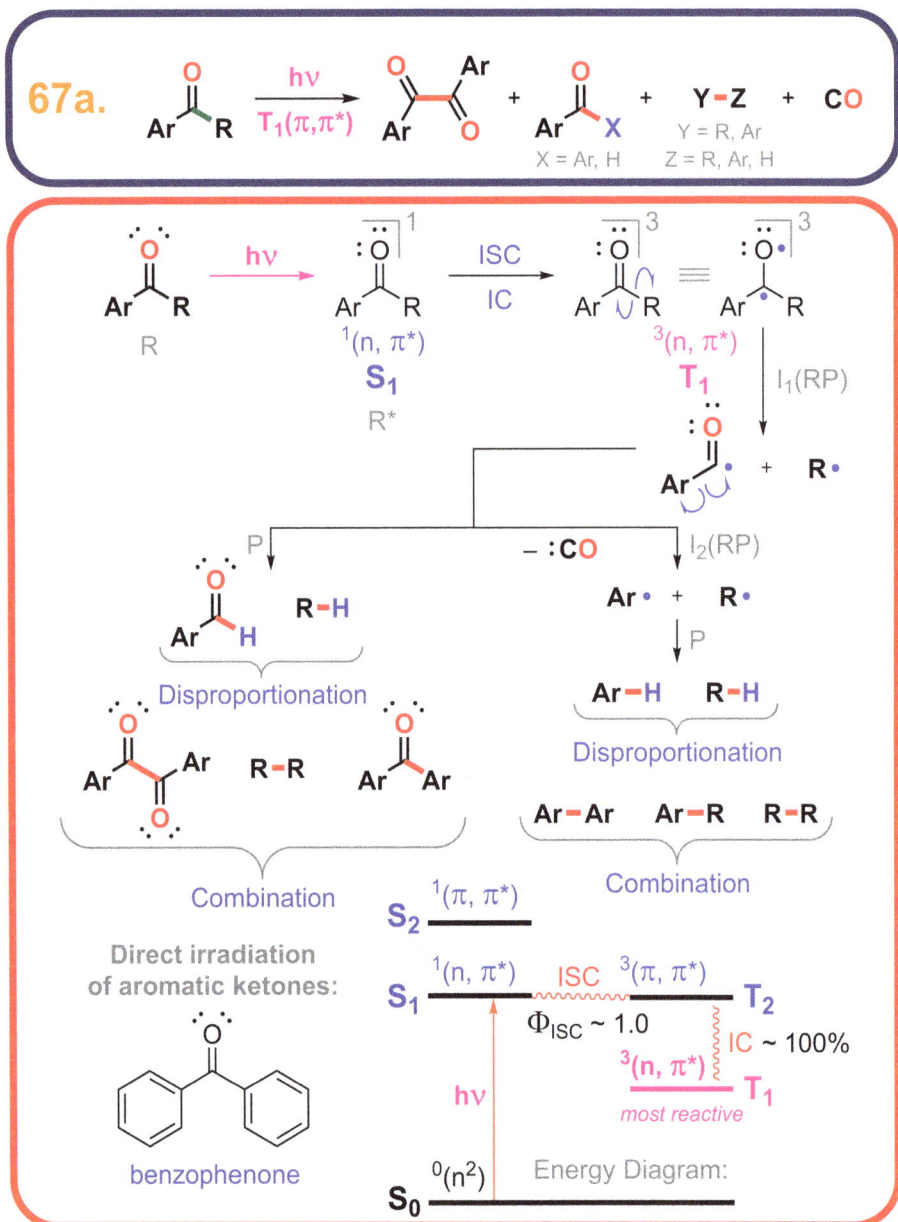

67a.

$$Ar \overset{O}{\underset{R}{\big|}} \quad \xrightarrow[T_1(\pi,\pi^*)]{h\nu} \quad \text{products} \quad + \quad Ar\overset{O}{\underset{X}{\big|}} \quad + \quad Y-Z \quad + \quad CO$$

X = Ar, H
Y = R, Ar
Z = R, Ar, H

Direct irradiation of aromatic ketones:

benzophenone

Energy Diagram:

Fig. 67.1: The **Norrish type I reaction** mechanism.[240]

https://doi.org/10.1515/9783110786835-067

67b.

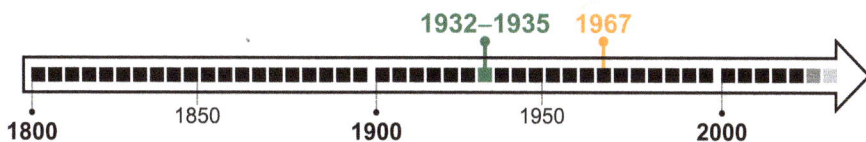

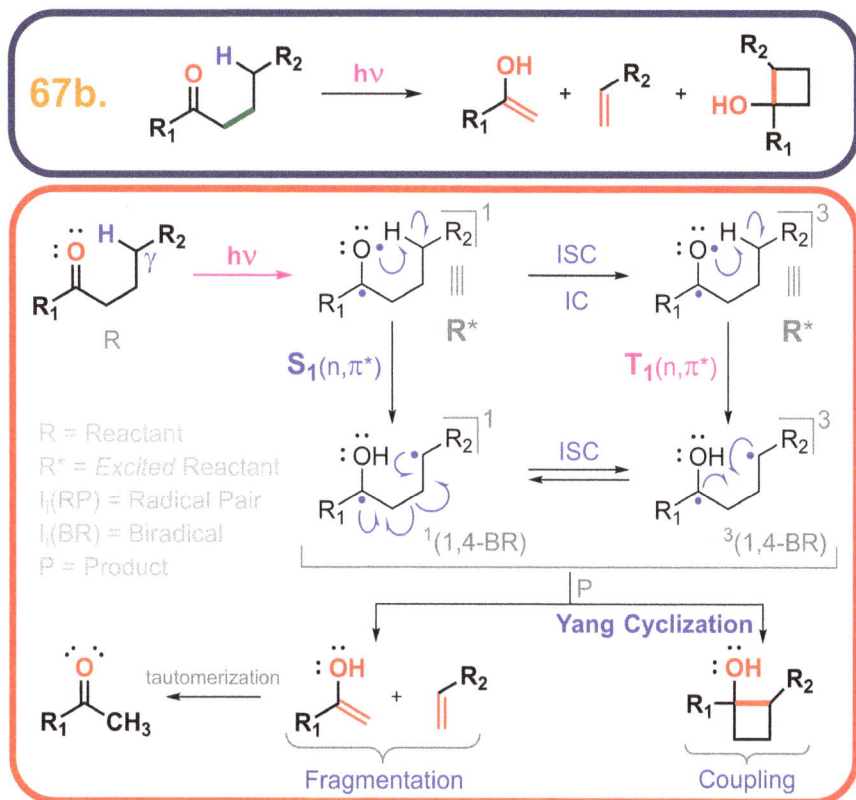

Fig. 67.2: The **Norrish type II reaction** mechanism.[241]

1932–1935 **1967**

1800   1850   1900   1950   2000

Fig. 67.3: The discovery of the **Norrish fragmentation**.[242]

---

240 The **Norrish type I reaction** is a photochemical decomposition (*α-cleavage*) of *aldehydes* and *ketones*. The products are formed after initial *fragmentation* and subsequent *disproportionation* or *(re) combination* of radical species. Upon *direct irradiation* of aromatic ketones (benzophenone), the reaction usually occurs from the <u>*triplet*</u> excited state $T_1 = {}^3(n, \pi^*)$ [2b].

241 The **Norrish type II reaction** is a photochemical intramolecular **γ-H abstraction**. The products are formed due to *fragmentation, (re)combination,* or the **Yang** *cyclization* of 1,4-biradicals. The reaction may occur from the <u>*singlet*</u> $S_1 = {}^1(n, \pi^*)$ or <u>*triplet*</u> excited state $T_1 = {}^3(n, \pi^*)$ [2b].

242 The **type I** and **II** reactions were likely first described between 1932 and 1935 [67a, 67b, 67c, 67d] or possibly earlier; see also [67e, 67f]. In **1967**, Ronald George Wreyford Norrish (jointly with Manfred Eigen and George Porter) received the Nobel Prize in Chemistry [67g].

68 Olefin (Alkene) Metathesis

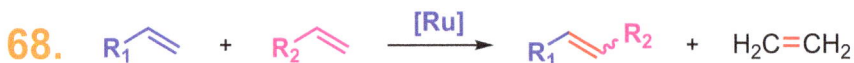

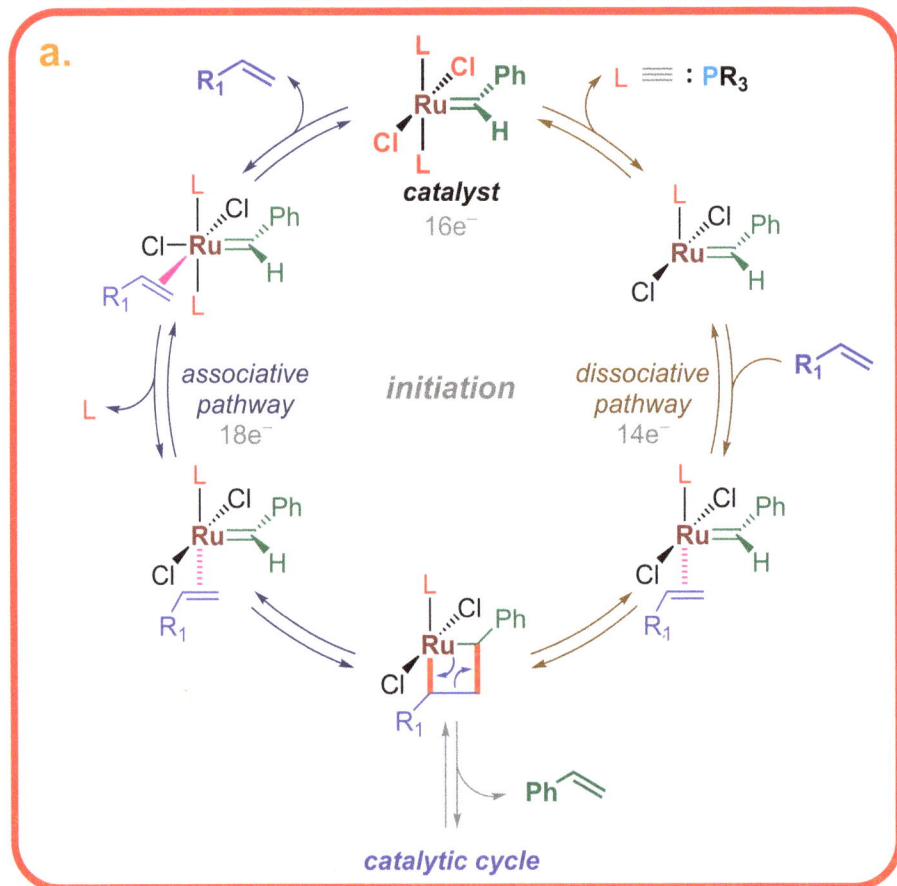

Fig. 68.1: The *olefin (alkene) metathesis* mechanism (initiation).[243]

---

243 The **Ru**-catalyzed olefin (alkene) metathesis mechanism starts with the *stable catalyst (16e⁻) initiation cycle (**a**)*: theoretically it can go either via a dissociative pathway (14e⁻), or an associative pathway (18e⁻), and an interchange pathway is not shown here [68a].

https://doi.org/10.1515/9783110786835-068

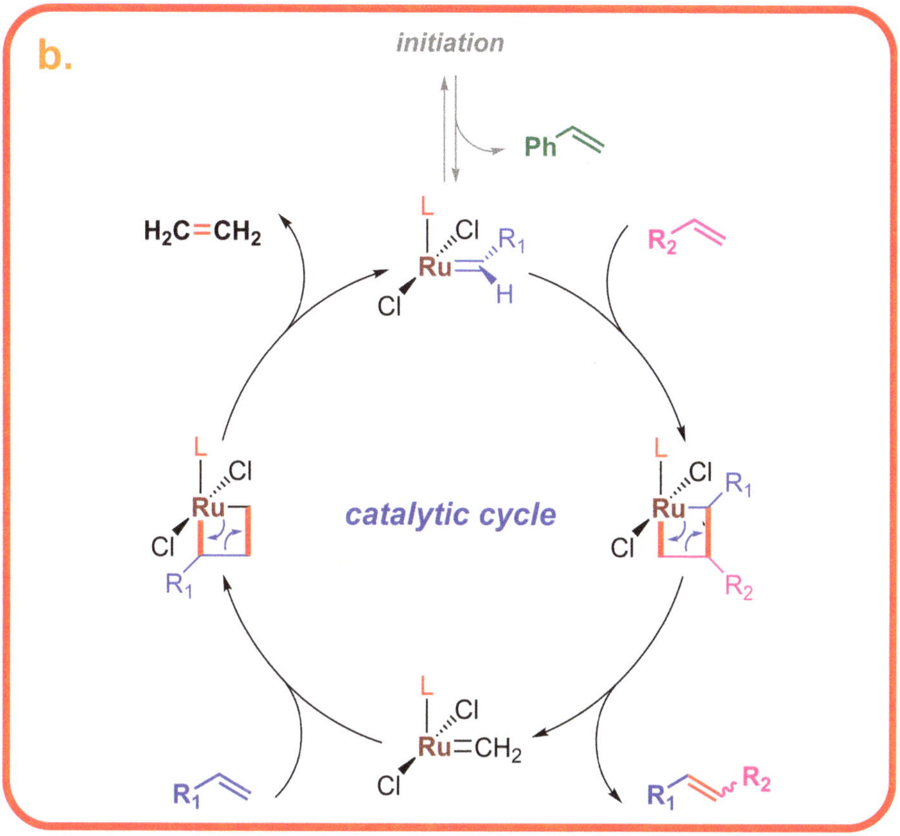

b.

Fig. 68.2: The *olefin (alkene) metathesis* mechanism (catalytic cycle).[244]

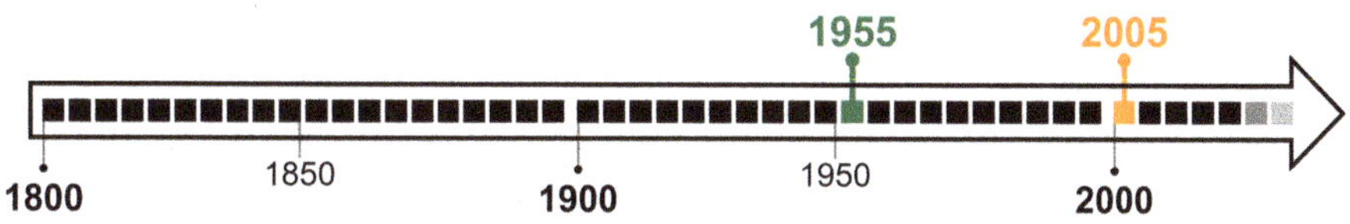

Fig. 68.3: The discovery of the *olefin metathesis*.[245]

---

244 After the loss of *styrene*, the *main catalytic cycle (b)* continues with the *"active"* catalyst. Please note that the mechanism is rather complex and varies significantly depending on the substrate and catalyst. For teaching purposes, a simplified and general example is shown.

245 The reaction was likely first described around **1955** [68b, 68c]. In **2005**, Yves Chauvin, Robert H. Grubbs, and Richard R. Schrock received the Nobel Prize in Chemistry for the development of the *metathesis* transformations [68d].

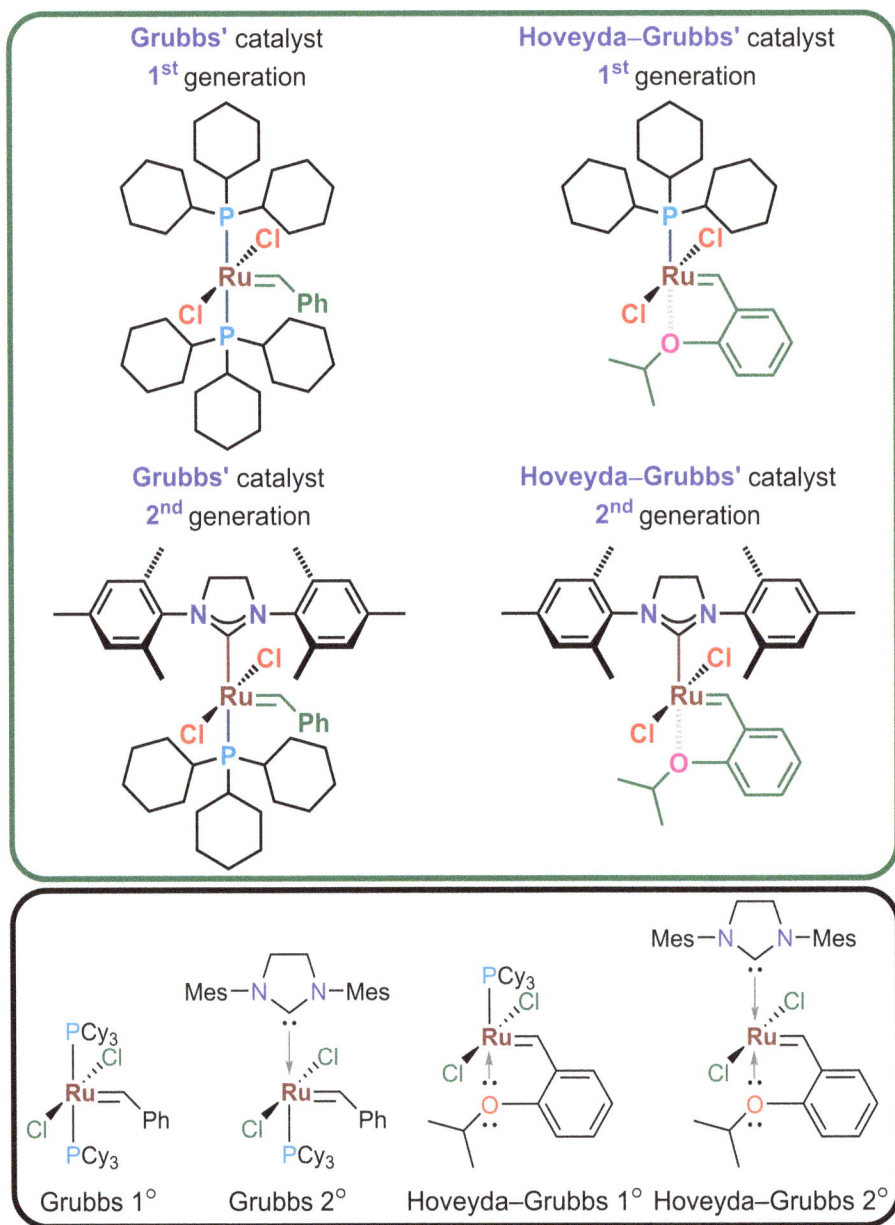

Fig. 68.4: The main *olefin (alkene) metathesis* catalysts.[246]

---

[246] The most common catalysts used in the ***Ru-catalyzed olefin metathesis*** are ***Grubbs' catalysts*** (first and second generation) [68e, 68f] and ***Hoveyda–Grubbs' catalysts*** (first and second generation) [68g].

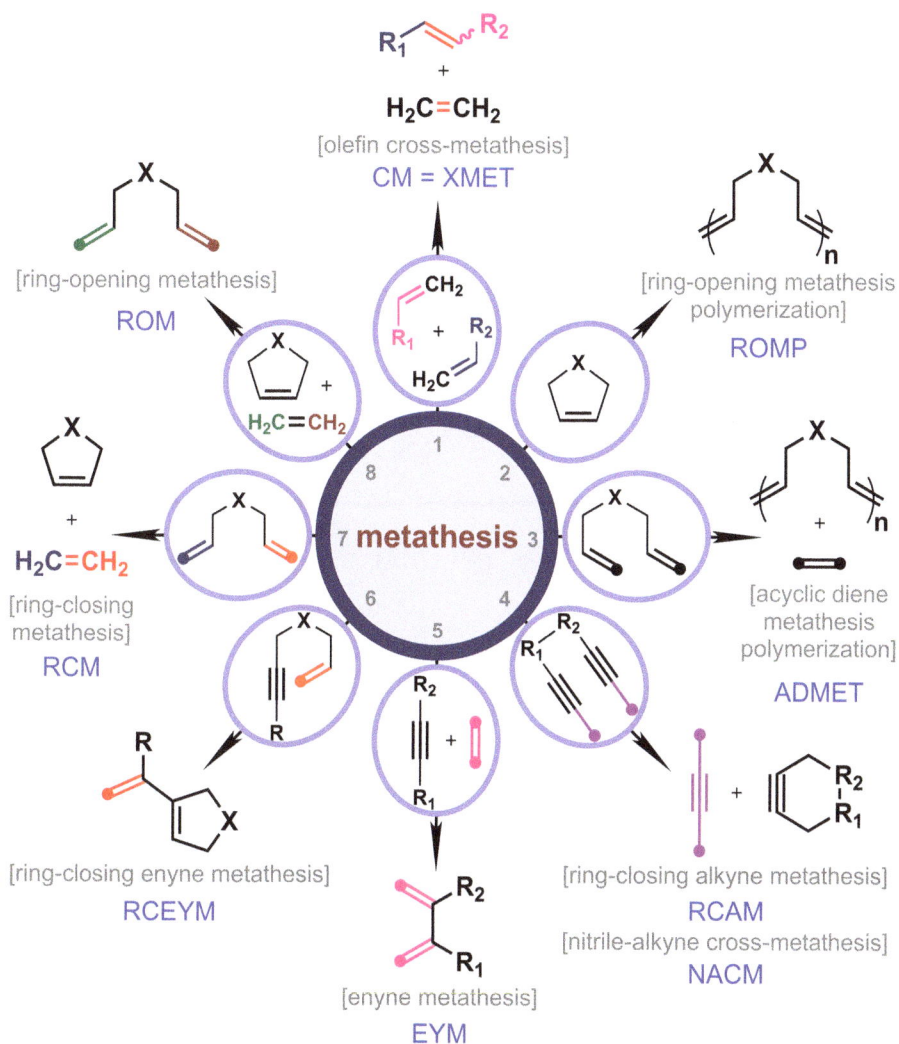

Fig. 68.5: Classification of *metathesis* reactions.[247]

---

247 The metathesis reactions can be classified as: 1. CM = XMET (olefin cross-metathesis); 2. ROMP (ring-opening metathesis polymerization); 3. ADMET (acyclic diene metathesis polymerization); 4. RCAM (ring-closing alkyne metathesis) and NACM (nitrile–alkyne cross-metathesis); 5. EYM (enyne metathesis); 6. RCEYM (ring-closing enyne metathesis); 7. **RCM** (ring-closing metathesis); and 8. ROM (ring-opening metathesis).

69 Oppenauer Oxidation

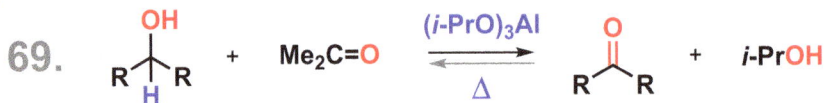

69.
$$\underset{\substack{R \;\; R \\ H}}{\overset{OH}{|}} + Me_2C=O \underset{\Delta}{\overset{(i\text{-PrO})_3Al}{\rightleftharpoons}} \underset{R \;\; R}{\overset{O}{\|}} + i\text{-PrOH}$$

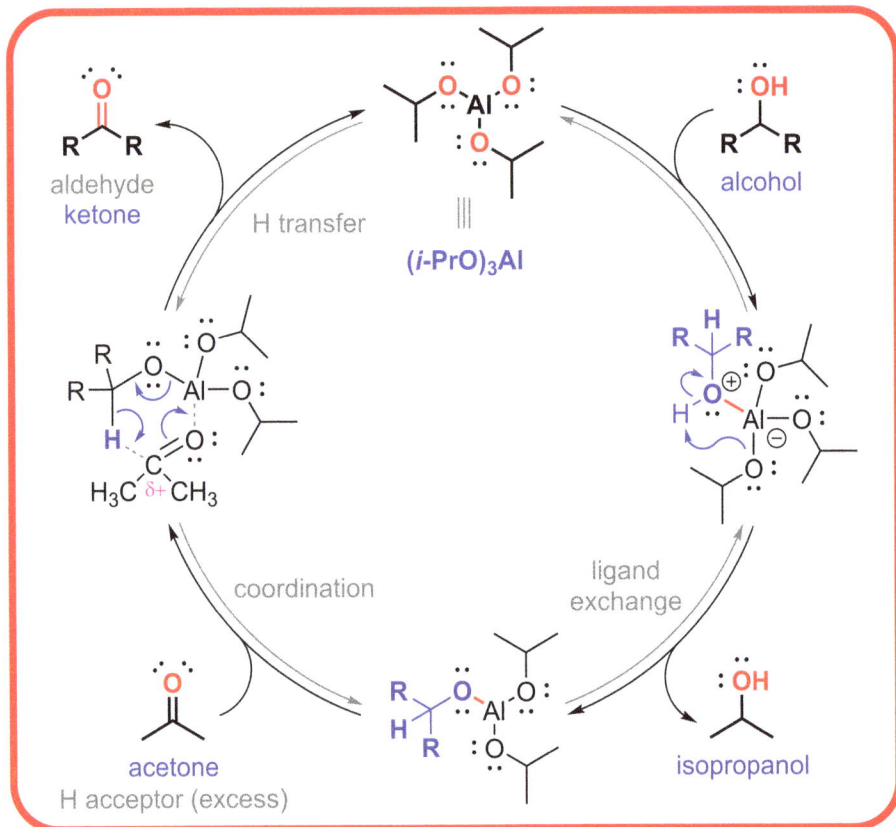

Fig. 69.1: The **Oppenauer** oxidation mechanism.[248]

---

248 The **Oppenauer** oxidation is reversible. The reversed reduction is called the **Meerwein–Ponndorf–Verley** (MPV) *reduction*. The equilibrium can be shifted towards *oxidation* by adding the excess of *acetone*.

https://doi.org/10.1515/9783110786835-069

Meerwein–Ponndorf–Verley Reduction

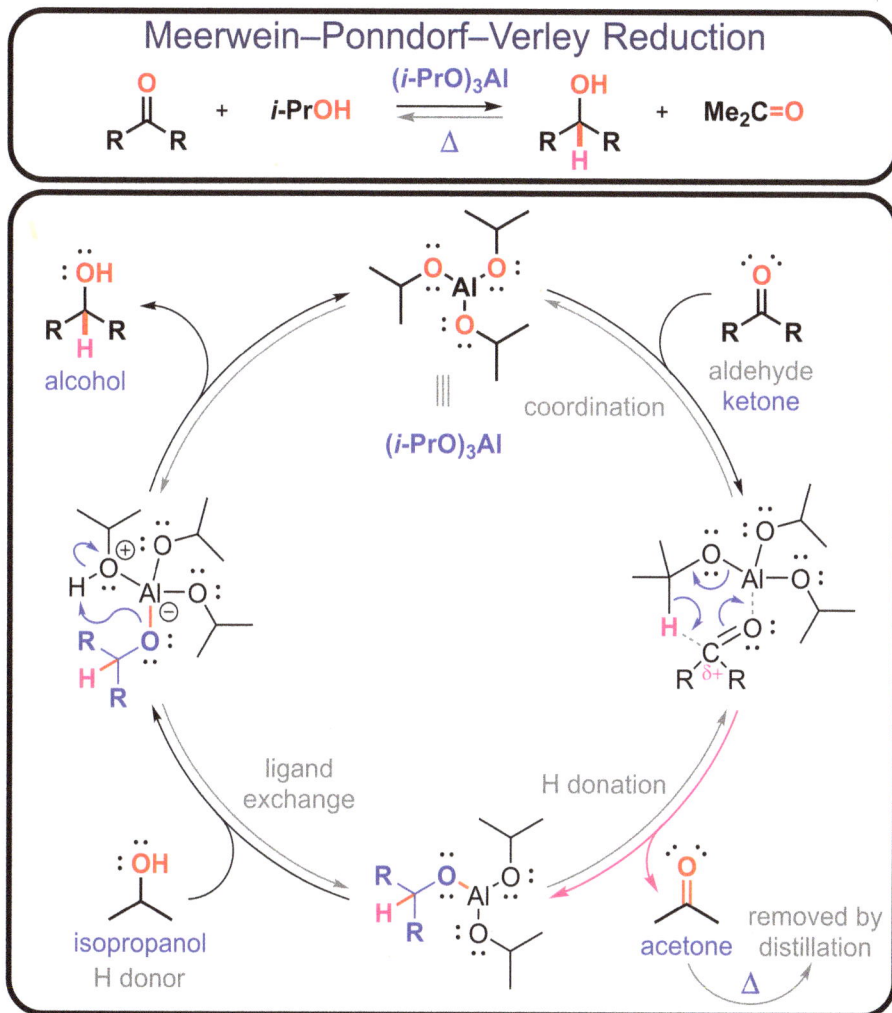

Fig. 69.2: The *Meerwein–Ponndorf–Verley* reaction mechanism.[249]

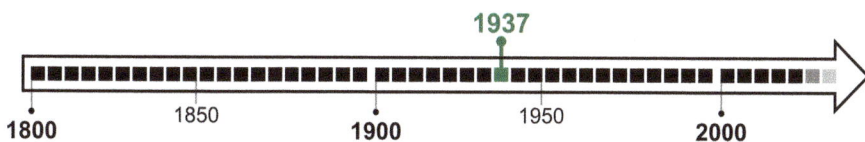

Fig. 69.3: The discovery of the *Oppenauer oxidation*.[250]

---

249 The **MPV** *reduction* is a reversed process of the **Oppenauer** *oxidation*. It is also covered in Chapter 58.
250 The reaction was likely first described around 1937 [69].

70 Ozonolysis

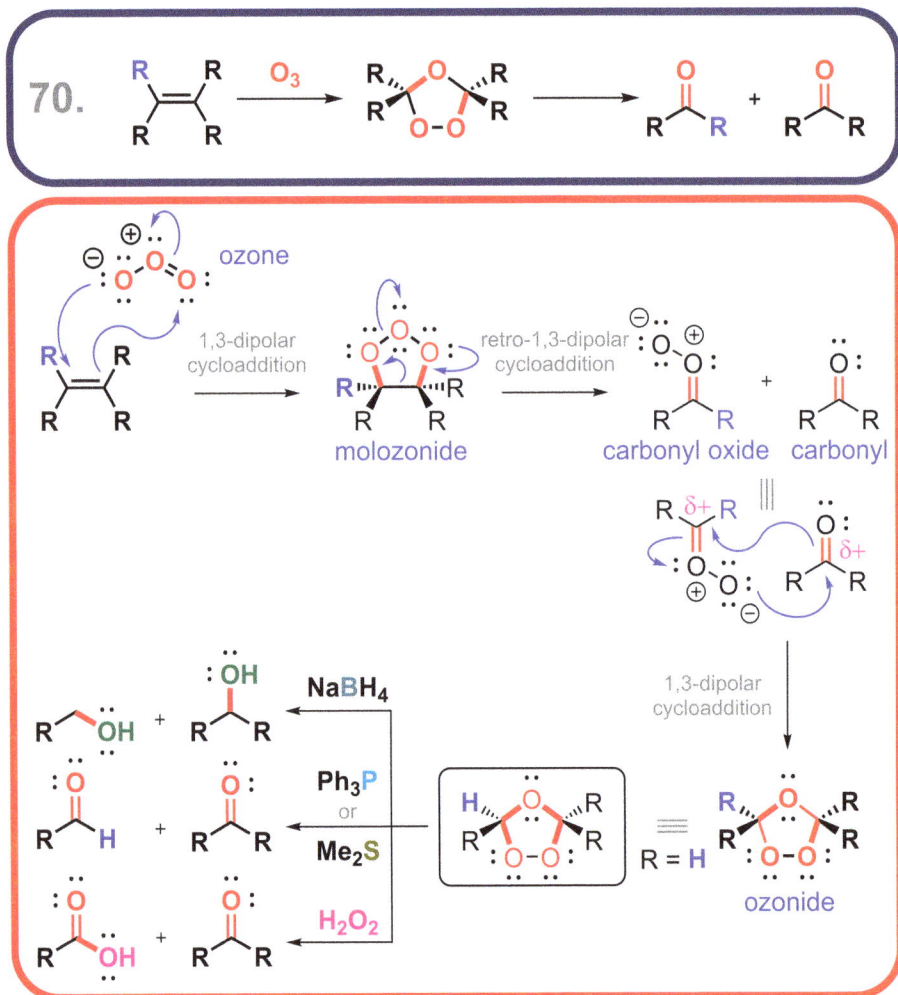

Fig. 70.1: The *ozonolysis* mechanism (the ***Criegee*** mechanism).[251]

_____

251 The *ozonolysis* mechanism was first proposed by Criegee [70a, 70b, 70c]; thus, it is often called the ***Criegee*** *mechanism* (it is different from *the **Criegee** oxidation* covered in Chapter 29). Formally, the first step of *ozonolysis* is a *1,3-dipolar cycloaddition reaction* or a (3 + 2)-*cycloadditon reaction*.

https://doi.org/10.1515/9783110786835-070

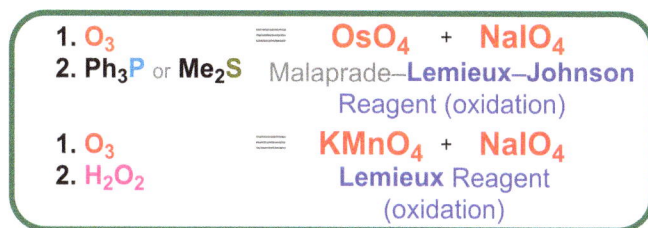

Fig. 70.2: Alternative to the *ozonolysis* reaction conditions.[252]

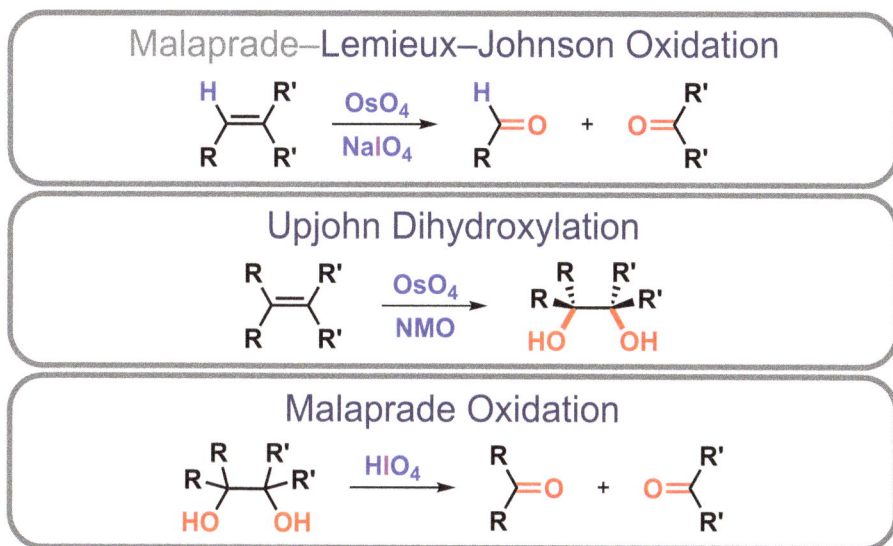

Malaprade–Lemieux–Johnson Oxidation

Upjohn Dihydroxylation

Malaprade Oxidation

Fig. 70.3: Reactions related to the *ozonolysis*.[253]

---

252 The *Malaprade–Lemieux–Johnson* reagent [70d] is an alternative to the use of *ozone* [70e], followed by Ph_3P or Me_2S to form *aldehydes* and *ketones*. The **Lemieux** reagent [70f] is an alternative to the use of *ozone*, followed by H_2O_2, to form *carboxylic acids* and *ketones*.

253 The *Malaprade–Lemieux–Johnson* reaction (oxidation) is an alternative to the *ozonolysis* reaction under Ph_3P or Me_2S conditions. The **Upjohn** dihydroxylation (covered in Chapter 93) followed by the **Malaprade** oxidation (covered in Chapter 29) can be also used as an alternative to *ozonolysis*.

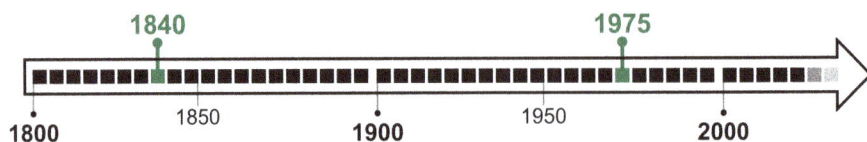

Fig. 70.4: The discovery of the *ozonolysis*.[254]

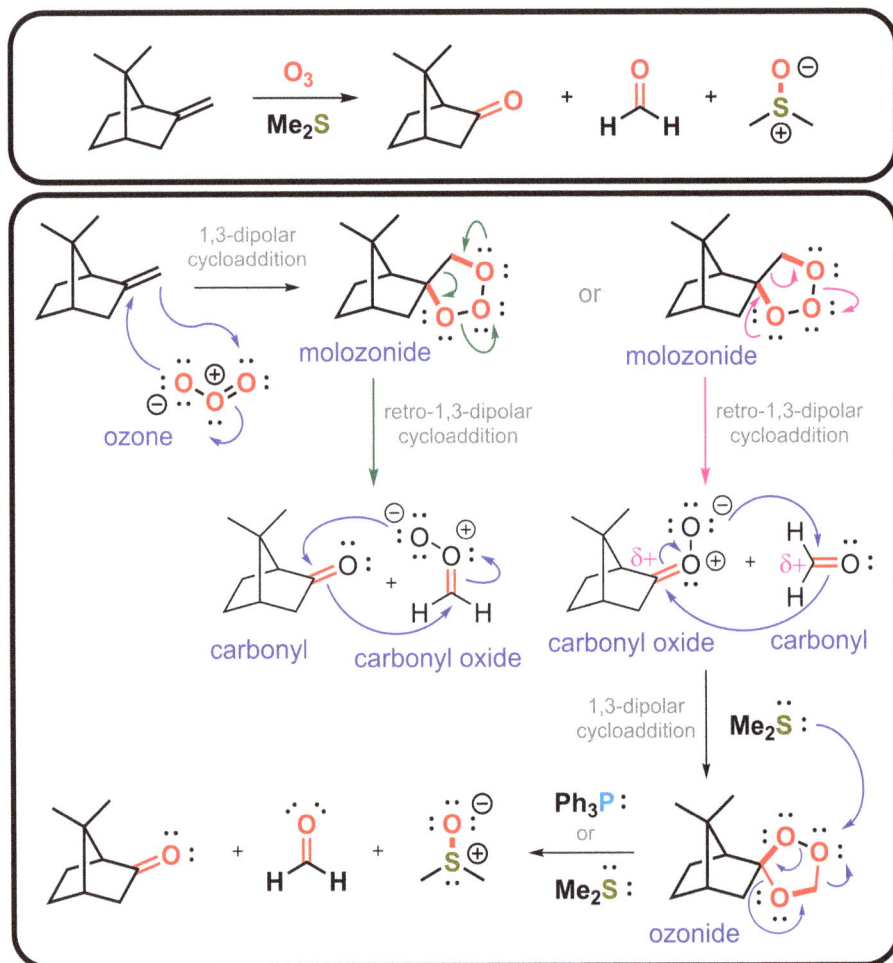

Fig. 70.5: The *ozonolysis* reaction mechanism of (–)-α-fenchene.[255]

254 The reaction was likely first described around 1840 [70g], and the mechanism was proposed around 1975 [70b, 70c].

255 An example of the *ozonolysis* reaction mechanism of (1*R*,4*S*)-7,7-dimethyl-2-methylenebicyclo [2.2.1]heptane (also known as (–)-α-fenchene).

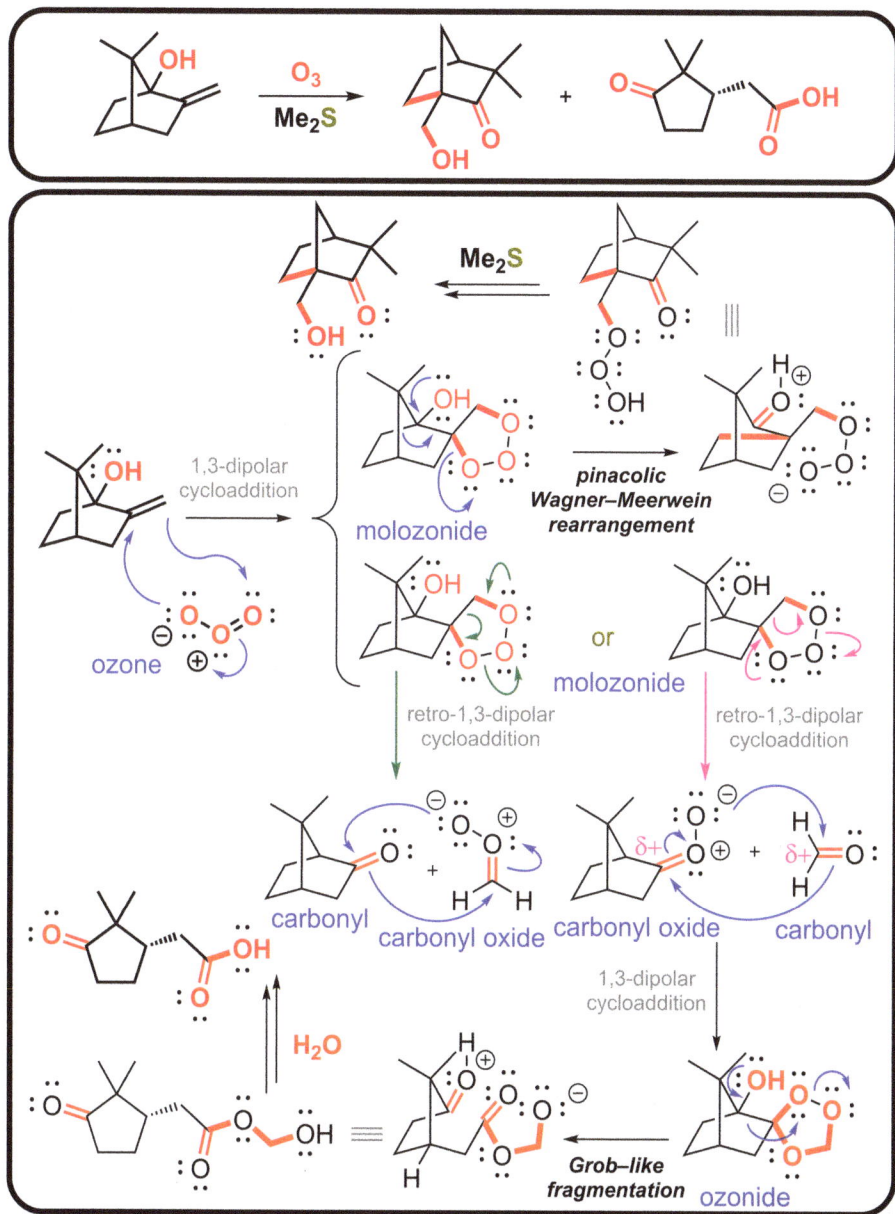

Fig. 70.6: An anomalous (interrupted) *ozonolysis* reaction mechanism.[256]

---

256 An example of an anomalous (interrupted) *ozonolysis* reaction mechanism of (1S,4S)-7,7-dimethyl-2-methylenebicyclo[2.2.1]heptan-1-ol yielding two unexpected products [70h].

71 Paal–Knorr Syntheses

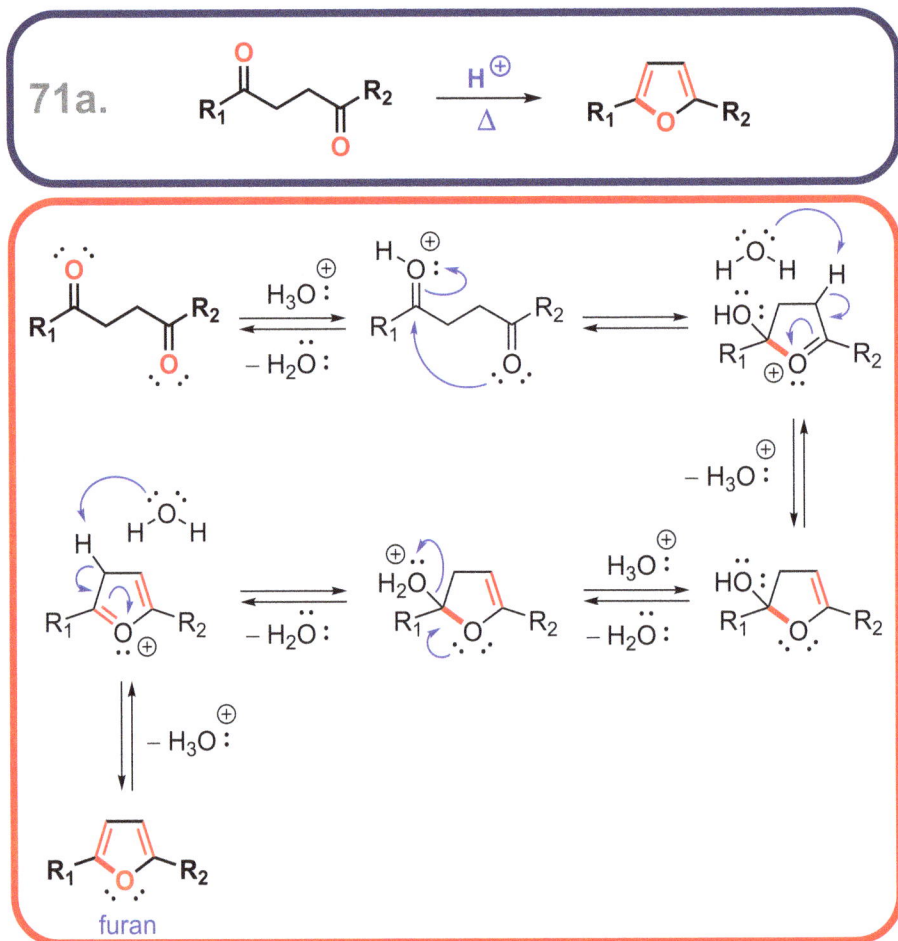

Fig. 71.1: The **Paal–Knorr** *furan synthesis* mechanism.[257]

---

[257] The **Paal–Knorr** *synthesis* is a reaction that was initially proposed for the synthesis of *furans* and *pyrroles*: the **Paal–Knorr** *furan synthesis*.

https://doi.org/10.1515/9783110786835-071

71b.

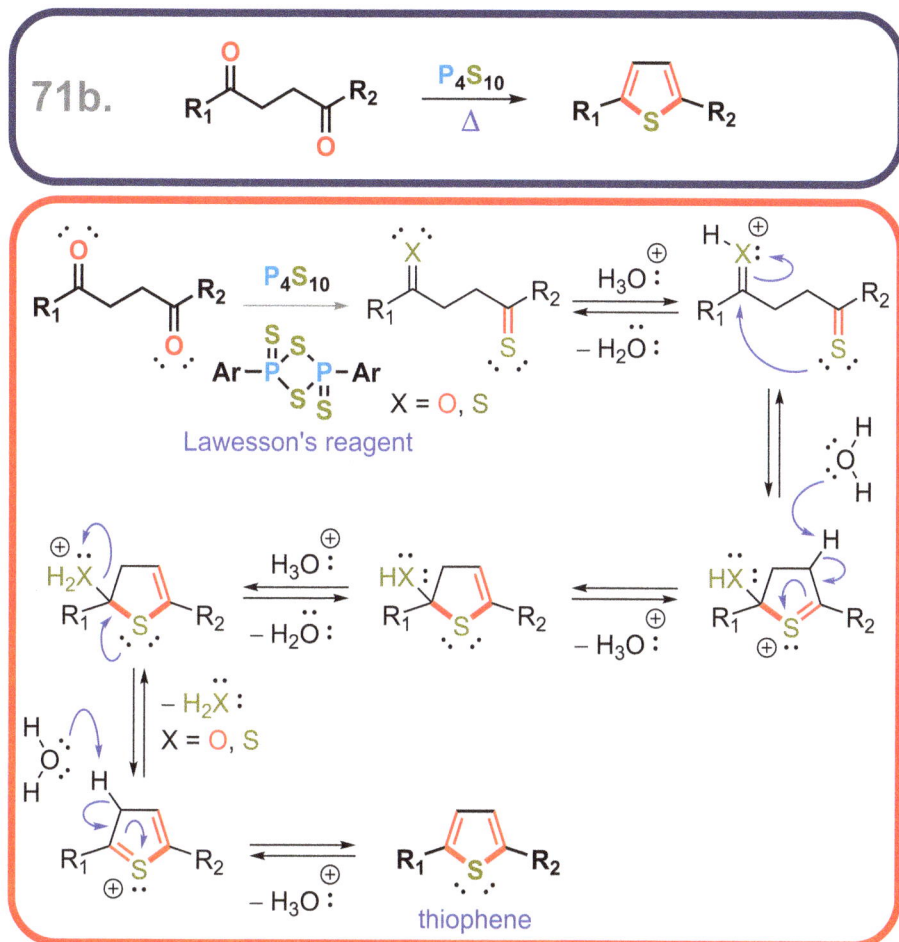

Fig. 71.2: The **Paal–Knorr** *thiophene synthesis* mechanism.[258]

---

258 The **Paal–Knorr** *thiophenes synthesis* was adopted for the preparation of *thiophenes*, for example, by using **Lawesson's** *reagent* [71a].

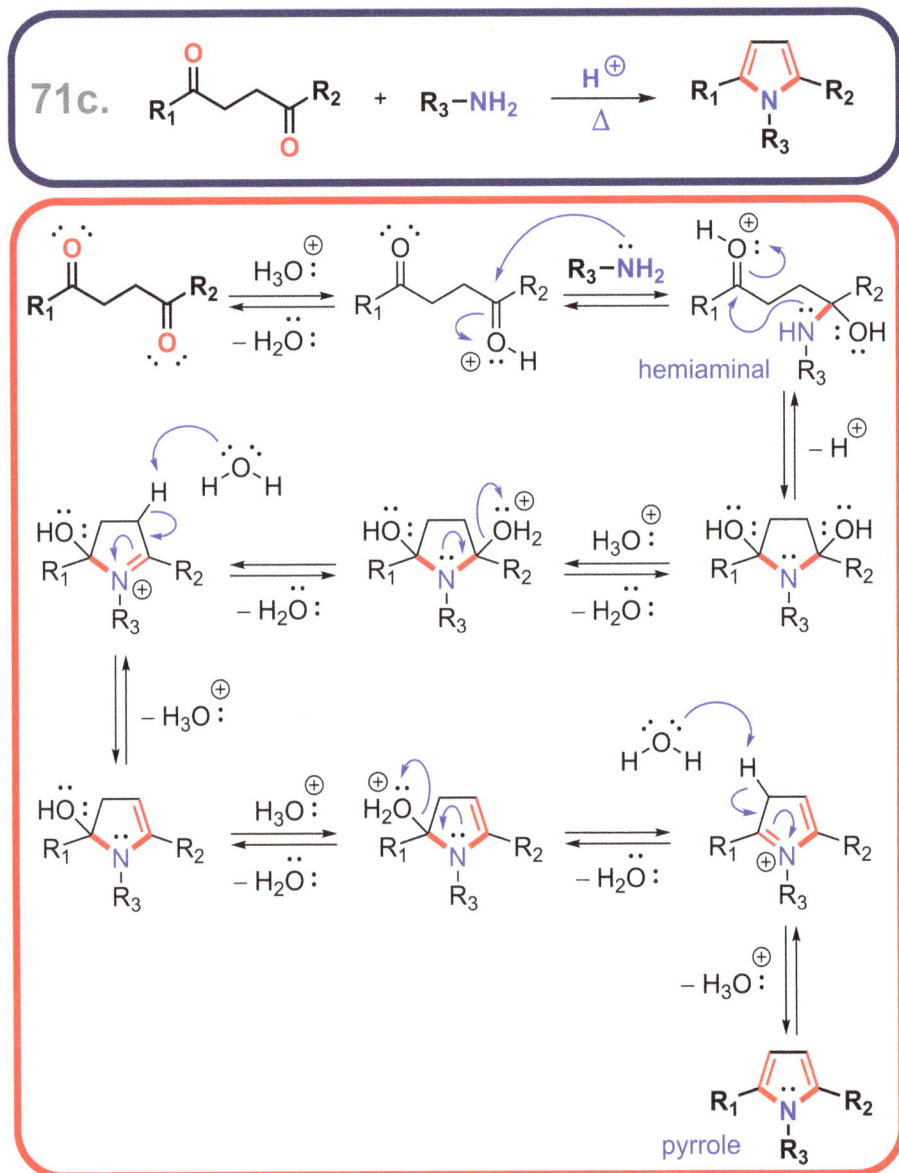

259 The **Paal–Knorr** *pyrrole synthesis* is a reaction that was initially proposed for the synthesis of *pyrroles*. It should not be confused with the **Knorr** *pyrrole synthesis* (not shown).

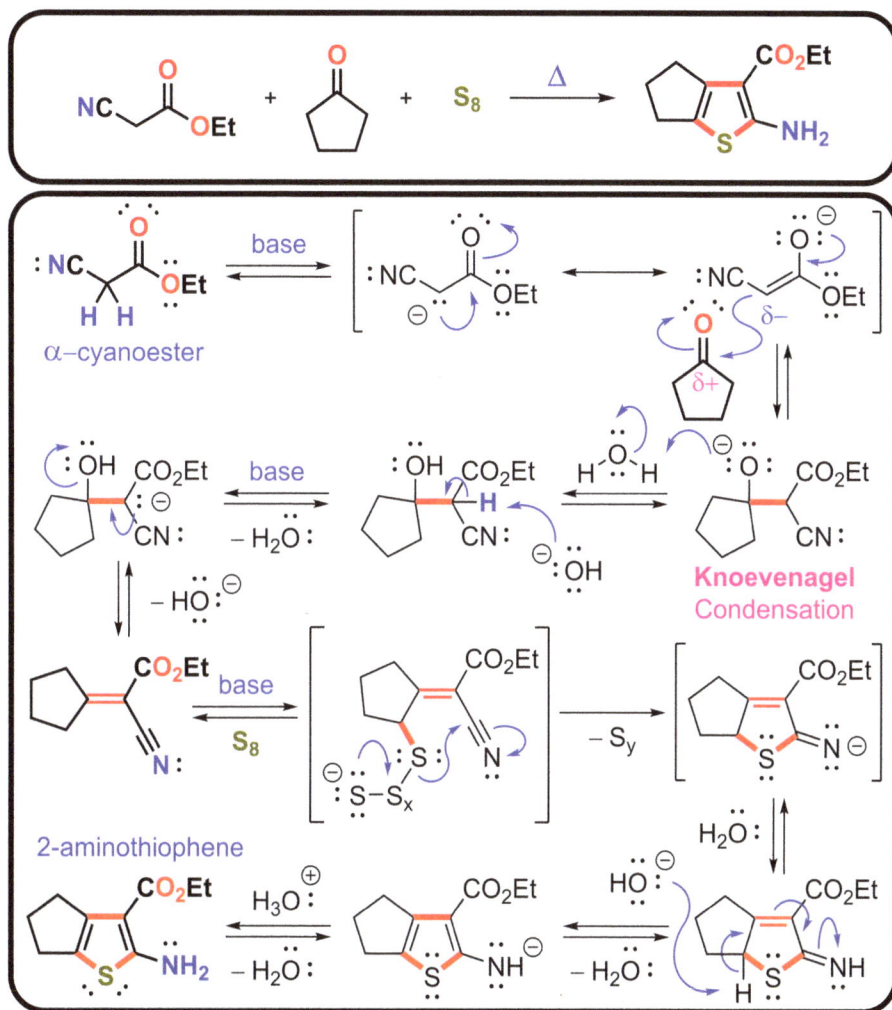

Fig. 71.4: Related to the **Paal–Knorr** synthesis: the **Gewald** condensation.[260]

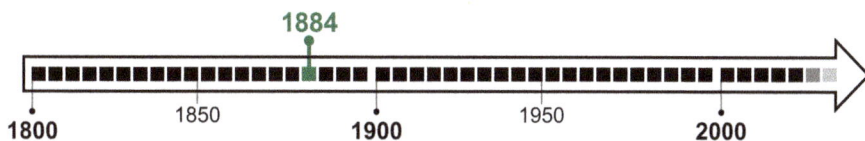

Fig. 71.5: The discovery of the **Paal–Knorr** syntheses.[261]

---

260 Thiophenes (2-aminothiophenes) can be prepared via the **Gewald** condensation (see Chapter 41).
261 The reaction was likely first described around 1884 [71b, 71c].

72 Paternò–Büchi Reaction

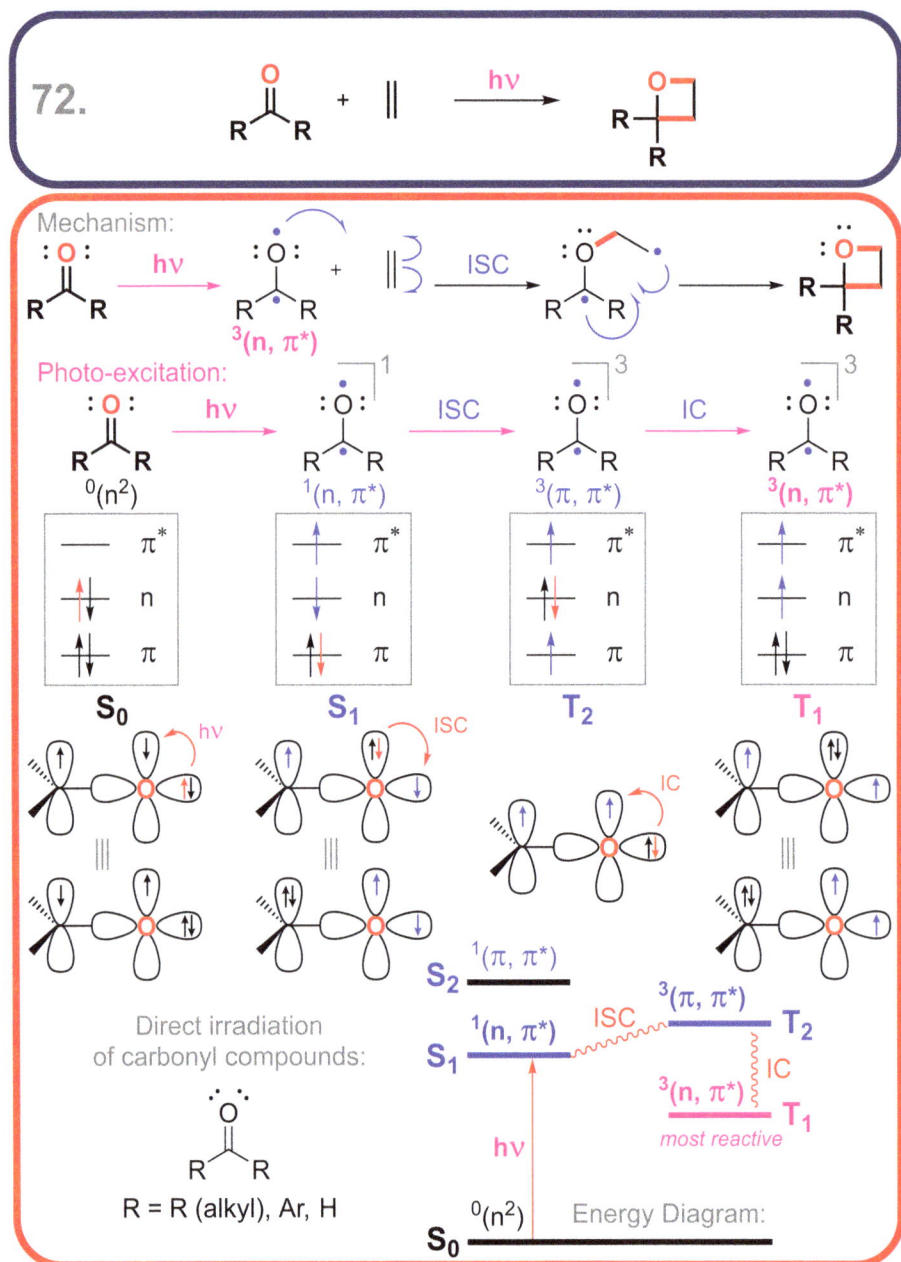

Fig. 72.1: The *Paternò–Büchi* reaction mechanism.[262]

https://doi.org/10.1515/9783110786835-072

	Δ	$h\nu$
4n	✗	✓
4n+2	✓	✗

Norrish Type II Fragmentation:

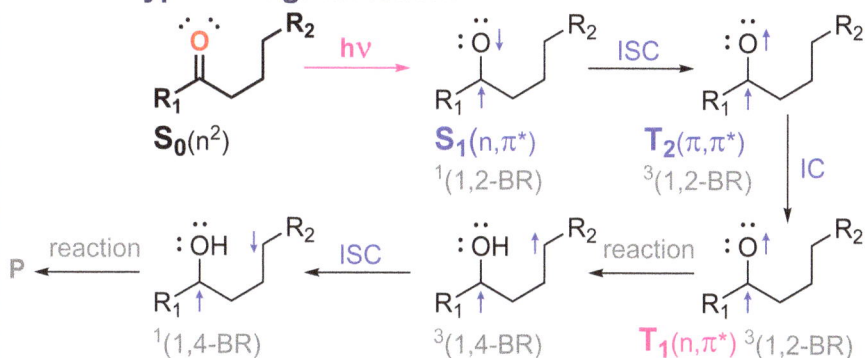

$S_0(n^2)$ →$h\nu$→ $S_1(n,\pi^*)$ $^1(1,2\text{-BR})$ →ISC→ $T_2(\pi,\pi^*)$ $^3(1,2\text{-BR})$ | IC

P ←reaction← $^1(1,4\text{-BR})$ ←ISC← $^3(1,4\text{-BR})$ ←reaction← $T_1(n,\pi^*)$ $^3(1,2\text{-BR})$

Paternò–Büchi Reaction:

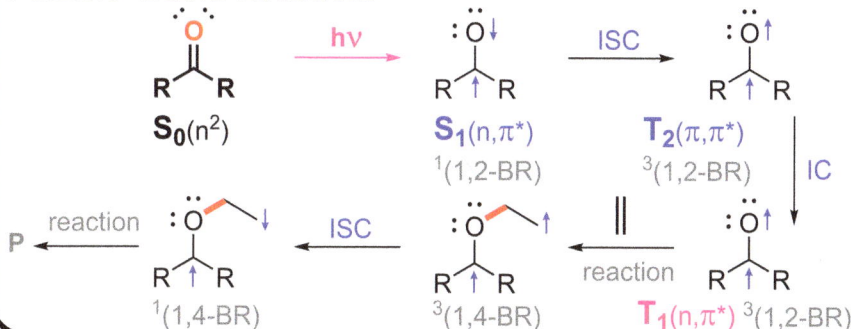

$S_0(n^2)$ →$h\nu$→ $S_1(n,\pi^*)$ $^1(1,2\text{-BR})$ →ISC→ $T_2(\pi,\pi^*)$ $^3(1,2\text{-BR})$ | IC

P ←reaction← $^1(1,4\text{-BR})$ ←ISC← $^3(1,4\text{-BR})$ ←reaction (=)← $T_1(n,\pi^*)$ $^3(1,2\text{-BR})$

Fig. 72.2: The *Norrish type II reaction* vs the *Paternò–Büchi reaction* mechanism.[263]

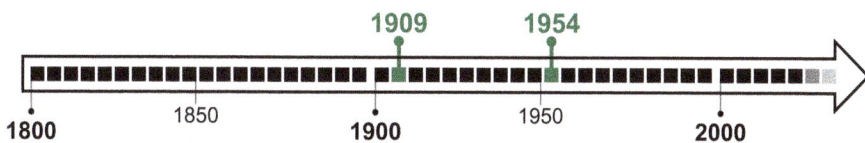

1909 **1954**

1800   1850   1900   1950   2000

Fig. 72.3: The discovery of the *Paternò–Büchi reaction*.[264]

---

262 The *Paternò–Büchi reaction* is a photochemical $[2_\pi + 2_\pi]$ or **[2 + 2]**-*cycloaddition reaction*. The *Woodward–Hoffmann rules* [64a, 64b, 64c]: this reaction ($4n = 4\pi$) is <u>not</u> allowed at the ground state = under thermal conditions (Δ) but <u>allowed</u> at the excited state = under photochemical conditions ($h\nu$) [2b].
263 Compare the mechanistic similarities between the *Norrish type II reaction* (covered in Chapter 67) and the *Paternò–Büchi cycloaddition reaction* [2b].
264 The reaction was likely described by Paternò around 1909 [72a] and by Büchi in 1954 [72b].

73 Pauson–Khand Reaction

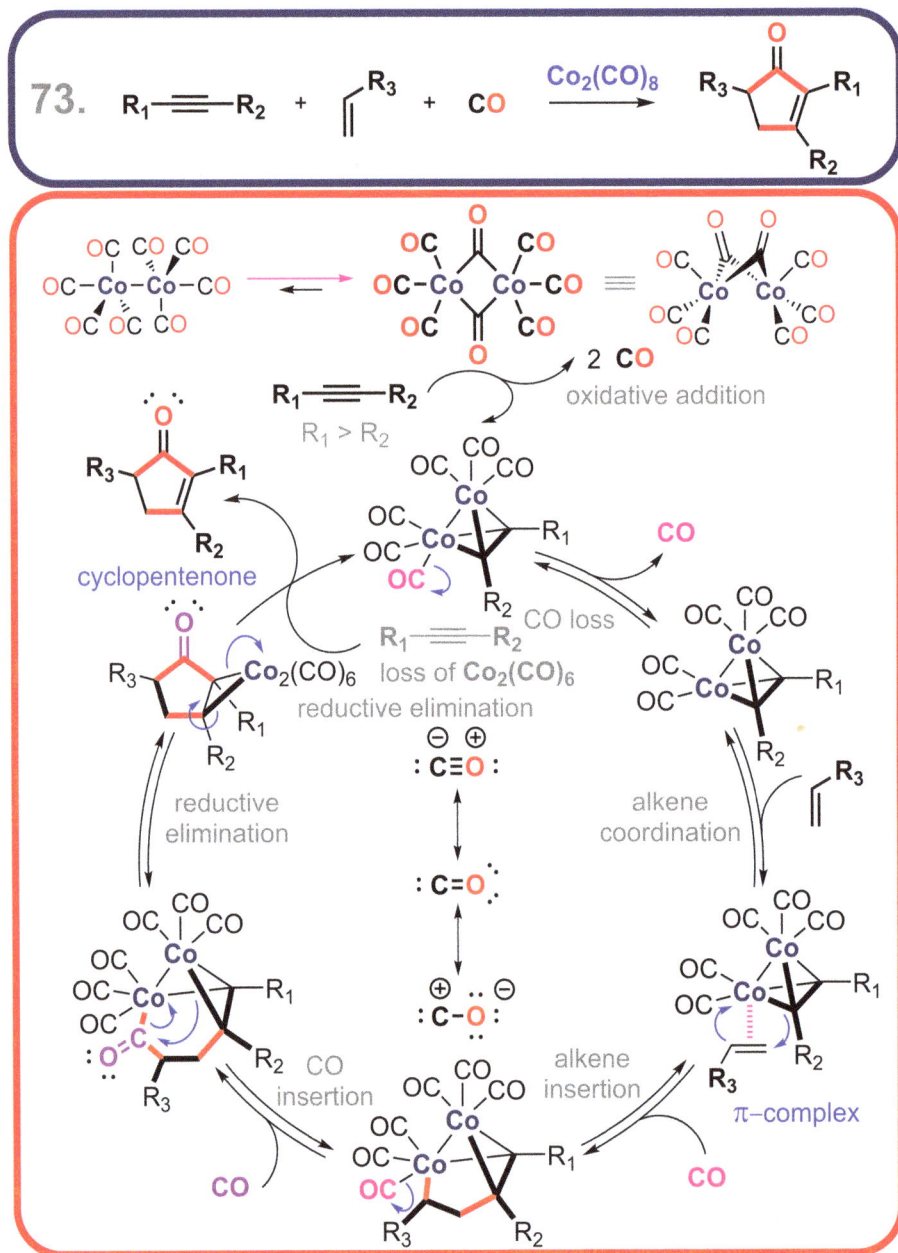

$$R_1\!\!=\!\!\!=\!\!\!R_2 \;+\; \overset{R_3}{\diagup\!\!\!\diagdown} \;+\; CO \;\;\xrightarrow{Co_2(CO)_8}\;\; $$

Fig. 73.1: The *Pauson–Khand reaction* mechanism.[265]

https://doi.org/10.1515/9783110786835-073

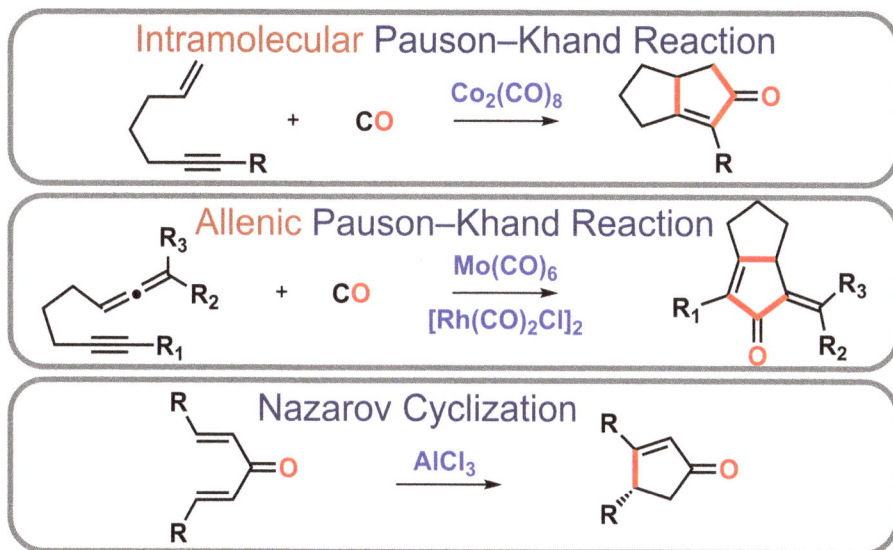

Fig. 73.2: Variations of the **Pauson–Khand** reaction.[266]

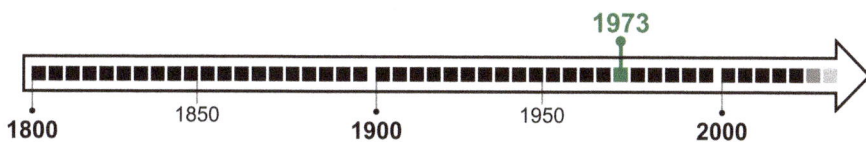

Fig. 73.3: The discovery of the **Pauson–Khand** reaction.[267]

---

265 The **Pauson–Khand** reaction is a **Co**-catalyzed (2 + 2 + 1)-cycloaddition reaction.
266 There are several variations of this reaction: the *intramolecular* **Pauson–Khand** reaction, the *allenic* **Pauson–Khand** reaction, and others (not shown) [73a]. Other metals can catalyze it: **Mo**, **Rh**, etc. The **Nazarov** cyclization undergoes a different [4π] **conrotatory electrocyclization** mechanism (Chapter 64), but it also yields *cyclopentenones.*
267 The reaction was likely first described around 1973 [73b, 73c, 73d].

74 Peptide (Amide) Coupling

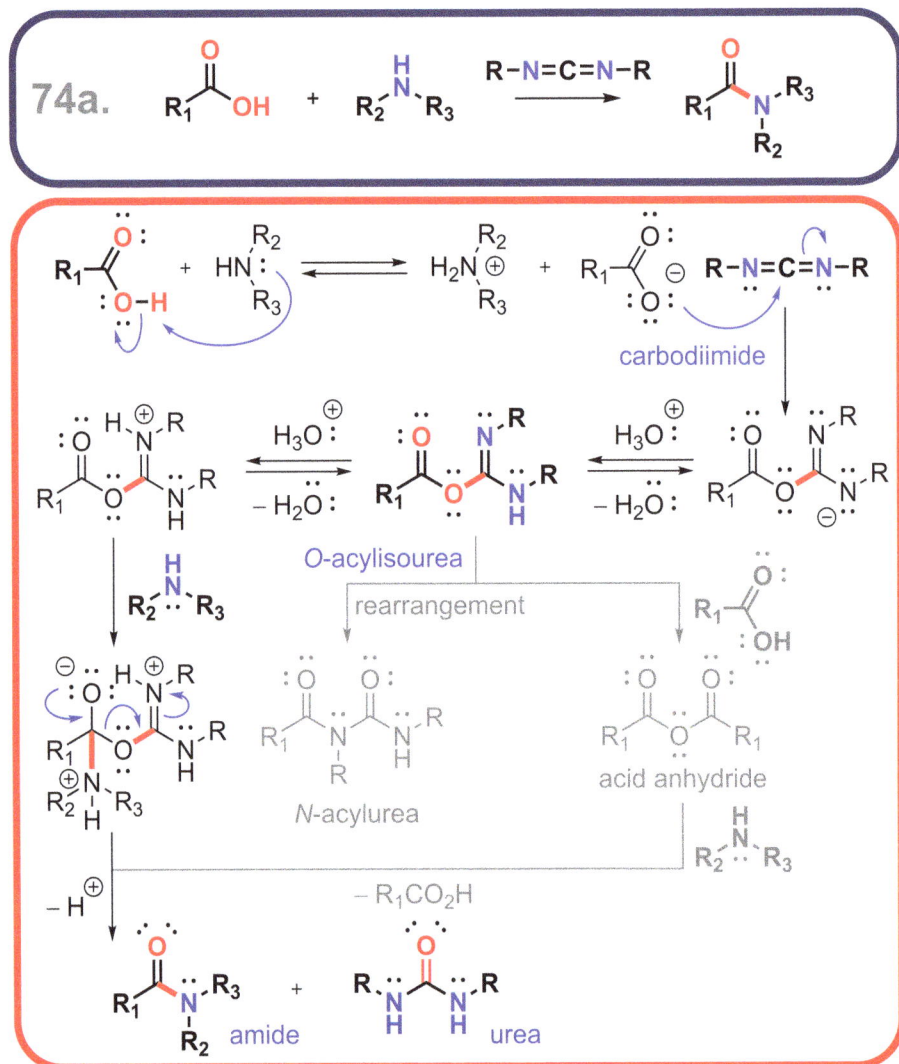

74a.

Fig. 74.1: The *peptide (amide) coupling* (DCC) mechanism.[268]

268 The *peptide (amide) coupling* mechanism is based on the use of *carbodiimide* coupling reagents (DCC) [74a, 74b].

https://doi.org/10.1515/9783110786835-074

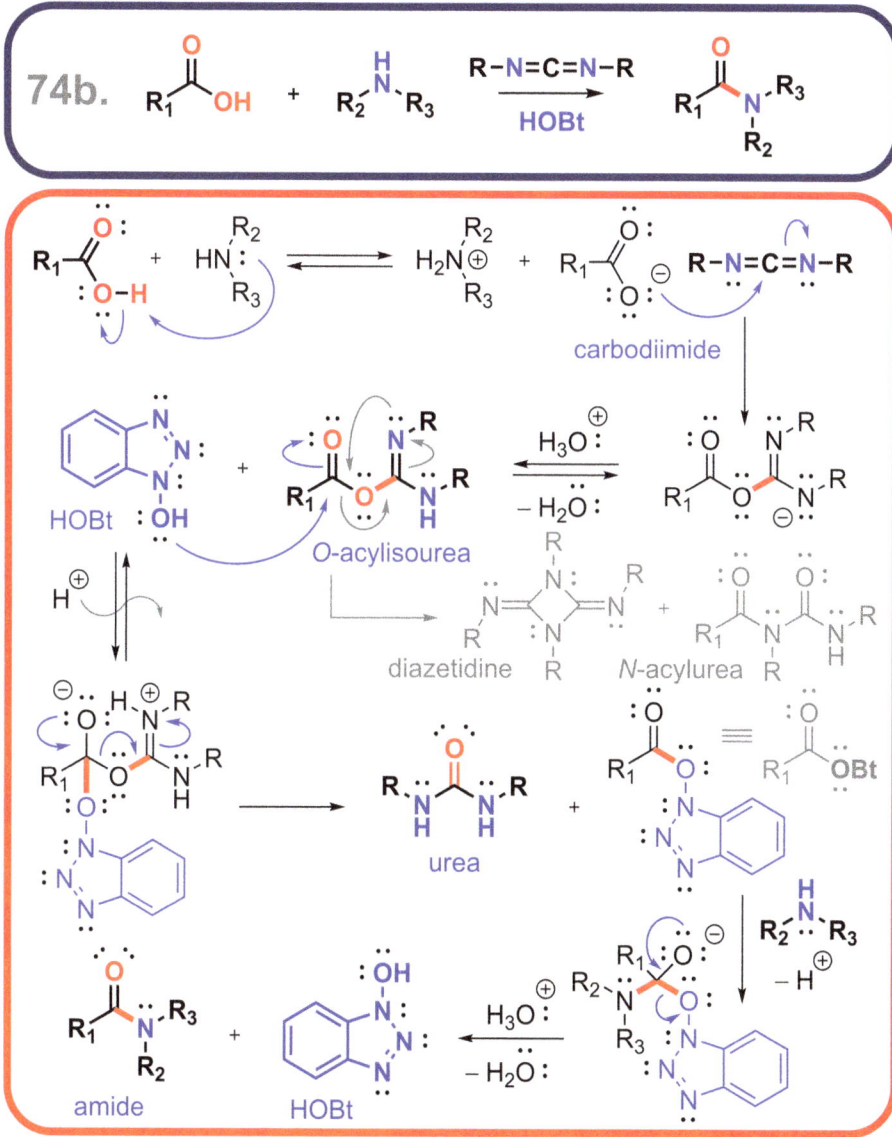

Fig. 74.2: The *peptide (amide) coupling* (DCC + HOBt) mechanism.[269]

---

269 The *peptide (amide) coupling* mechanism is based on the use of *carbodiimide* coupling reagents and *additives* (DCC and HOBt) [74a, 74b].

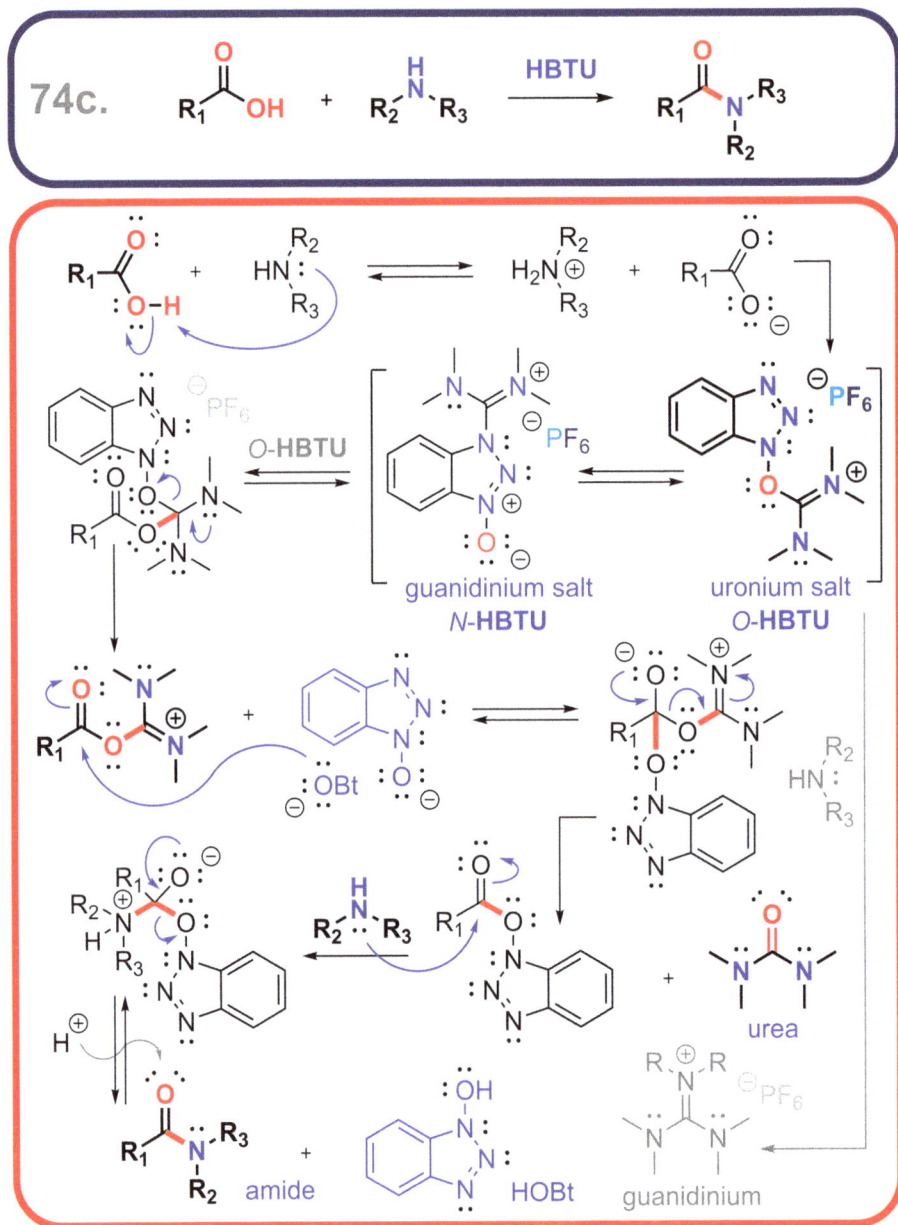

---

270 The *peptide (amide) coupling* mechanism is based on the use of *benzotriazole = guanidinium/uronium salts* coupling reagents (HBTU) [74c].

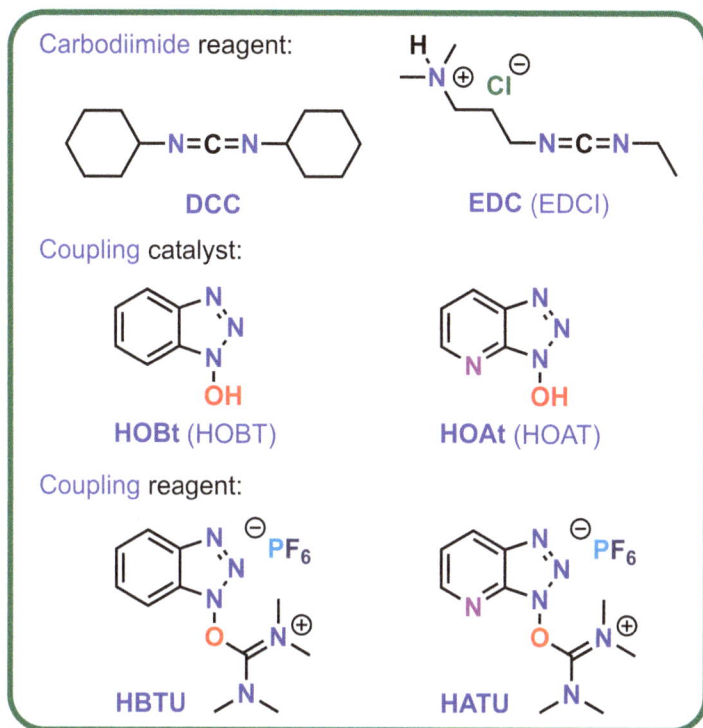

Fig. 74.4: The main *peptide (amide) coupling* reagents and catalysts.[271]

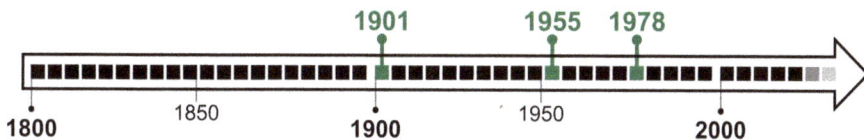

Fig. 74.5: The discovery of the *peptide (amide) coupling*.[272]

---

271 The most common reagents used in the *peptide (amide) coupling* or the *peptide synthesis* are the *carbodiimide reagents* (DCC [74d], EDC [74e], and many other); **guanidinium/uronium salts** (HBTU [74f], HATU [74g]; and many more like *phosphonium salts* PyBOP [74]). The most common additives (catalysts) used in the *peptide synthesis* are HOBt [74i] and HOAt, among others.

272 A. The *peptide (amide) coupling* reaction was likely first described around 1901 [74j]. B. DCC coupling reagent was likely first described around 1955 [74k]. C. HBTU coupling reagent was likely first described around 1978 [74l].

75 Pictet–Spengler Reaction

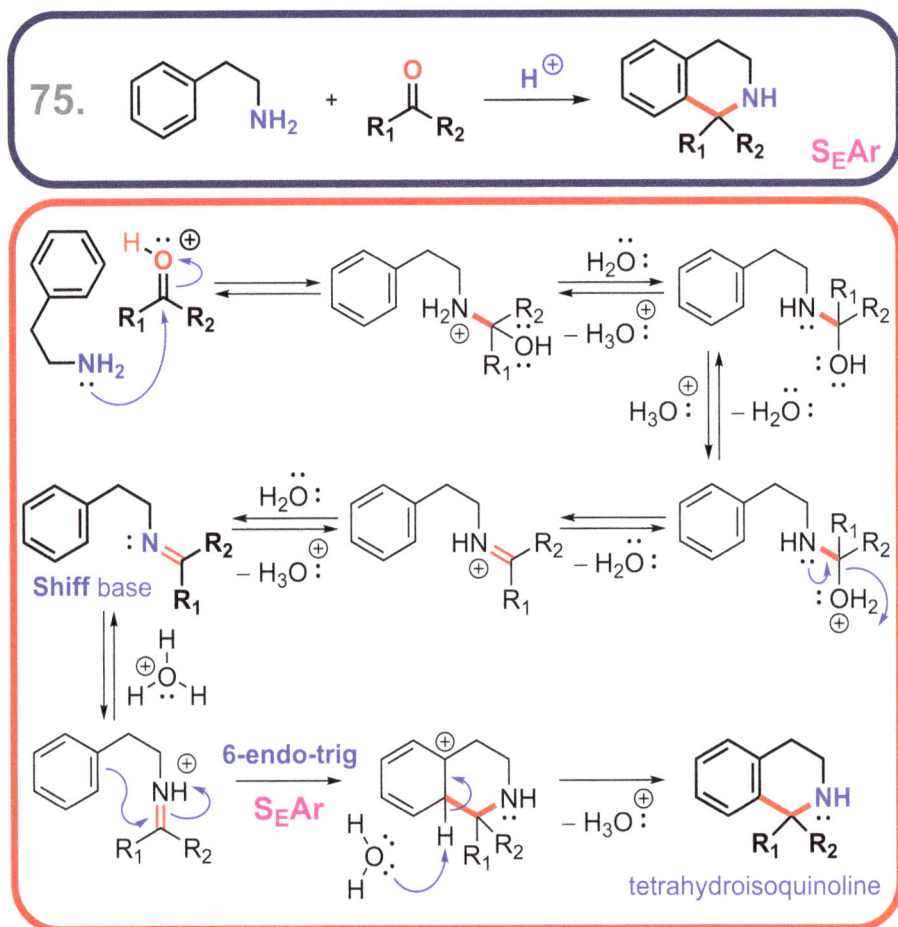

Fig. 75.1: The *Pictet–Spengler* reaction mechanism.[273]

273 The *Pictet–Spengler* reaction or the *Pictet–Spengler* condensation mechanism is a combination of the *Mannich* condensation = the *imine condensation (the Shiff base)* (see Chapter 56) and the **aromatic electrophilic substitution** (the **arenium ion** mechanism or S_EAr, which was covered in Chapter 3).

https://doi.org/10.1515/9783110786835-075

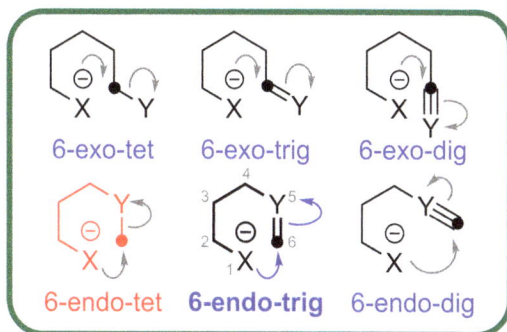

Fig. 75.2: **Baldwin's** rules.[274]

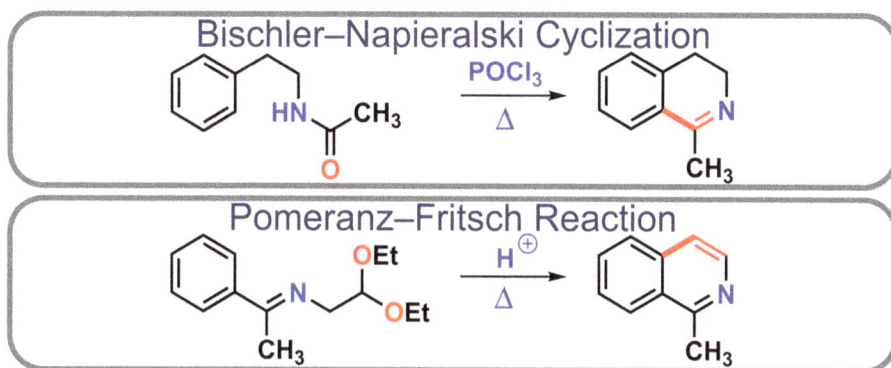

Fig. 75.3: Reactions related to the **Pictet–Spengler** reaction.[275]

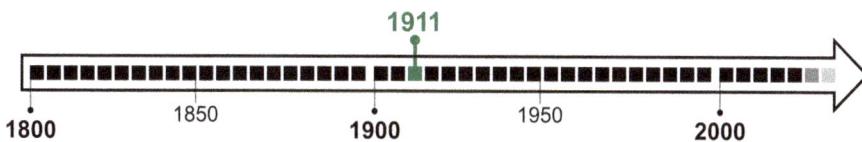

Fig. 75.4: The discovery of the **Pictet–Spengler** reaction.[276]

---

274 The cyclization (S_EAr) step is allowed according to *Baldwin's rules*: **6-endo-trig** [75a].

275 Several named reactions are related to the **Pictet–Spengler** reaction: the **Bischler–Napieralski** cyclization (Chapter 19) and the **Pomeranz–Fritsch** reaction [19a, 19b]. Both reactions yield isoquinolines.

276 The reaction was likely first described around 1911 [75b].

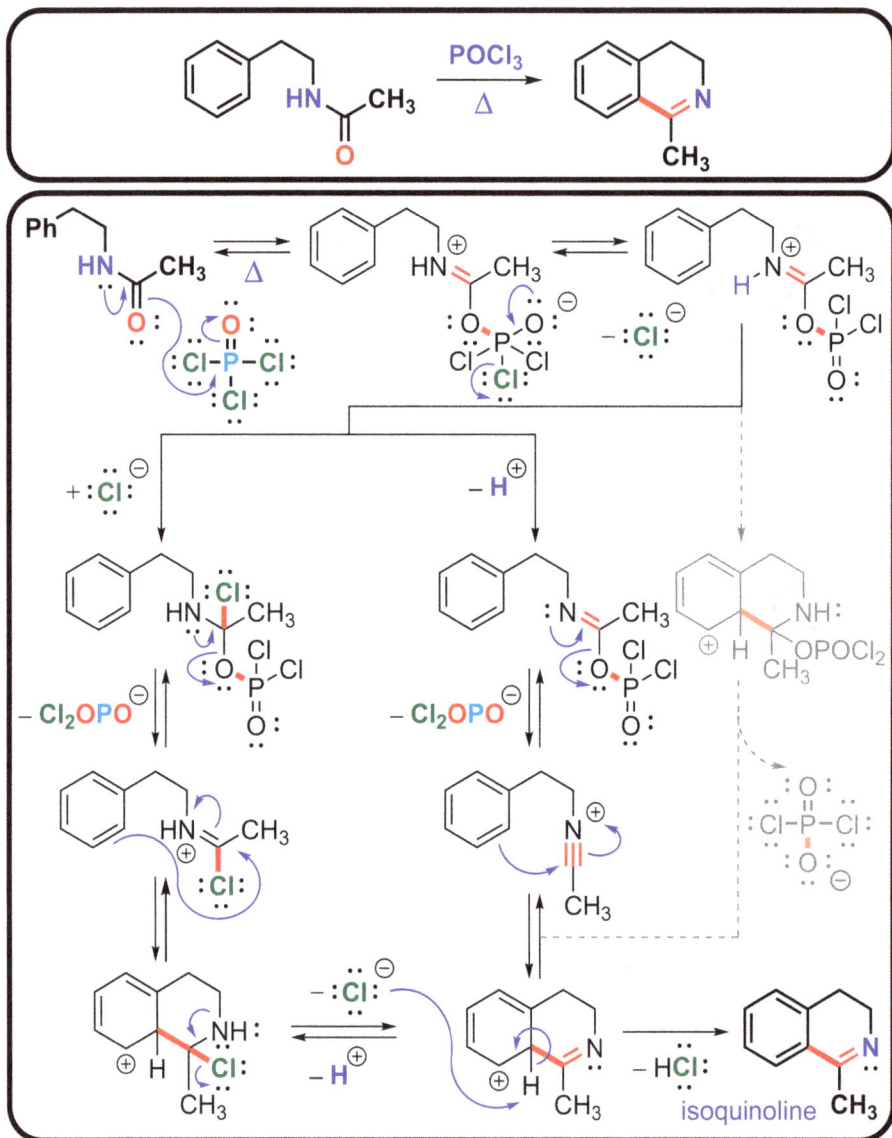

Fig. 75.5: The ***Bischler–Napieralski*** *cyclization* mechanism of *N*-phenethylacetamide.[277]

277 An example of the ***Bischler–Napieralski*** *cyclization* of *N*-phenethylacetamide yielding 1-methyl-3,4-dihydroisoquinoline.

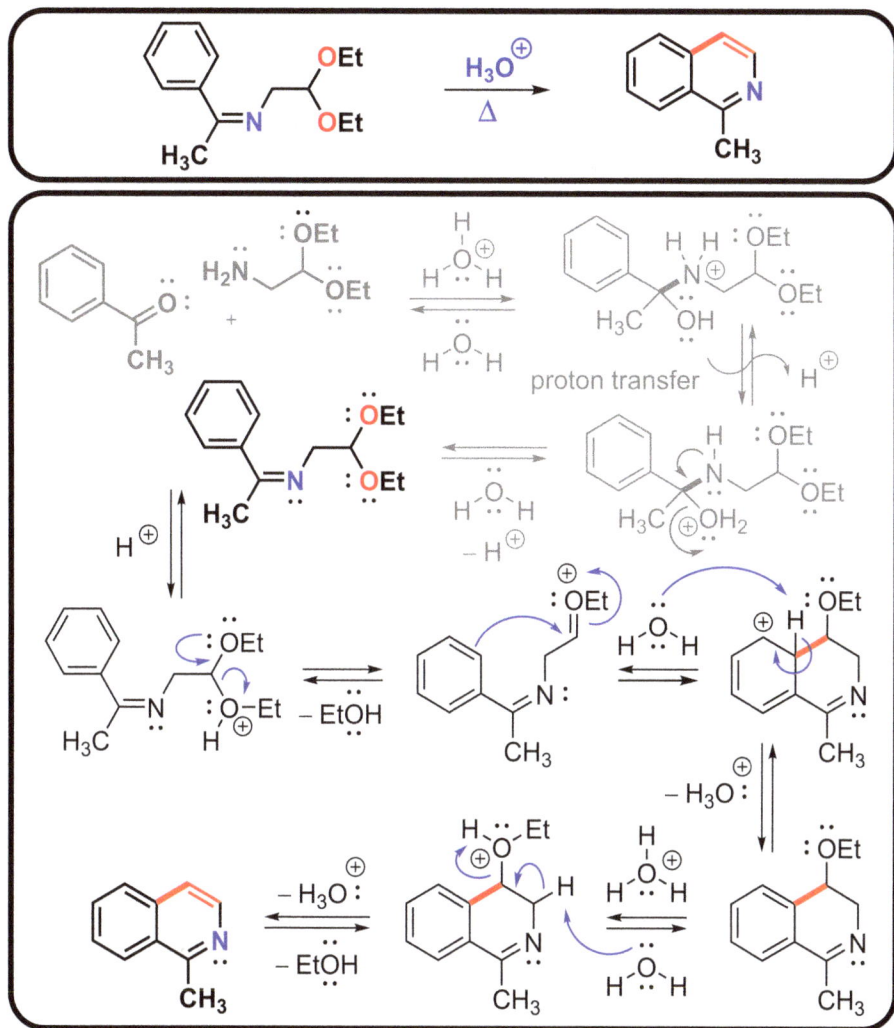

Fig. 75.6: The **Pomeranz–Fritsch** reaction mechanism.[278]

---

278 The **Pomeranz–Fritsch** reaction [19a, 19b] yielding 1-methylisoquinoline.

76 Pinacol–Pinacolone Rearrangement

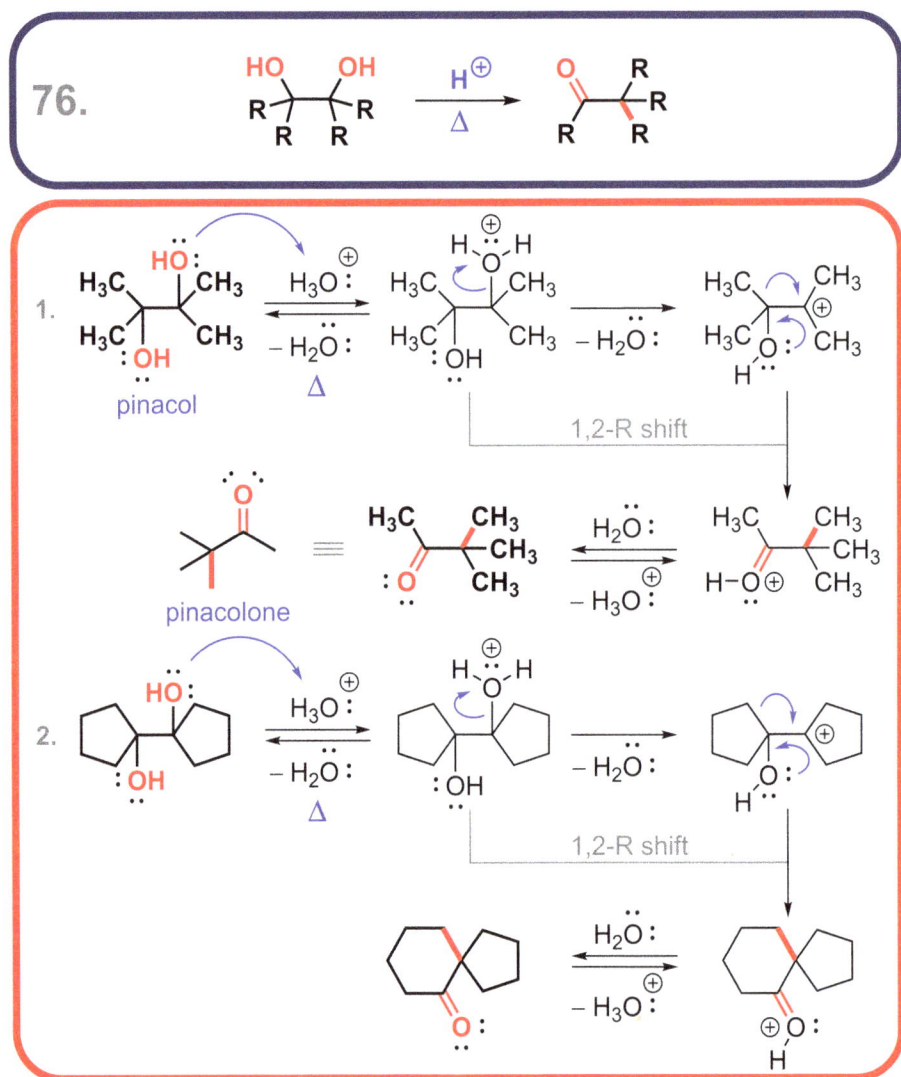

Fig. 76.1: The *pinacol–pinacolone rearrangement* mechanism.[279]

---

[279] The *pinacol–pinacolone rearrangement* or simply the *pinacol rearrangement* mechanism is distantly related to the **Wagner–Meerwein** *rearrangement* covered in Chapter 96. The *pinacol–pinacolone rearrangement* should not be confused with the *pinacol coupling* covered in Chapter 57. Please also note: *2,3-dimethylbutane-2,3-diol* is called **pinacol** and *3,3-dimethyl-2-butanone* is called **pinacolone**.

https://doi.org/10.1515/9783110786835-076

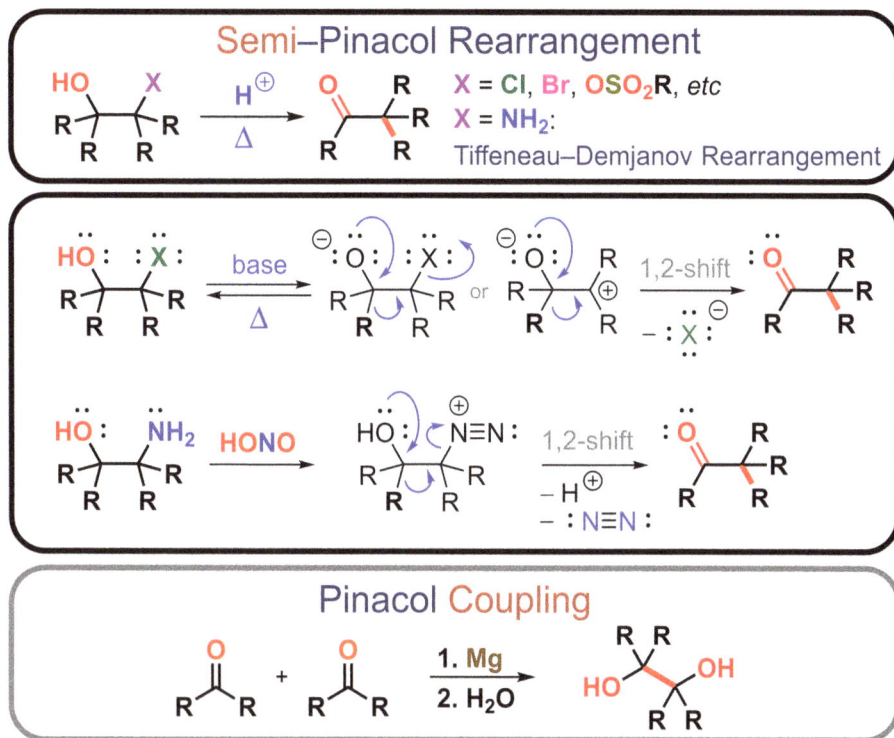

Fig. 76.2: The *semi-pinacol rearrangement* mechanism.[280]

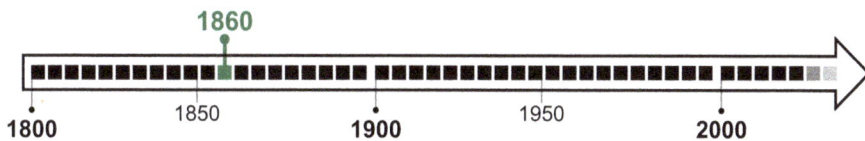

Fig. 76.3: The discovery of the *pinacol–pinacolone rearrangement*.[281]

---

280 The *semi-pinacol rearrangement* mechanism [1] is analogous to the *pinacol rearrangement*. It occurs in *α-substituted alcohols*. If X = NH$_2$, the reaction is called the **Tiffeneau–Demjanov** *rearrangement* [76a, 76b].

281 The reaction was likely first described around 1860 [76c].

77 Polonovski Reaction

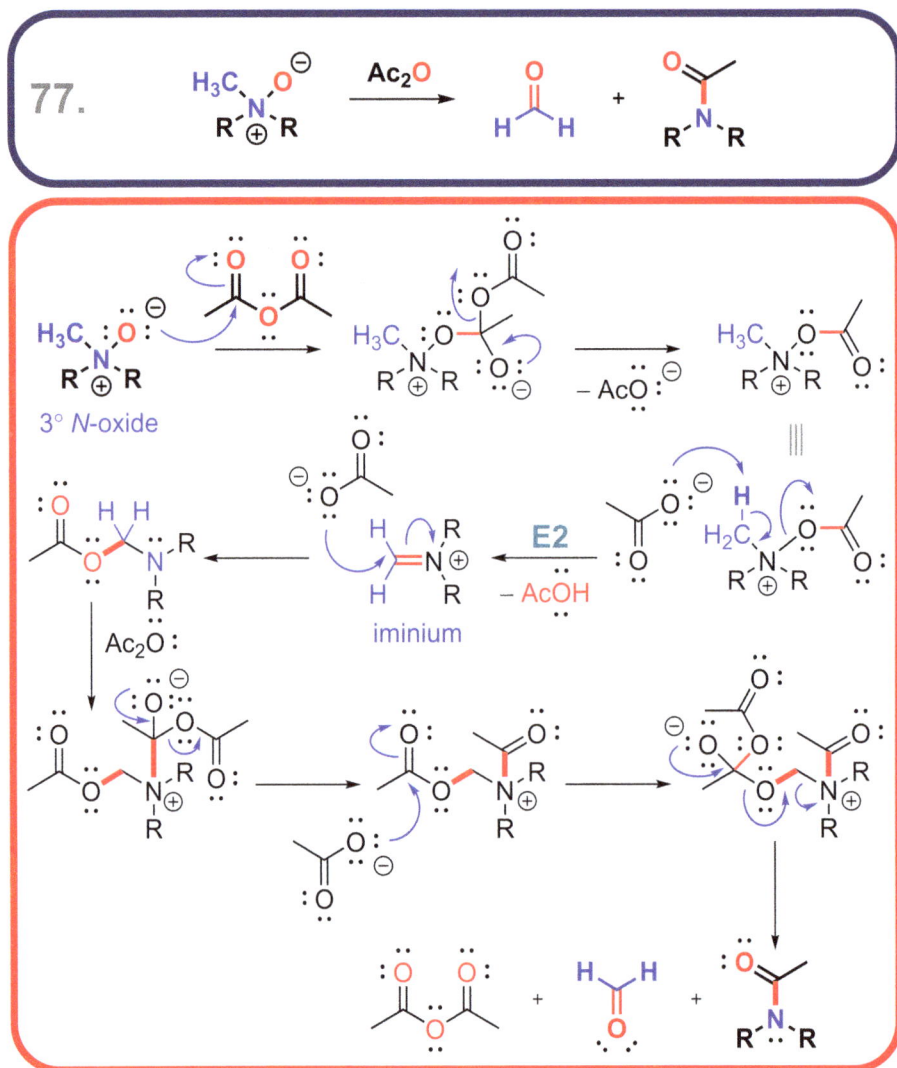

Fig. 77.1: The *Polonovski reaction* mechanism.[282]

---

[282] The *Polonovski reaction* can be called the *Polonovski rearrangement*. The key intermediate is an *iminium ion* (see the *Mannich* reaction in Chapter 56).

https://doi.org/10.1515/9783110786835-077

Polonovski–Potier Reaction

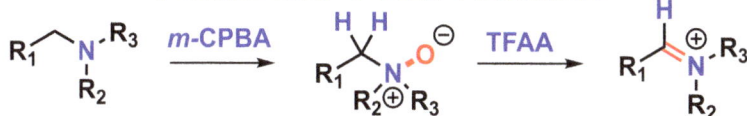

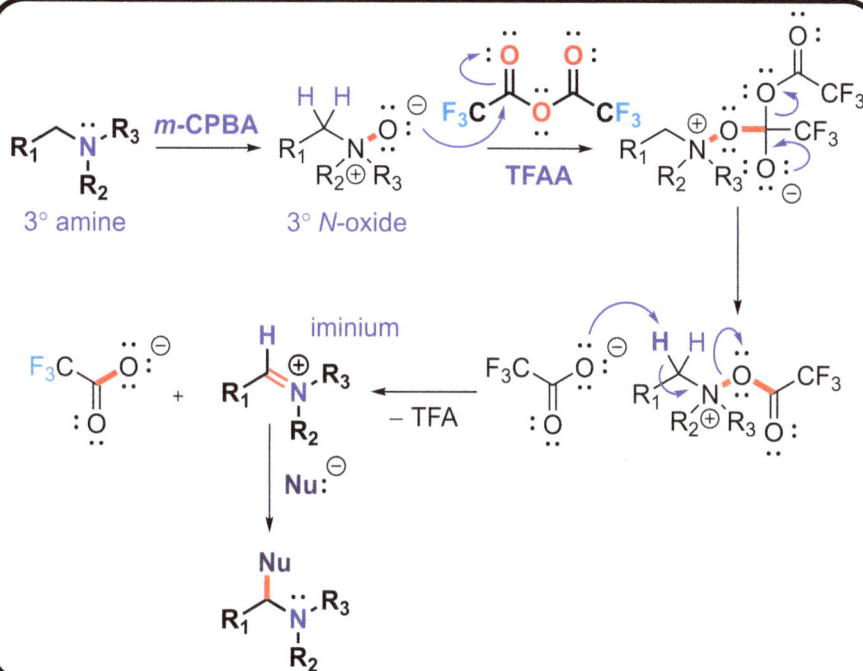

Fig. 77.2: The **Polonovski–Potier** reaction mechanism.[283]

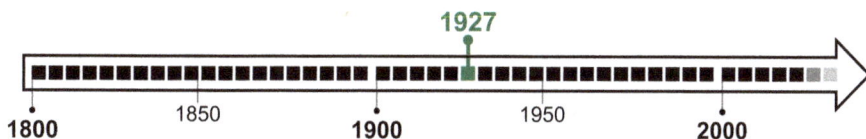

Fig. 77.3: The discovery of the **Polonovski** reaction.[284]

---

283 The **Polonovski–Potier** reaction is closely related [77a, 77b]. Trifluoroacetic anhydride (TFAA) is used instead of acetic anhydride, and the iminium ion can be trapped with various nucleophiles.
284 The reaction was likely first described around 1927 [77c].

78 Prilezhaev Epoxidation

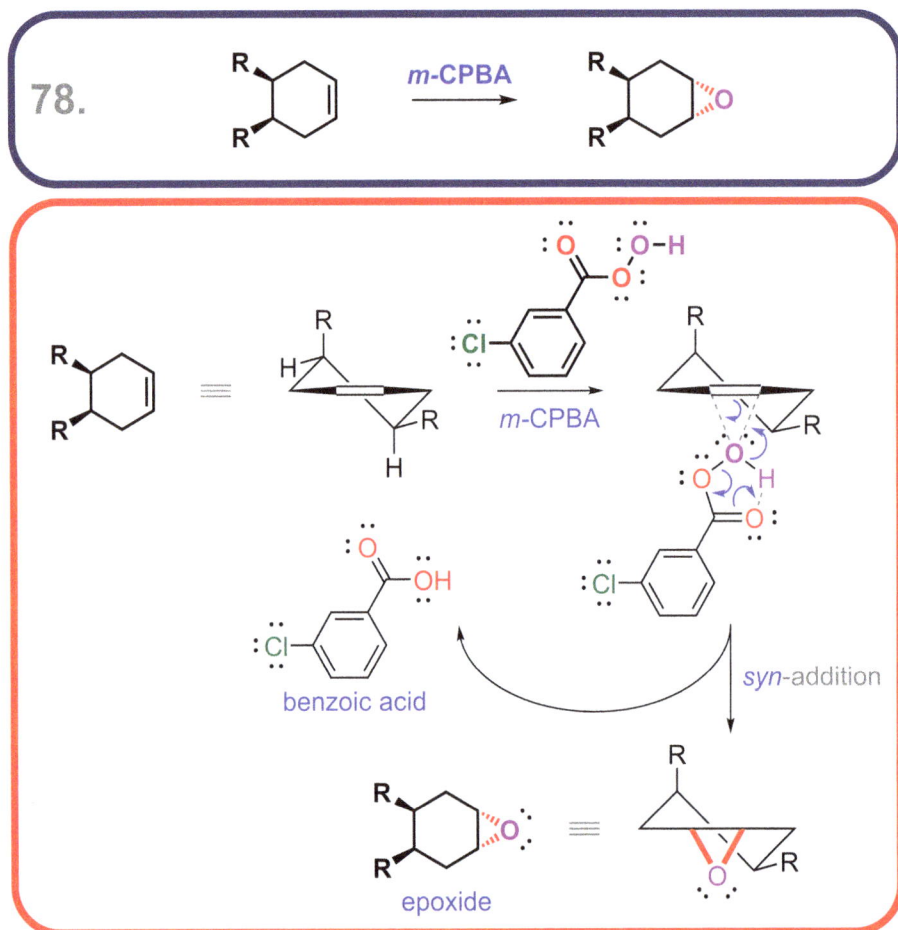

78.

Fig. 78.1: The *Prilezhaev epoxidation* mechanism.[285]

---

285 The *Prilezhaev reaction* (in Russian Прилежаев) is a type of epoxidation, and it is often called the *Prilezhaev epoxidation*.

https://doi.org/10.1515/9783110786835-078

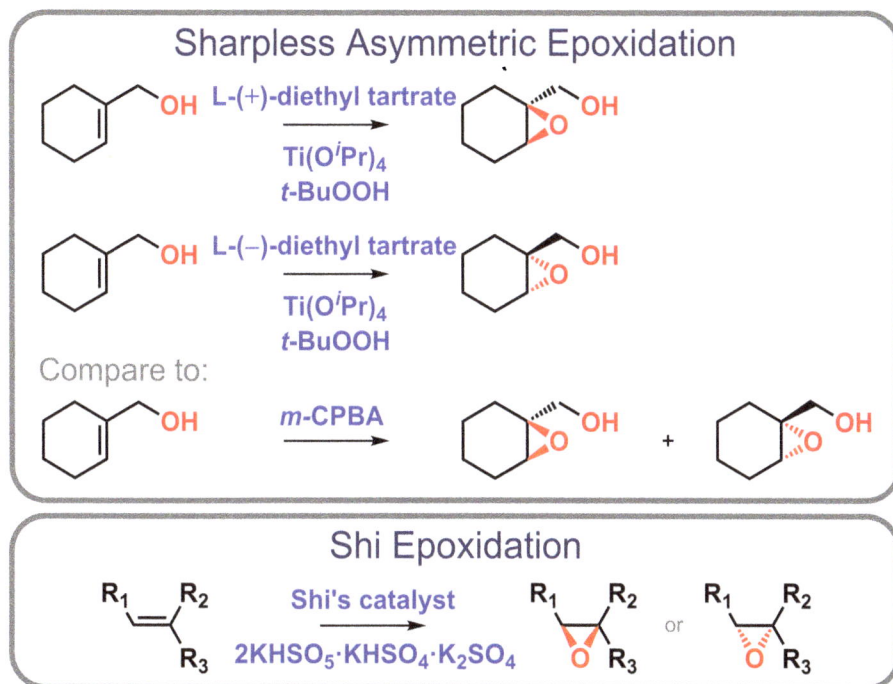

Fig. 78.2: Reactions related to the **Prilezhaev** epoxidation.[286]

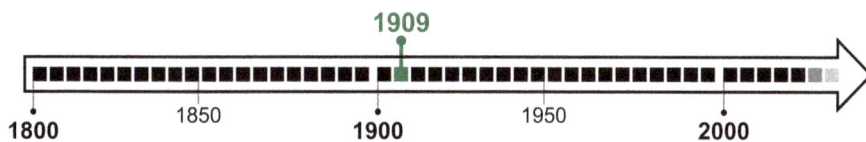

Fig. 78.3: The discovery of the **Prilezhaev** epoxidation.[287]

---

286 There are many ways to synthesize *epoxides*, such as: the **Sharpless** *asymmetric epoxidation* [78a] (compare to the **Prilezhaev** epoxidation where a mixture of enantiomers is formed); the **Shi** *asymmetric epoxidation* [78b]; and many more other examples (not shown) [1].
287 The reaction was likely first described around 1909 [78c].

79 Prins Reaction

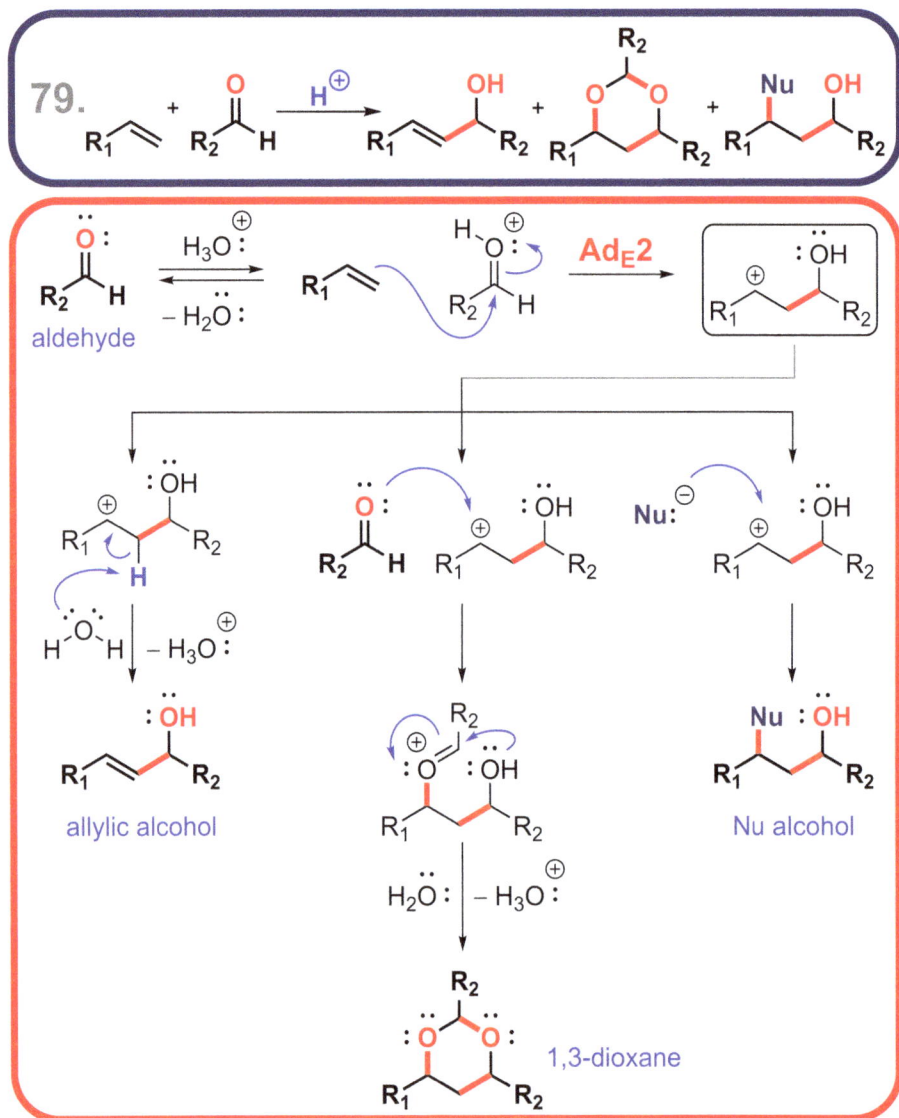

Fig. 79.1: The *Prins reaction* mechanism.[288]

---

288 The *Prins reaction* is a type of *condensation* with various possible products. Mechanistically (addition of a protonated *aldehyde* to an *alkene*), it is an example of the **electrophilic addition** covered in Chapter 1.

https://doi.org/10.1515/9783110786835-079

Aza–Prins Reaction

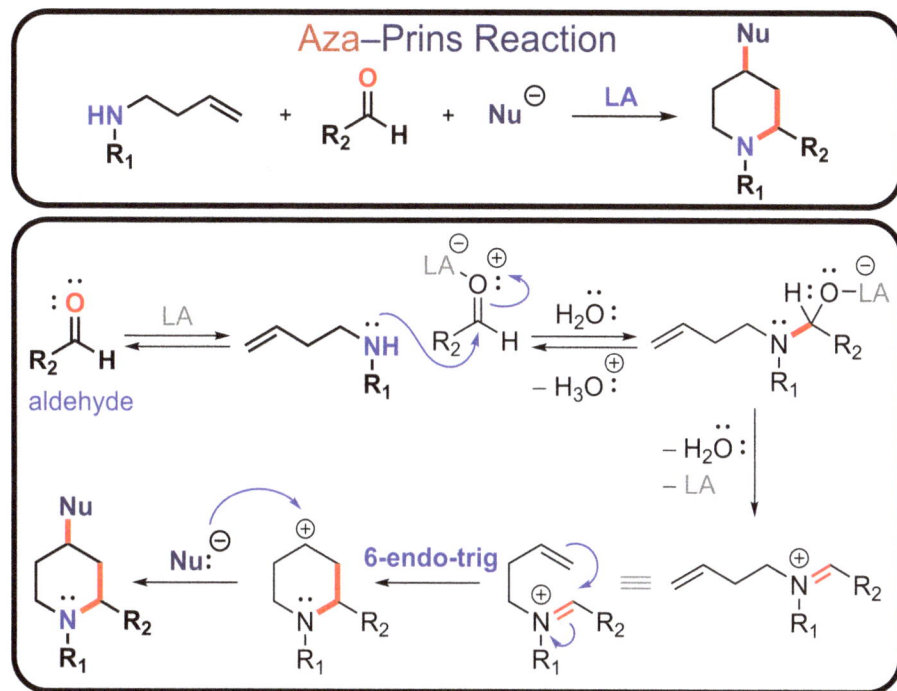

Fig. 79.2: The *aza-**Prins** reaction* mechanism.[289]

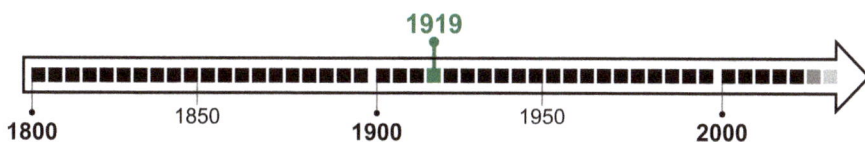

Fig. 79.3: The discovery of the ***Prins** reaction*.[290]

---

289 The *aza-**Prins** reaction* mechanism is related to the ***Prins** reaction* [79a, 79b]. It yields the *piperidine* core (see *Baldwin's rules* mentioned in Chapter 75: **6-endo-trig**). Other variations exist: for example, the ***Prins–pinacol** reaction* (not shown here) [79c].
290 The reaction was likely first described around 1919 [79d, 79e].

80 Pummerer Rearrangement

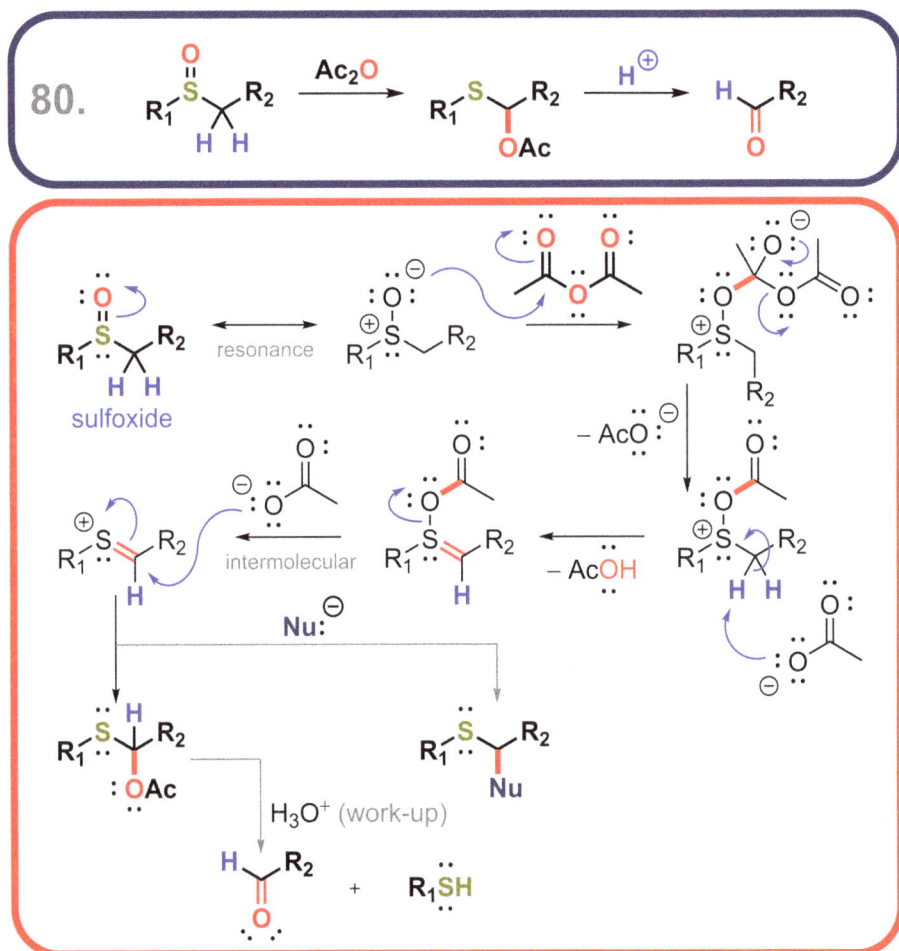

Fig. 80.1: The **Pummerer** *rearrangement* mechanism.[291]

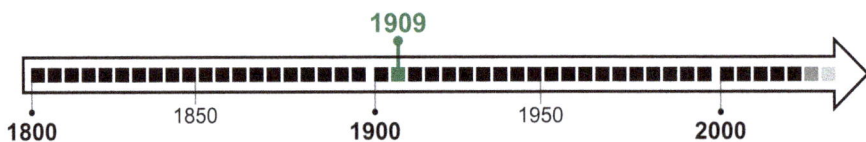

Fig. 80.2: The discovery of the **Pummerer** *rearrangement*.[292]

---

291 The **Pummerer** *rearrangement* can be called the **Pummerer** *fragmentation*.
292 The reaction was likely first described around 1909 [80a].

https://doi.org/10.1515/9783110786835-080

Polonovski Reaction

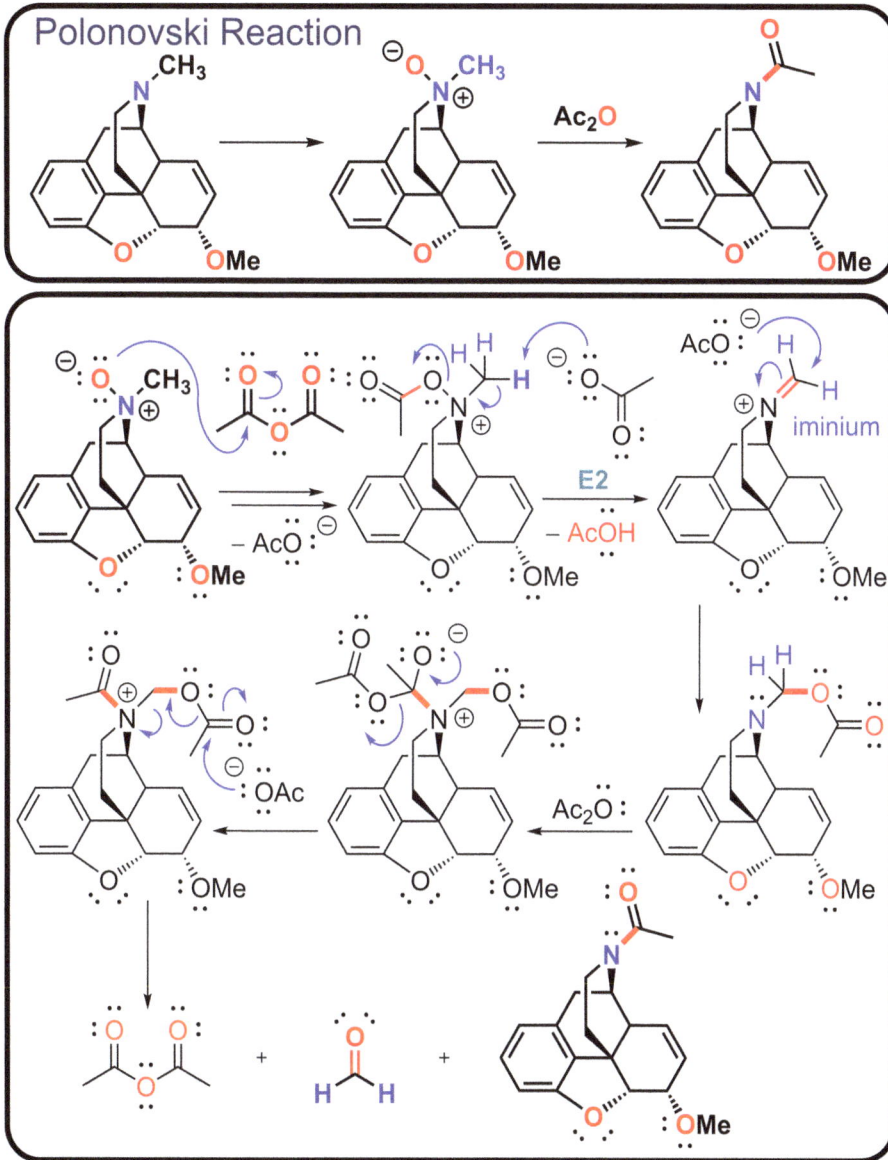

Fig. 80.3: The **Polonovski** reaction mechanism (N-demethylation).[293]

---

293 The **Polonovski** reaction mechanism (Chapter 77) is related to the **Pummerer** rearrangement. Here an amine oxide plays a similar role as a sulfoxide (in the **Pummerer** rearrangement) [80b, 80c].

81 Ramberg–Bäcklund Rearrangement

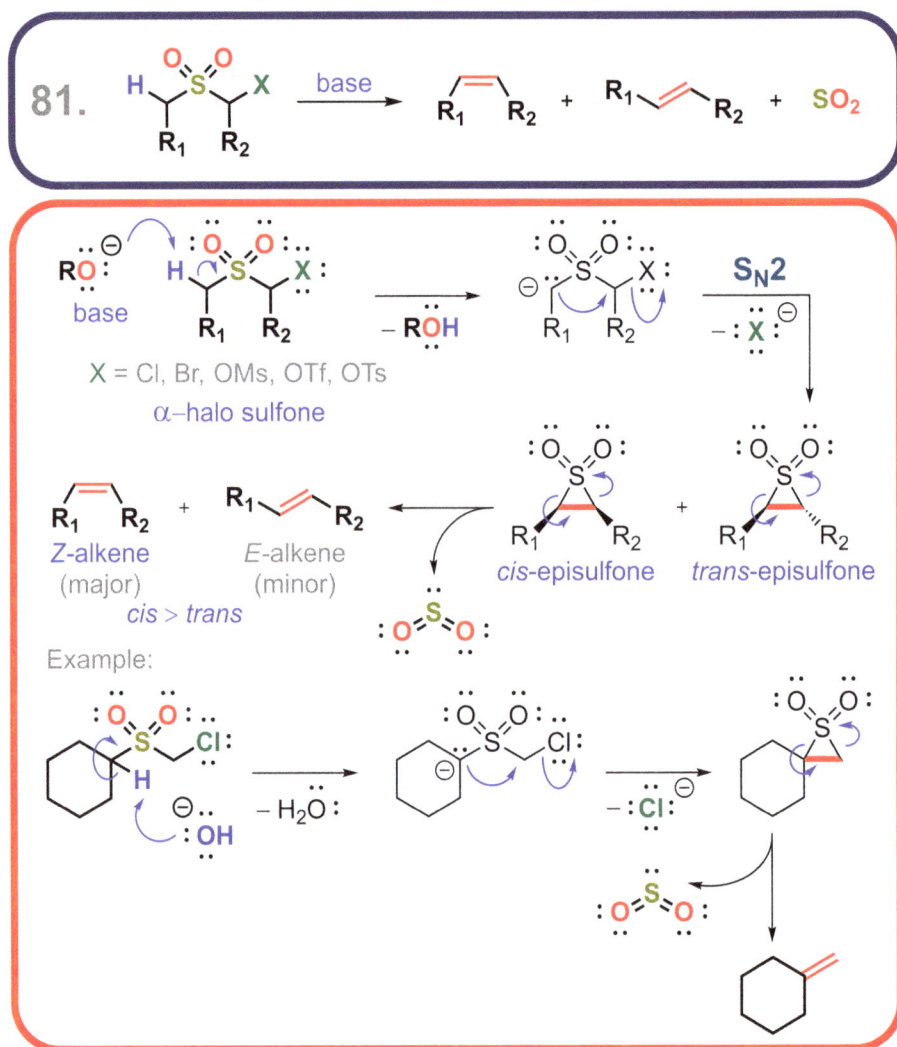

81.

X = Cl, Br, OMs, OTf, OTs

α–halo sulfone

Z-alkene (major)
E-alkene (minor)
cis > trans

cis-episulfone
trans-episulfone

Example:

Fig. 81.1: The *Ramberg–Bäcklund* rearrangement mechanism.[294]

---

294 The *Ramberg–Bäcklund* rearrangement or the *Ramberg–Bäcklund* reaction mechanism is a combination of the bimolecular **nucleophilic substitution** (S_N2), covered in Chapter 2, and subsequent concerted **elimination** (*cheletropic elimination reaction*) [1a] and [81a].

https://doi.org/10.1515/9783110786835-081

Ramberg–Bäcklund Variation

Favorskii Rearrangement

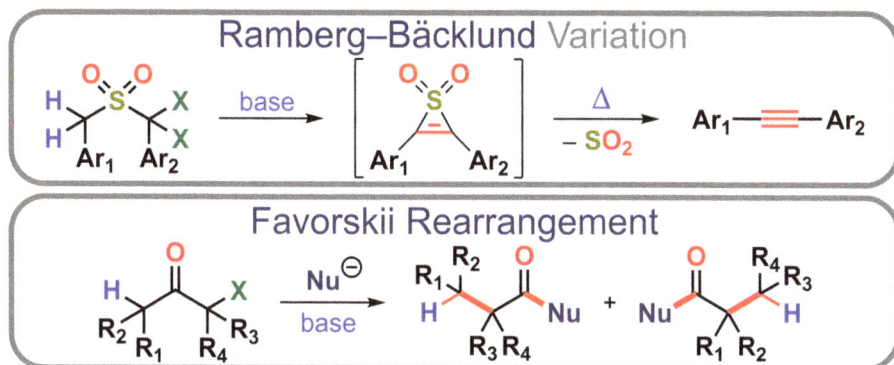

Fig. 81.2: Reactions related to the *Ramberg–Bäcklund* rearrangement.[295]

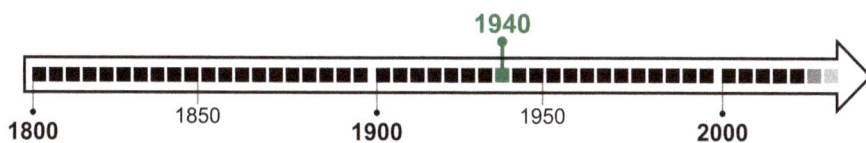

Fig. 81.3: The discovery of the *Ramberg–Bäcklund* rearrangement.[296]

---

295 There are several variations of the *Ramberg–Bäcklund* rearrangement; for example, the formation of *alkynes* instead of *alkenes* [81b] and [1a]. The S_N2 step in the *Favorskii* rearrangement (covered in Chapter 37) is related to the *Ramberg–Bäcklund* rearrangement.
296 The reaction was likely first described around 1940 [81c].

82 Reformatsky Reaction

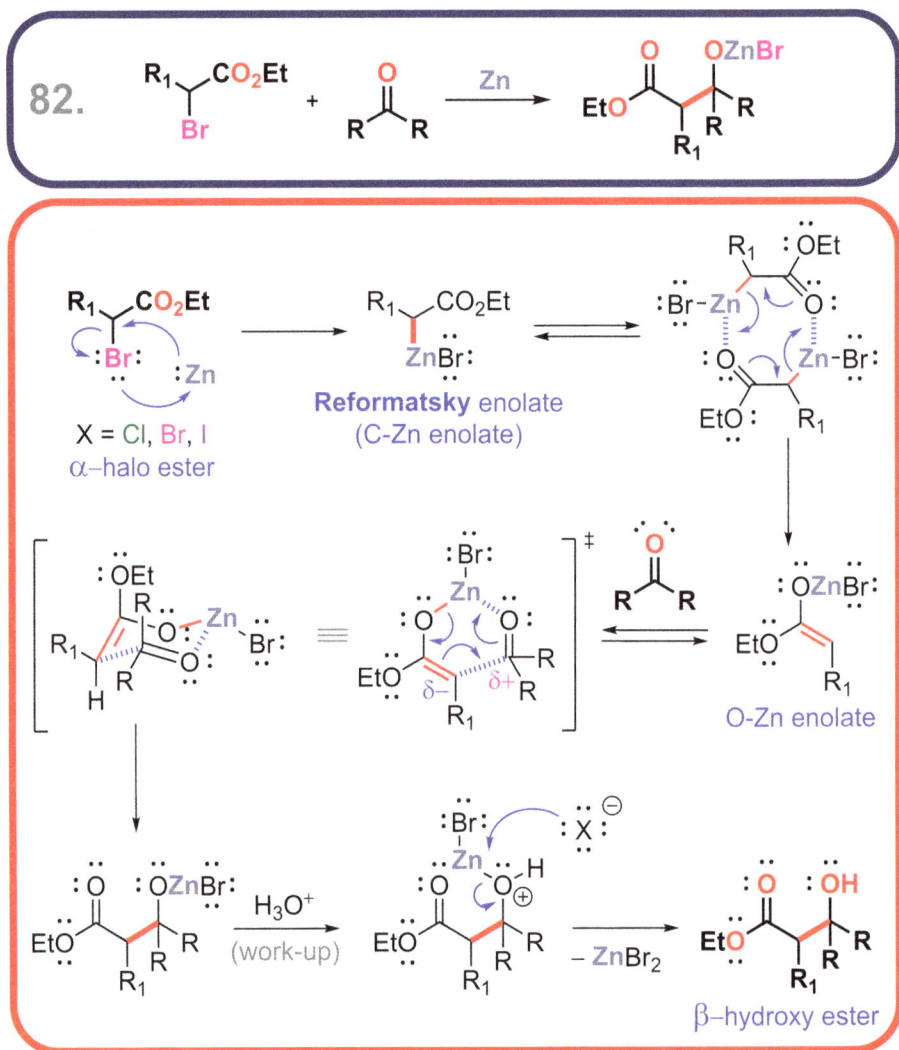

82. R_1—CO_2Et with Br + R—C(=O)—R →(Zn)→ EtO—C(=O)—C(R)(R_1)—OZnBr

82. $R_1\text{—CHBr—}CO_2Et$ + $R_2C{=}O$ $\xrightarrow{\text{Zn}}$ product

- α–halo ester, X = Cl, Br, I
- **Reformatsky** enolate (C-Zn enolate)
- O-Zn enolate
- β–hydroxy ester
- H$_3$O$^+$ (work-up)
- $-$ ZnBr$_2$

Fig. 82.1: The **Reformatsky** reaction mechanism.[297]

---

297 The **Reformatsky** reaction (condensation) (also spelled Reformatskii, and in Russian Сергей Николаевич Реформатский or С. Н. Реформатский) mechanistically is much like the aldol condensation reaction (see Chapter 83).

https://doi.org/10.1515/9783110786835-082

Blaise Reaction

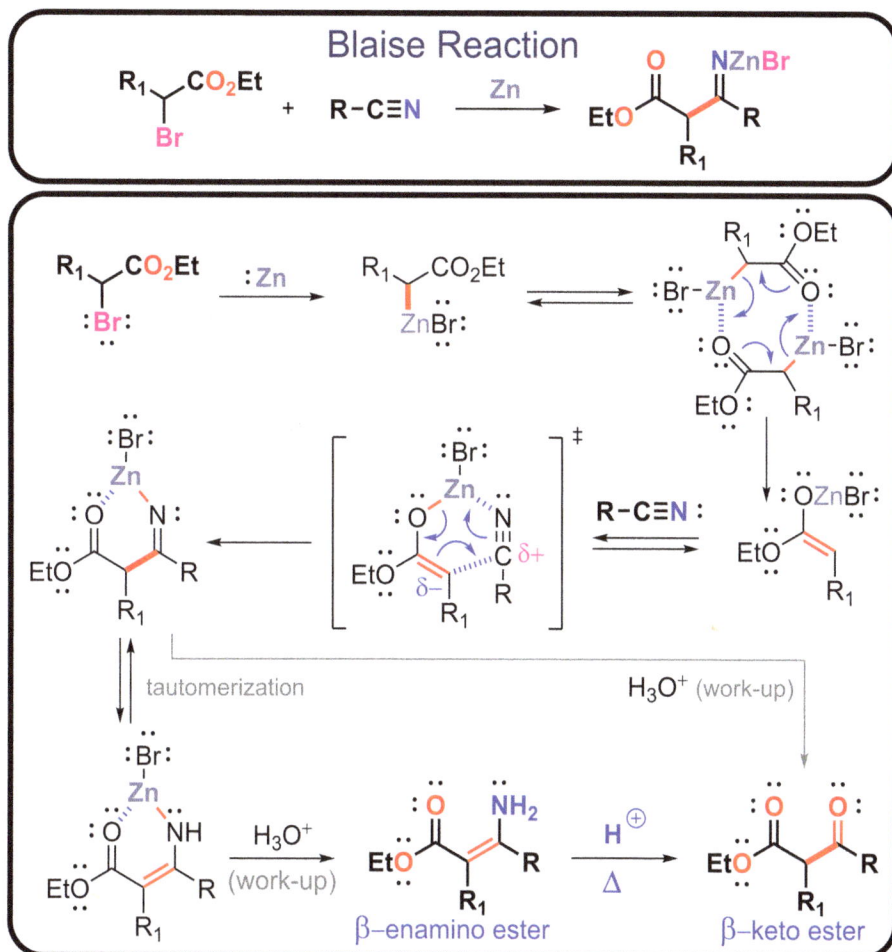

Fig. 82.2: The **Blaise** reaction mechanism.[298]

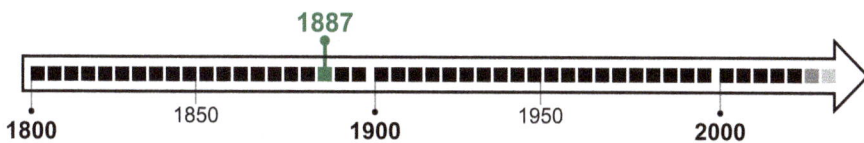

Fig. 82.3: The discovery of the **Reformatsky** reaction.[299]

---

298 The **Blaise** reaction is a variation of the **Reformatsky** reaction [82a, 82b]. Here, the preformed **Reformatsky** enolate (C-Zn or O-Zn enolate) reacts with a *nitrile* instead of an *aldehyde* or *ketone*.
299 The reaction was likely first described around 1887 [82].

83 Robinson Annulation

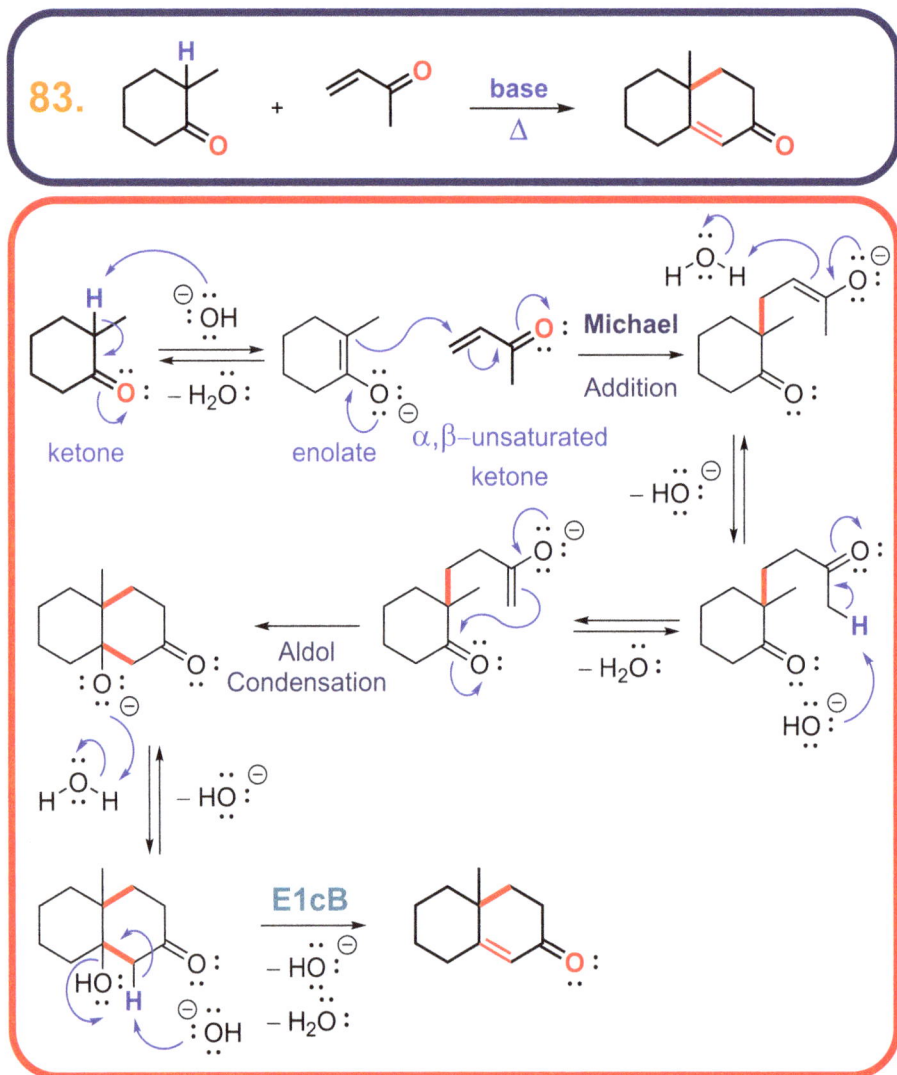

Fig. 83.1: The *Robinson* annulation mechanism.[300]

---

300 The *Robinson* annulation mechanism is a cascade of the *Michael* conjugate addition (see Chapter 59), followed by the *aldol condensation*, and finally **E1cB** elimination (see Chapter 6).

https://doi.org/10.1515/9783110786835-083

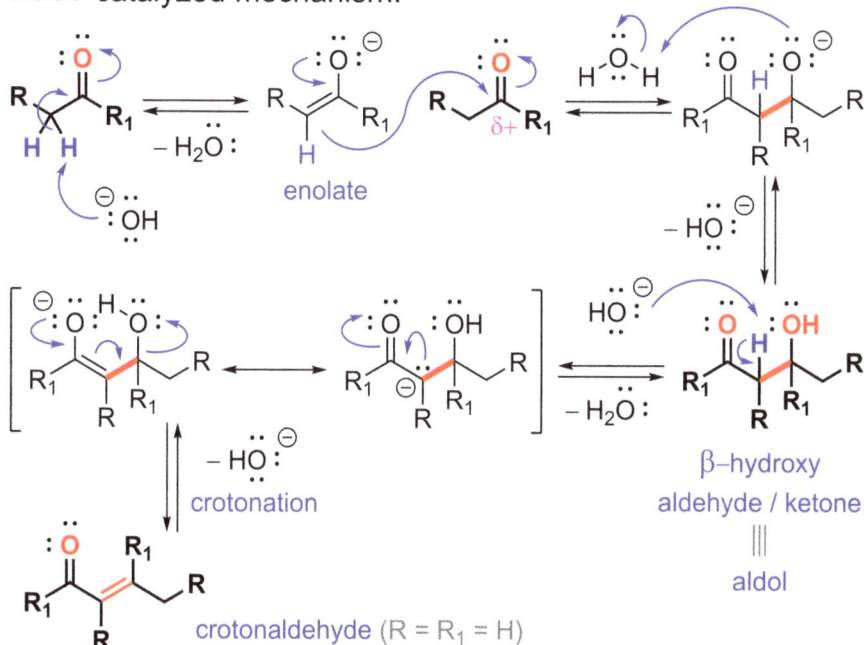

Fig. 83.2: The *aldol condensation* mechanism.[301]

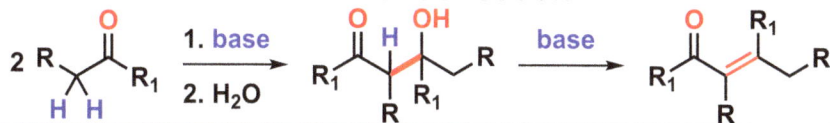

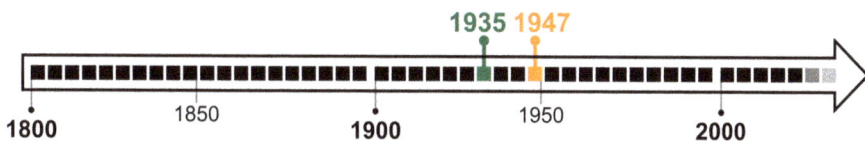

Fig. 83.3: The discovery of the **Robinson** annulation.[302]

---

301 The *base-catalyzed aldol condensation* can yield β-hydroxy aldehydes **(aldols)**. The *aldols* can undergo an elimination and yield *crotonaldehydes* (the *croton condensation* = *crotonation*) [1].
302 The reaction was likely first described around 1935 [83a]. In **1947**, Sir Robert Robinson received the Nobel Prize in Chemistry for his work related to alkaloids [83b].

84 Shapiro Reaction

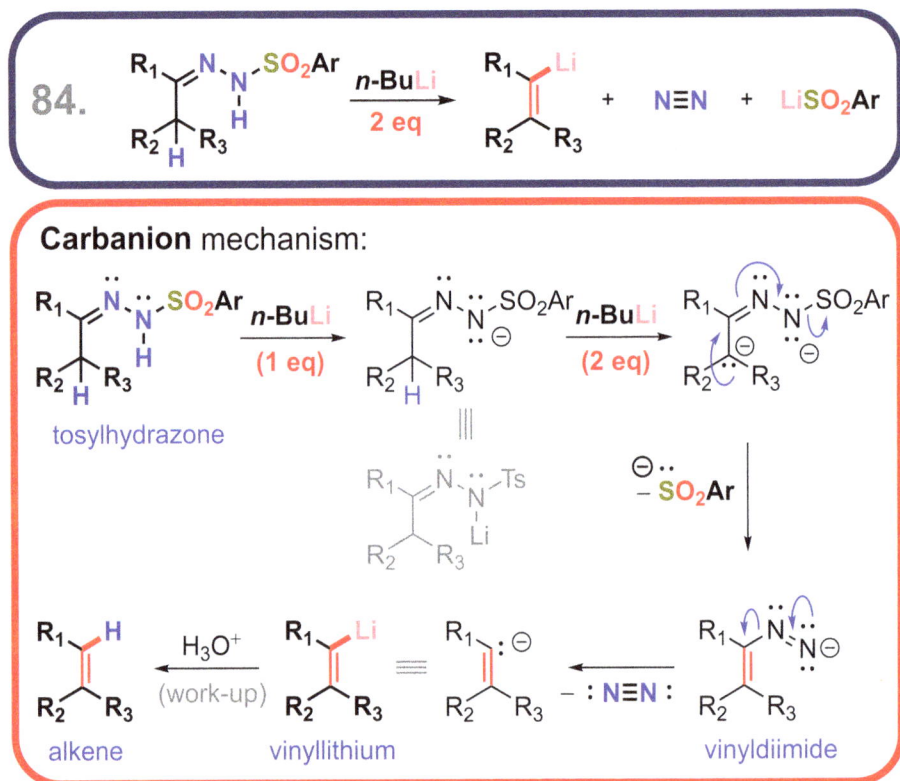

84. R_1—C(=N—N(H)—SO$_2$Ar)—CH(R_2)(R_3)$\xrightarrow[\text{2 eq}]{\textit{n}\text{-BuLi}}$ alkene R_1,Li / R_2,R_3 + N≡N + LiSO$_2$Ar

Carbanion mechanism:

tosylhydrazone $\xrightarrow[\text{(1 eq)}]{\textit{n}\text{-BuLi}}$ $\xrightarrow[\text{(2 eq)}]{\textit{n}\text{-BuLi}}$

$\xrightarrow{-\ :\overset{\ominus}{SO_2Ar}}$

alkene R_1,H / R_2,R_3 $\xleftarrow[\text{(work-up)}]{H_3O^+}$ vinyllithium R_1,Li / R_2,R_3 ≡ R_1,:$^\ominus$ / R_2,R_3 $\xleftarrow{-\ :N≡N:}$ vinyldiimide R_1,N=N$^\ominus$ / R_2,R_3

Fig. 84.1: The **Shapiro** *reaction* mechanism.[303]

---

303 The **Shapiro** *reaction* is a type of **elimination** reaction that undergoes the *carbanion* mechanism.

https://doi.org/10.1515/9783110786835-084

Bamford–Stevens Reaction

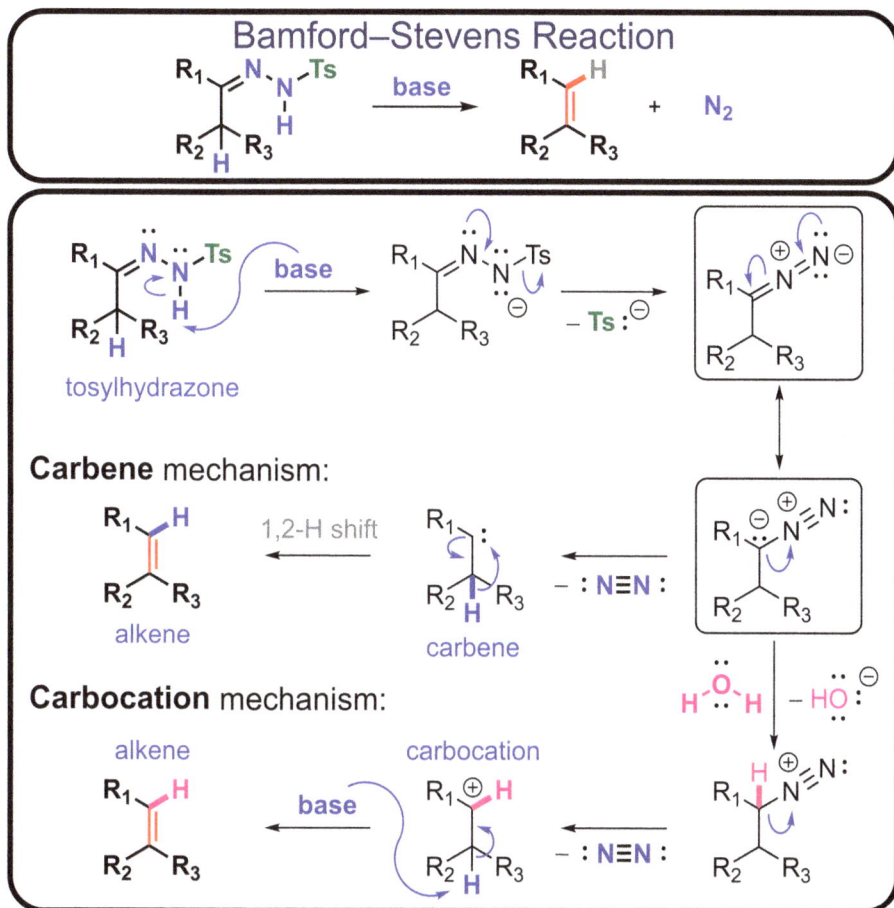

Fig. 84.2: The **Bamford–Stevens** reaction mechanism.[304]

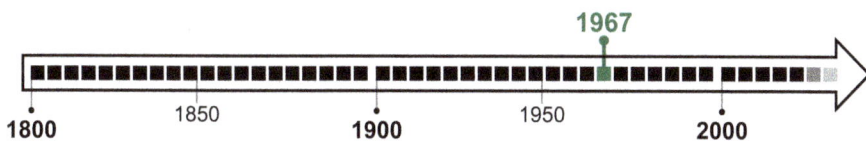

Fig. 84.3: The discovery of the **Shapiro reaction**.[305]

---

304 The **Bamford–Stevens reaction** is a more general variation of the **Shapiro reaction**. Two mechanisms are possible: the *carbene* mechanism and the *carbocation* (*carbenium ion*) mechanism. [84a].
305 The reaction was likely first described around 1967 [84b]; see also [84c, 84d].

85 Sonogashira Cross-Coupling

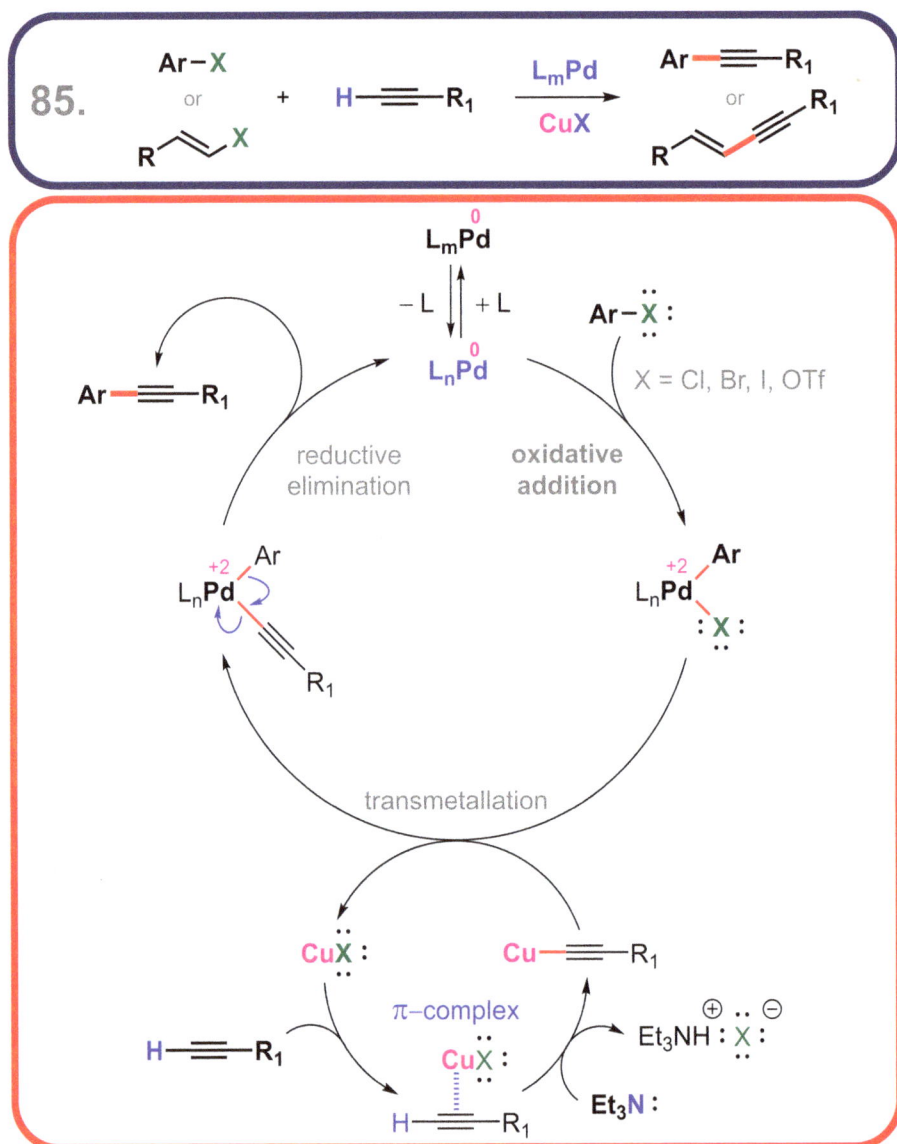

Fig. 85.1: The *Sonogashira* cross-coupling mechanism.[306]

---

306 The *Sonogashira* cross-coupling is a type of mixed *Pd-catalyzed* and *Cu-co-catalyzed* cross-coupling reaction (C–C bond formation using *aryl halides* and *terminal alkynes*). For teaching purposes, a simplified and general mechanism (with two catalytic cycles using **Pd** and **Cu**) is shown.

https://doi.org/10.1515/9783110786835-085

Castro–Stephens Coupling

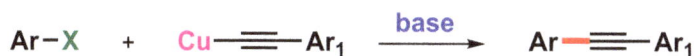

$$Ar-X \quad + \quad Cu\!\!\!=\!\!\!=\!\!\!-Ar_1 \quad \xrightarrow{\text{base}} \quad Ar\!-\!\!\!=\!\!\!=\!\!\!-Ar_1$$

Suzuki Cross Coupling

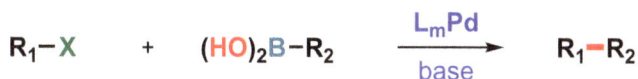

$$R_1-X \quad + \quad (HO)_2B-R_2 \quad \xrightarrow[\text{base}]{L_mPd} \quad R_1\!-\!R_2$$

Stille Cross Coupling

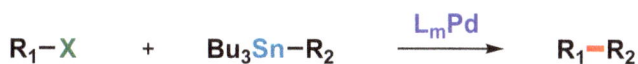

$$R_1-X \quad + \quad Bu_3Sn-R_2 \quad \xrightarrow{L_mPd} \quad R_1\!-\!R_2$$

Negishi Cross Coupling

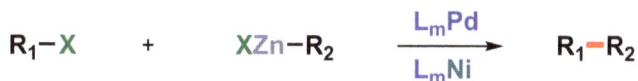

$$R_1-X \quad + \quad XZn-R_2 \quad \xrightarrow[L_mNi]{L_mPd} \quad R_1\!-\!R_2$$

Kumada Cross Coupling

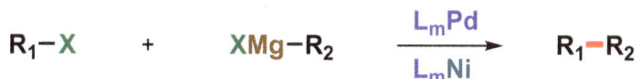

$$R_1-X \quad + \quad XMg-R_2 \quad \xrightarrow[L_mNi]{L_mPd} \quad R_1\!-\!R_2$$

Fig. 85.2: Reactions related to the **Sonogashira** cross-coupling.[307]

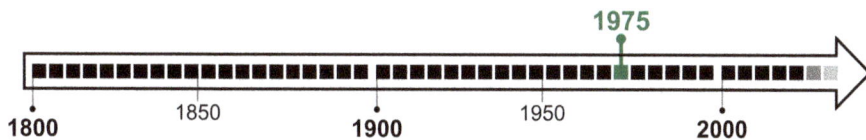

1975

1800   1850   1900   1950   2000

Fig. 85.3: The discovery of the **Sonogashira** cross-coupling.[308]

---

307 The **Castro–Stephens** cross-coupling is **Cu**-catalyzed and closely related (C–C bond formation using *aryl halides* and pre-formed or *in situ* generated *copper(I) acetylides*) [85a]. Other cross-coupling reactions are also related to the **Sonogashira** cross-coupling: the **Suzuki** (Chapter 89), the **Stille** (Chapter 88), the **Negishi** (Chapter 66), and the **Kumada** cross-coupling (Chapter 53).
308 The reaction was likely first described around 1975 [85b].

86 Staudinger Reaction

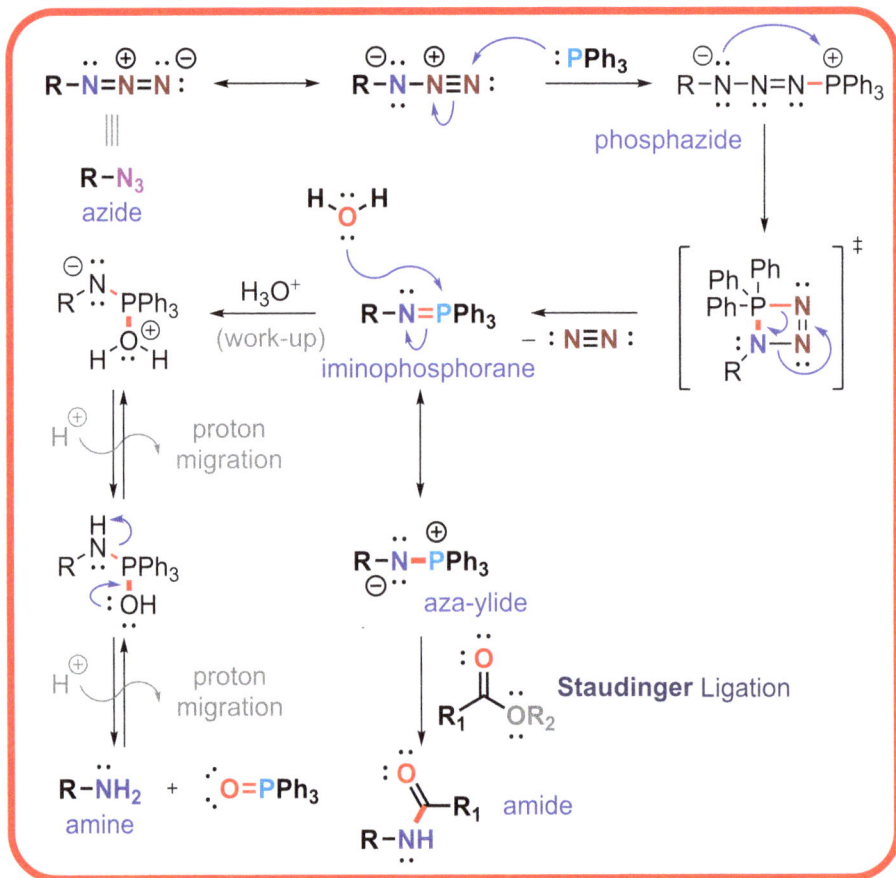

86. $R-N=N=N^{\oplus\ominus}$ + Ph_3P $\longrightarrow$ $R-NH_2$ + $N\equiv N$ + $O=PPh_3$

$R-N=N=N$ (azide) $\longleftrightarrow$ $R-\overset{\ominus}{N}-N\equiv N$ $\xrightarrow{:PPh_3}$ $R-N-N=N-PPh_3$ phosphazide

$R-N_3$
azide

$R-\overset{\ominus}{N}-PPh_3$ ... $\xrightarrow[\text{(work-up)}]{H_3O^+}$ $R-N=PPh_3$ iminophosphorane $\xleftarrow{-:N\equiv N:}$ $\left[\begin{array}{c} Ph\ Ph \\ Ph-P=N \\ N-N \\ R \end{array} \right]^{\ddagger}$

H_2O

proton migration

$R-\overset{H}{\underset{N}{N}}-PPh_3$: OH

proton migration

$R-\overset{\oplus}{N}-PPh_3$ aza-ylide

Staudinger Ligation

R_1 OR$_2$

$R-NH_2$ + $O=PPh_3$
amine

$R-NH$ R_1 amide

Fig. 86.1: The *Staudinger* reaction mechanism.[309]

---

309 The *Staudinger* reaction (reduction) is a reduction of *azides* to primary amines using *triphenylphosphine*. It should not be confused with the *Staudinger* synthesis or the *Staudinger* ketene cycloaddition reaction (for example, formation of β-lactams) [86a, 86b].

https://doi.org/10.1515/9783110786835-086

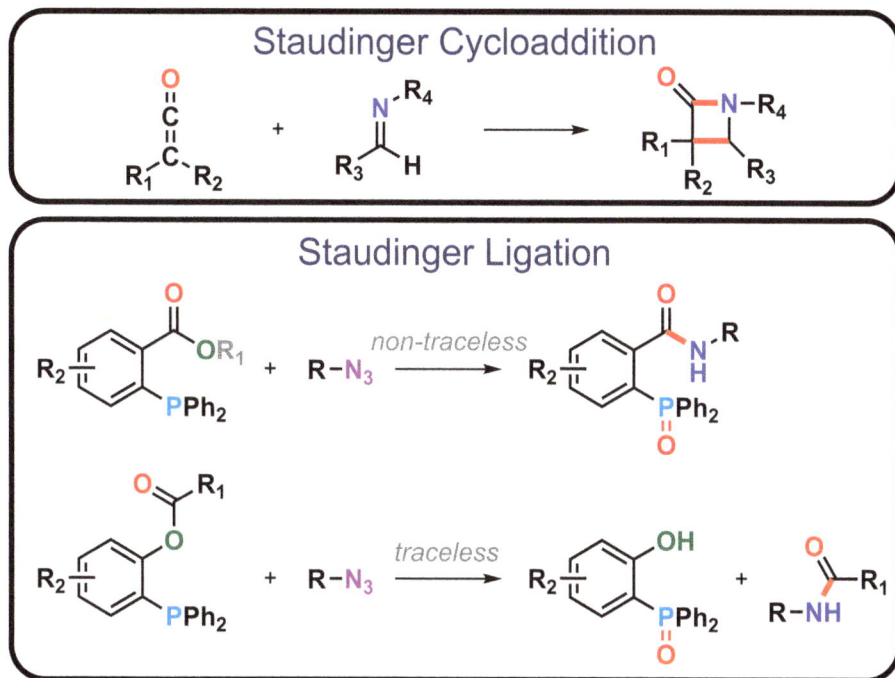

Fig. 86.2: The **Staudinger** *cycloaddition* and *ligation*.[310]

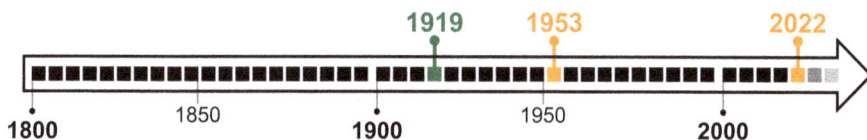

Fig. 86.3: The discovery of the **Staudinger** *reaction*.[311]

---

310 The **Staudinger** *ligation* [86c, 86d] is a modification of the **Staudinger** *reaction*: in this case, the generated *aza-ylide* is trapped with an *ester* to form an *amide* bond. There are two general types: *non-traceless* and *traceless* **Staudinger** *ligation* [86e].

311 The reaction was likely first described around 1919 [86f]. In **1953**, Hermann Staudinger received the Nobel Prize in Chemistry for his work in macromolecular chemistry [86g]. In **2022**, Carolyn R. Bertozzi, Morten Meldal, and K. Barry Sharpless received the Nobel Prize in Chemistry for the development of click chemistry and bioorthogonal chemistry [30g, 30h].

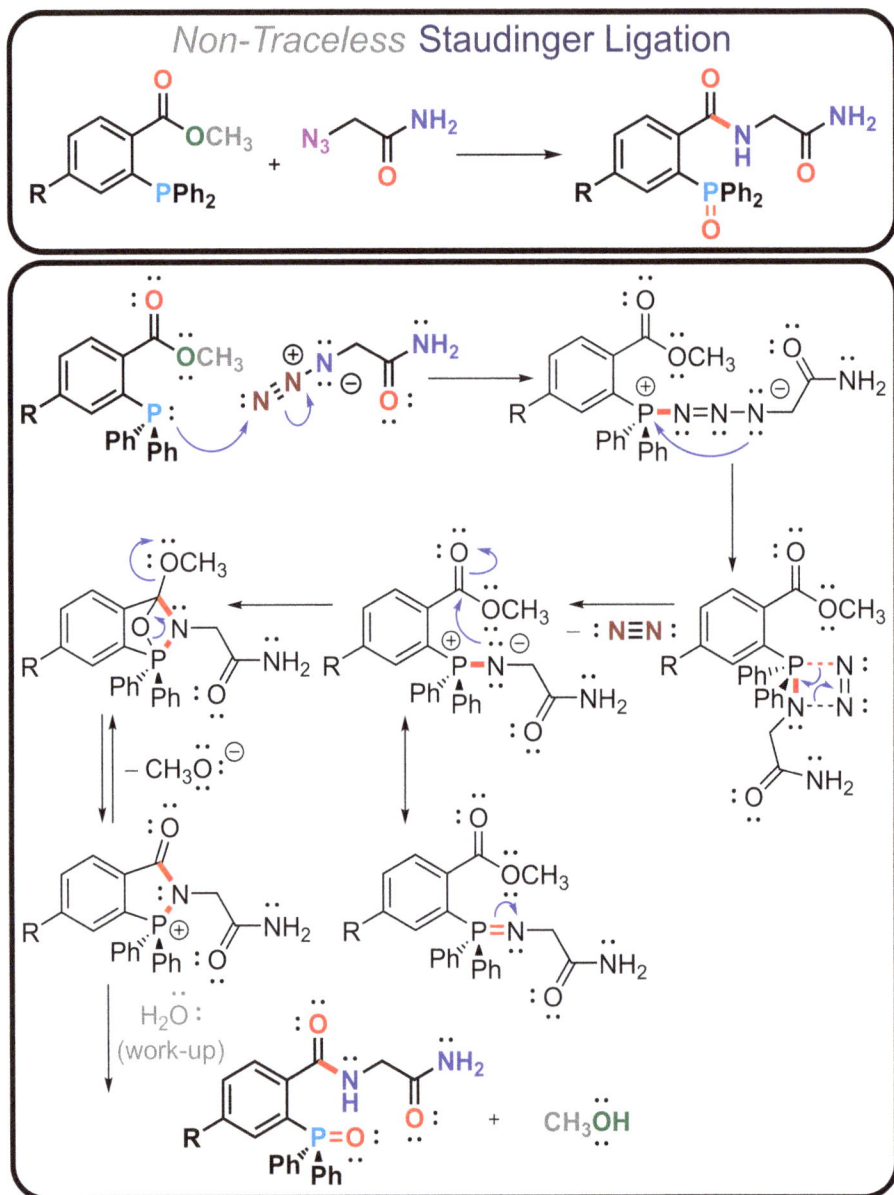

Fig. 86.4: The *non-traceless* **Staudinger** *ligation* mechanism.[312]

312 Types of the **Staudinger** *ligation*: *non-traceless* and *traceless* **Staudinger** *ligation* [86d, 86e, 86h].

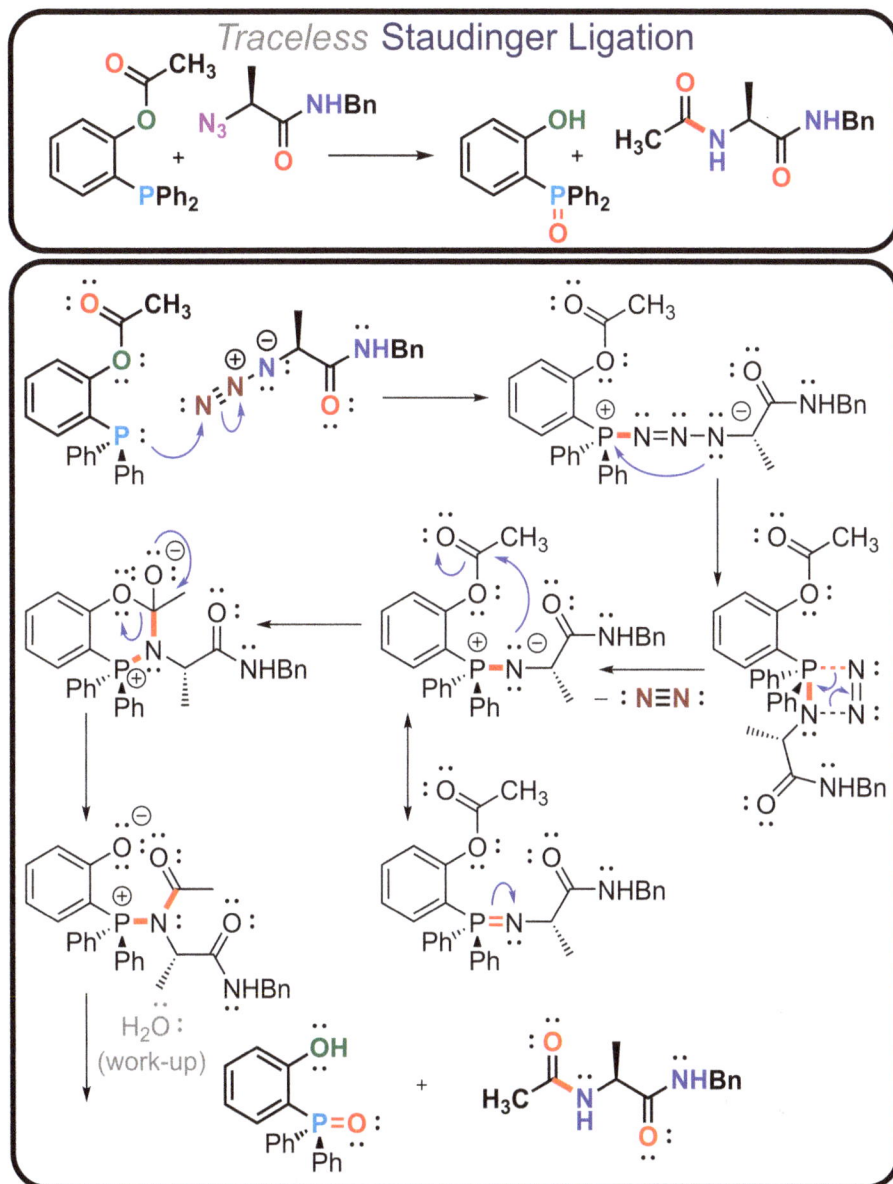

Fig. 86.5: The *traceless Staudinger ligation* mechanism.[313]

---

313 Types of the **Staudinger** ligation: *non-traceless* and *traceless* **Staudinger** *ligation* [86d, 86e, 86h].

87 Steglich Esterification

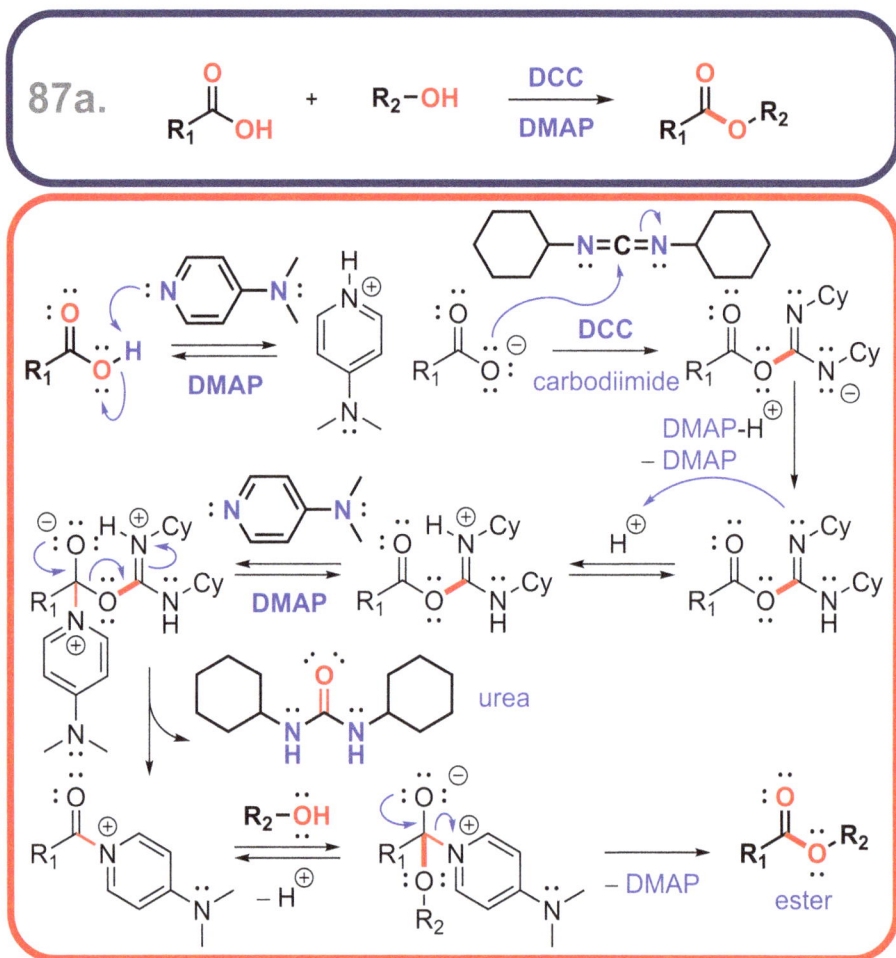

87a.

R_1—COOH + R_2—OH $\xrightarrow[\text{DMAP}]{\text{DCC}}$ R_1—C(O)—O—R_2

Fig. 87.1: The *Steglich* esterification mechanism (DCC + DMAP).[314]

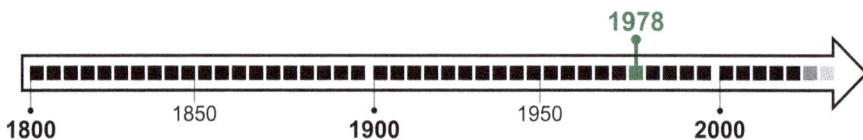

1978

1800   1850   1900   1950   2000

Fig. 87.2: The discovery of the *Steglich* esterification.[315]

---

314 The *Steglich* esterification is an *ester coupling reaction* (compare to the *peptide (amide) coupling*

https://doi.org/10.1515/9783110786835-087

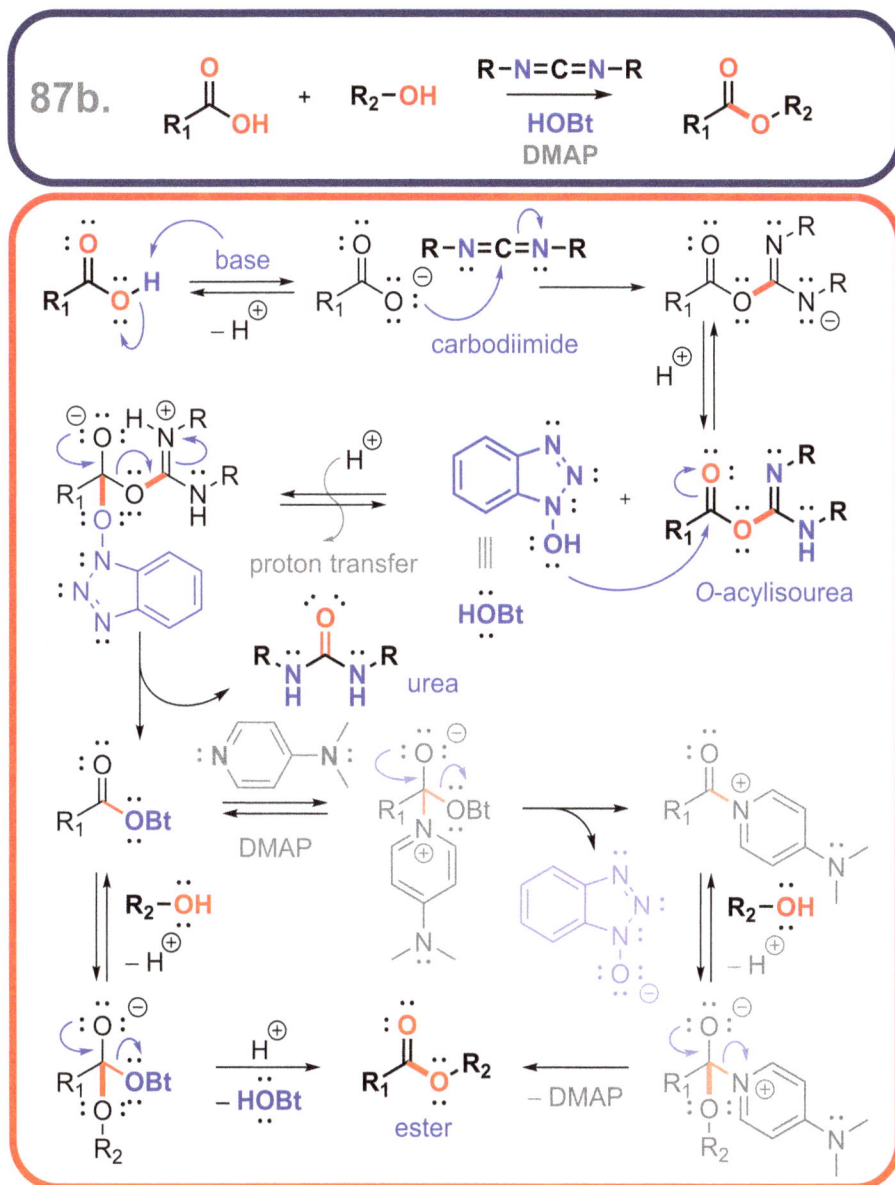

Fig. 87.3: The **Steglich** *esterification* mechanism (DCC + HOBt + DMAP).[316]

mechanism in Chapter 74 or the **Fischer** *esterification*, Fig. 61.5). The mechanism involves the use of *carbodiimide* coupling reagents (DCC) and DMAP catalyst [87a].

315 The reaction was likely first described around 1978 [87b].

316 The **Steglich** *esterification* can be carried out with DCC in the presence of other *peptide (amide) coupling additives* (for example, HOBt) with or without DMAP catalyst.

88 Stille Cross-Coupling

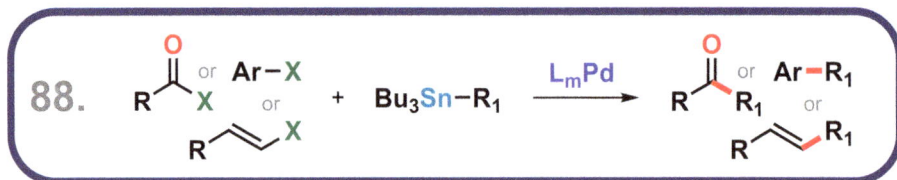

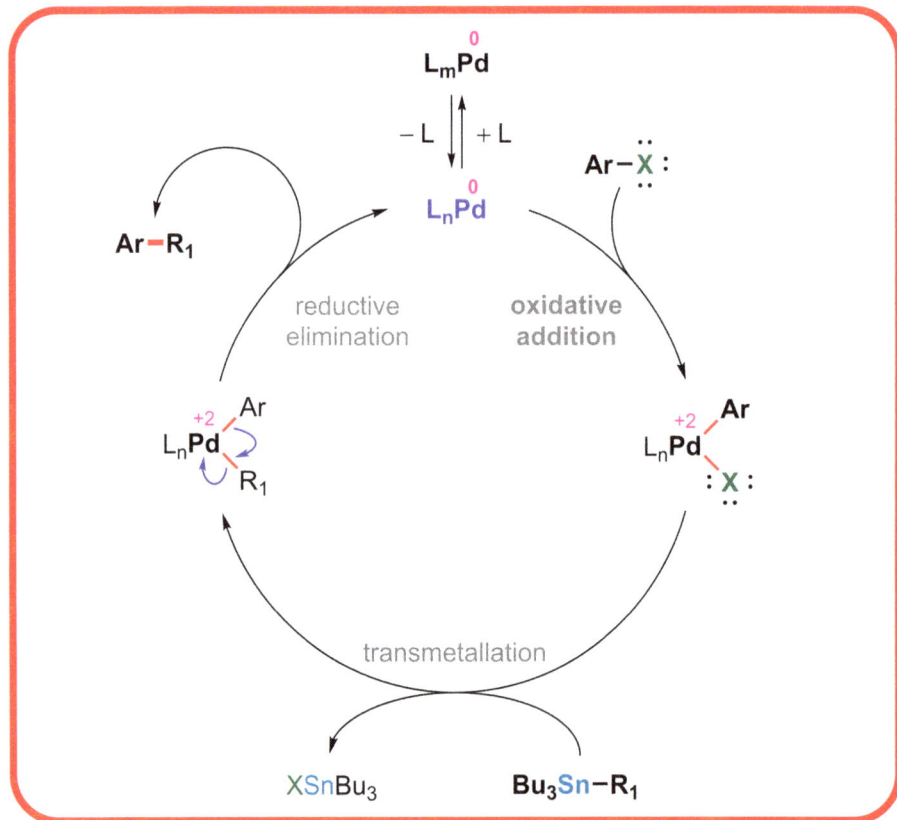

88. $R-\overset{\overset{\displaystyle O}{\|}}{\underset{}{C}}-X$ or **Ar–X** or $R{\diagdown}\!{=}\!{\diagup}^X$ + **Bu$_3$Sn–R$_1$** $\xrightarrow{\text{L}_m\text{Pd}}$ $R-\overset{\overset{\displaystyle O}{\|}}{\underset{}{C}}-R_1$ or **Ar–R$_1$** or $R{\diagdown}\!{=}\!{\diagup}^{R_1}$

Fig. 88.1: The *Stille cross-coupling* mechanism.[317]

---

317 The *Stille* cross-coupling or the *Migita–Kosugi–Stille* cross-coupling is a versatile type of *Pd-catalyzed cross-coupling* reaction (C–C bond formation using *aryl halides* or other *electrophiles* and *organotin compounds = organostannanes*). For teaching purposes, a simplified and general mechanism is shown.

https://doi.org/10.1515/9783110786835-088

Carbonylative Stille Cross Coupling

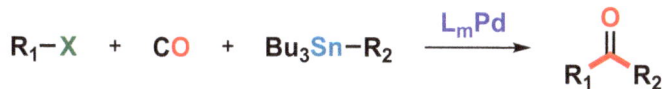

$$R_1-X \ + \ CO \ + \ Bu_3Sn-R_2 \ \xrightarrow{\ L_mPd\ } \ R_1 \overset{O}{\underset{}{\diagup\!\!\!\!\diagdown}} R_2$$

Fukuyama Cross Coupling

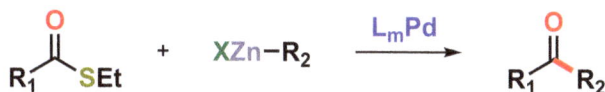

$$R_1 \overset{O}{\underset{}{\diagup\!\!\!\!\diagdown}} SEt \ + \ XZn-R_2 \ \xrightarrow{\ L_mPd\ } \ R_1 \overset{O}{\underset{}{\diagup\!\!\!\!\diagdown}} R_2$$

Fig. 88.2: Reactions related to the *Stille* cross-coupling.[318]

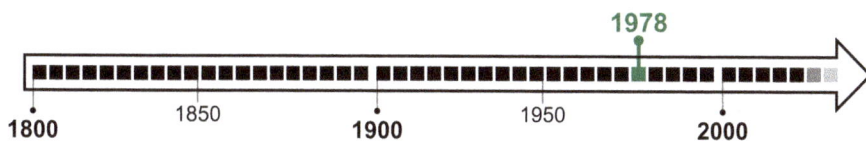

Fig. 88.3: The discovery of the *Stille* cross-coupling.[319]

---

318 The *carbonylative Stille cross-coupling* is related to the *Stille cross-coupling*. It is a method to form *ketones* (two C–C bond formations using *aryl halides* or other *electrophiles, organostannanes*, and *carbon monoxide*) [88a]. Ketones can also be formed via the *Fukuyama cross-coupling* (C–C bond formation using *thioesters* and *organozinc compounds*) [88b] or the *Liebeskind–Srogl cross-coupling* covered in Chapter 55 (C–C bond formation using *thioesters* and *boronic acids*).

319 The reaction was likely first described around 1978 [88c, 88d].

89 Suzuki Cross-Coupling

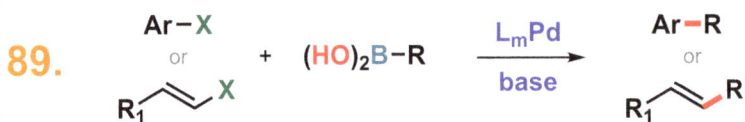

89. $Ar-X$ or $\underset{R_1}{\diagup}X$ + $(HO)_2B-R$ $\xrightarrow[\text{base}]{L_mPd}$ $Ar-R$ or $\underset{R_1}{\diagup}R$

1. Oxo-Pd pathway (a)

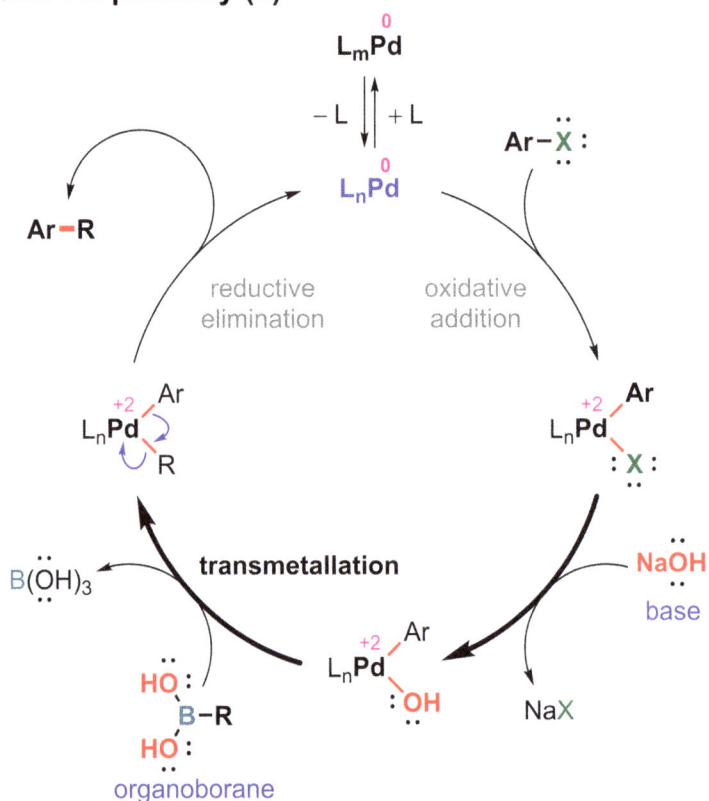

Fig. 89.1: The *Suzuki* cross-coupling mechanism (oxo-**Pd** pathway (**a**)).[320]

---

320 The *Suzuki* cross-coupling or the *Suzuki–Miyaura* cross-coupling is a type of *Pd*-catalyzed cross-coupling reaction (C–C bond formation using *aryl halides* and *organoboronic acids*). It is one of the most important reactions in synthetic organic and medicinal chemistry. The *oxo-Pd pathway* (**a**) is the preferred mechanism [89a].

https://doi.org/10.1515/9783110786835-089

2. Boronate pathway (b)

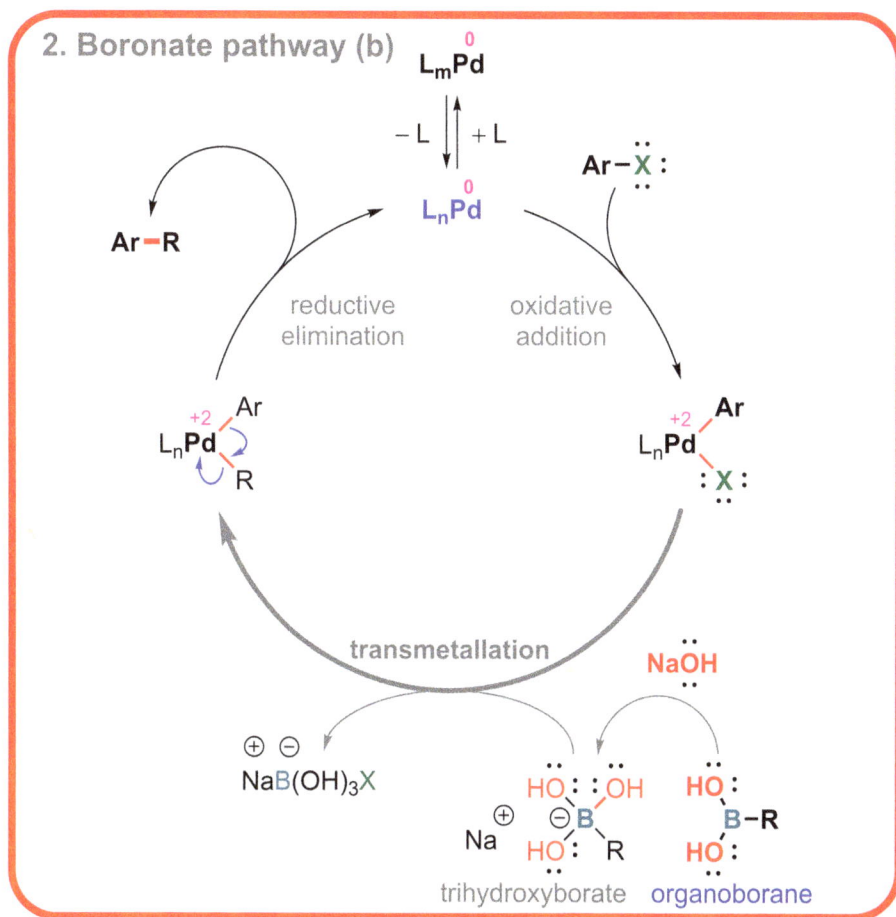

Fig. 89.2: The **Suzuki cross-coupling** mechanism (boronate pathway (b)).[321]

Fig. 89.3: The discovery of the **Suzuki cross-coupling**.[322]

---

321 The reaction mechanism can be also explained by the *boronate pathway* (b). For teaching purposes, a simplified and general mechanism is shown [89b].

322 The reaction was likely first described around 1979 [89c, 89d]. In **2010**, Akira Suzuki (jointly with Richard F. Heck and Ei-ichi Negishi) received the Nobel Prize in Chemistry for the development of **Pd**-catalyzed cross-coupling reactions [46c].

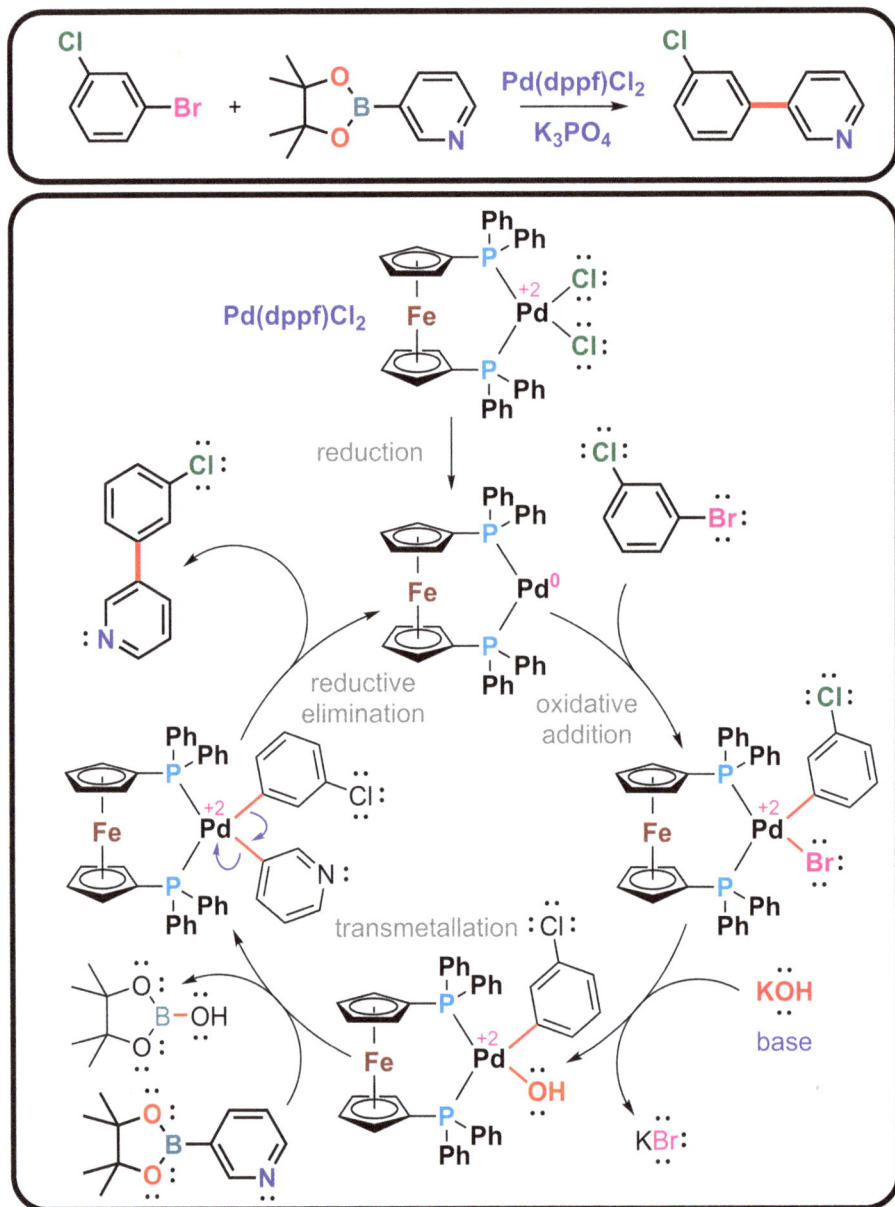

Fig. 89.4: The *Suzuki cross-coupling* mechanism catalyzed by **Pd(dppf)Cl₂**.[323]

---

323 [1,1'-Bis(diphenylphosphino)ferrocene]dichloropalladium(II) or **Pd(dppf)Cl₂** is one of the most common **Pd** catalysts [89e, 89g].

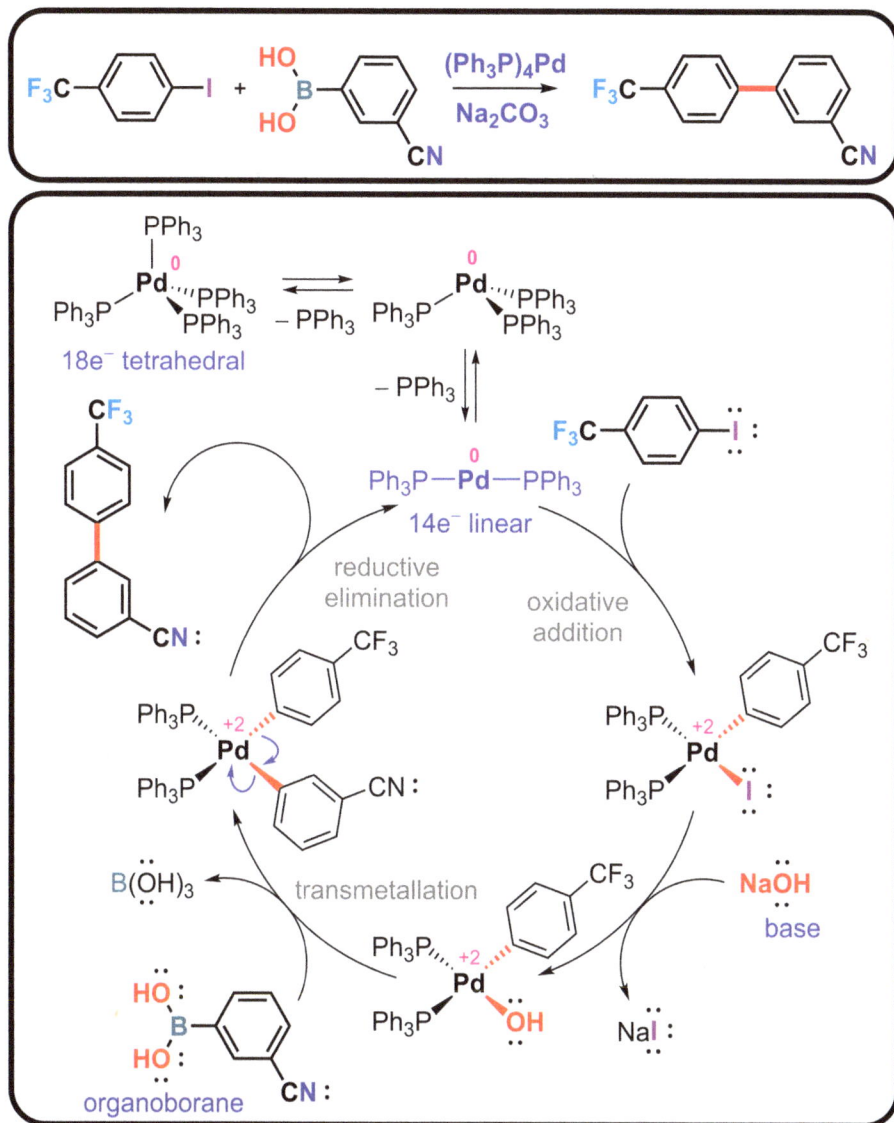

Fig. 89.5: The *Suzuki cross-coupling* mechanism catalyzed by **Pd(PPh₃)₄**.[324]

---

324 Tetrakis(triphenylphosphine)palladium(0) or **Pd(PPh₃)₄** is one of the most common **Pd** catalysts [89f, 89g].

90 Swern Oxidation

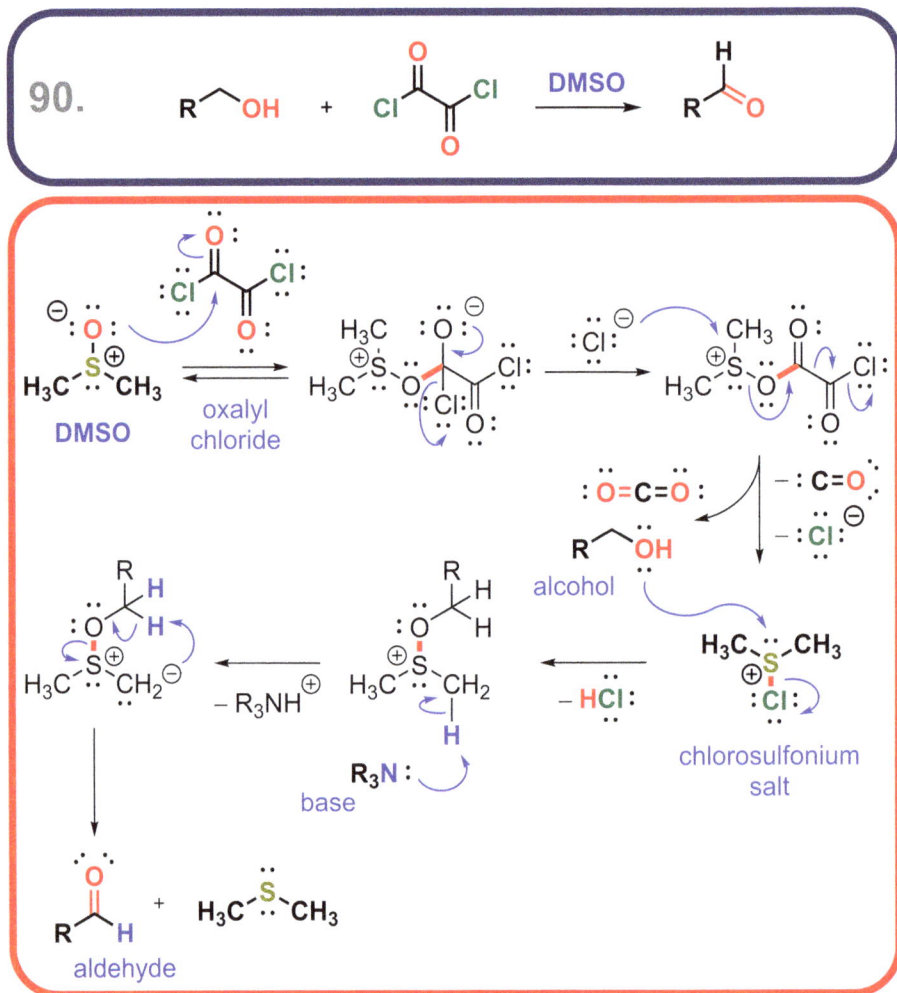

Fig. 90.1: The *Swern oxidation* mechanism.[325]

---

325 The *Swern oxidation* is one of the most important reactions in synthetic organic and medicinal chemistry.

https://doi.org/10.1515/9783110786835-090

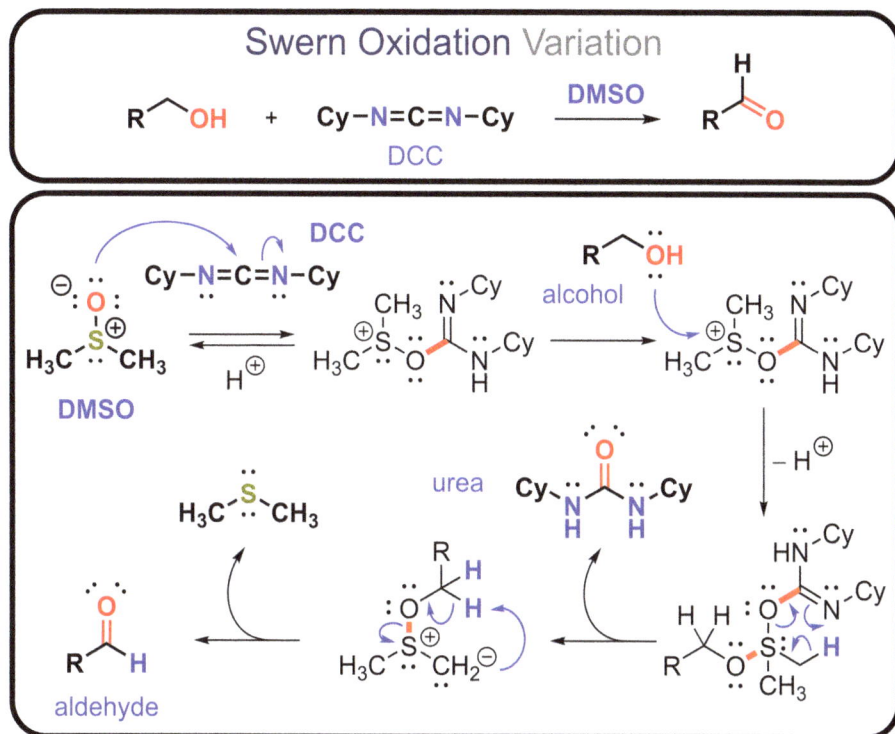

Fig. 90.2: The **Swern oxidation** variation mechanism (DCC + DMSO).[326]

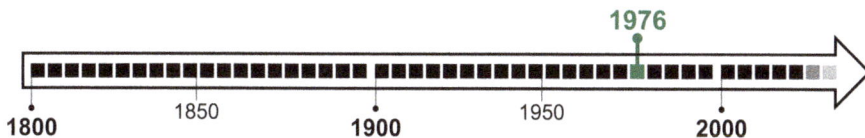

Fig. 90.3: The discovery of the **Swern oxidation**.[327]

---

326 There are numerous variations of the **Swern oxidation**: the **Swern variation** using TFAA and DMSO [90a] or *carbodiimide reagent* (DCC) and DMSO [90b]. Several important named oxidation reactions yield *ketones* from *alcohols*: the **Dess–Martin** oxidation (Chapter 33) and the **Jones** oxidation (Chapter 51).

327 The reaction was likely first described around 1976 [90a]; see also [90c, 90d].

91 Ugi Reaction

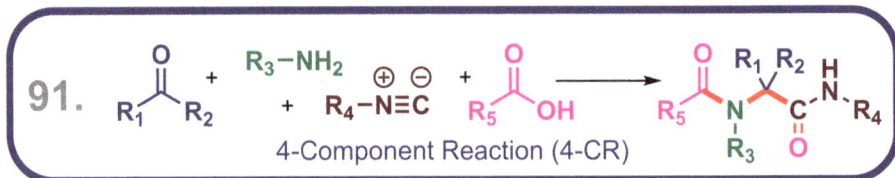

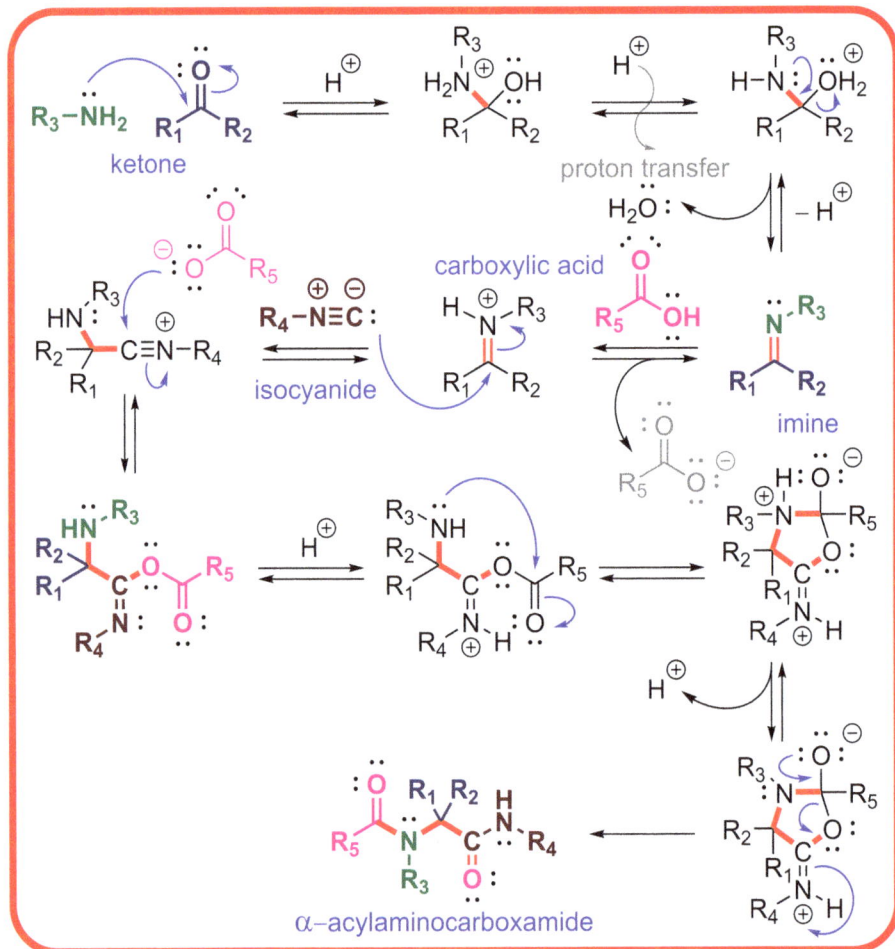

$$91.$$

4-Component Reaction (4-CR)

Fig. 91.1: The **Ugi reaction** mechanism.[328]

328 The **Ugi reaction** or the **Ugi condensation** is a type of multi-component reaction (MCR): a four-component reaction (4-CR).

https://doi.org/10.1515/9783110786835-091

Passerini Reaction

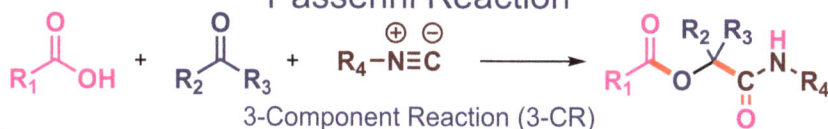

3-Component Reaction (3-CR)

1. Concerted Mechanism:

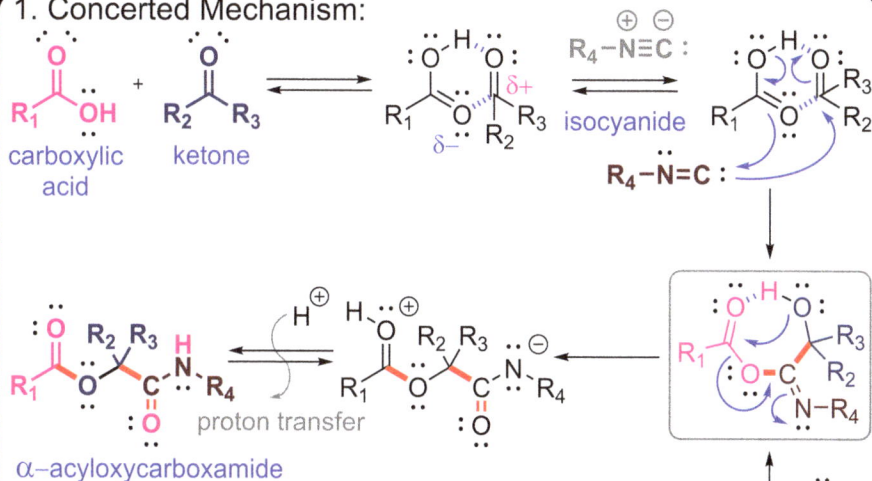

carboxylic acid   ketone   isocyanide

$R_4-N=C:$

α–acyloxycarboxamide

proton transfer

2. Ionic Mechanism:

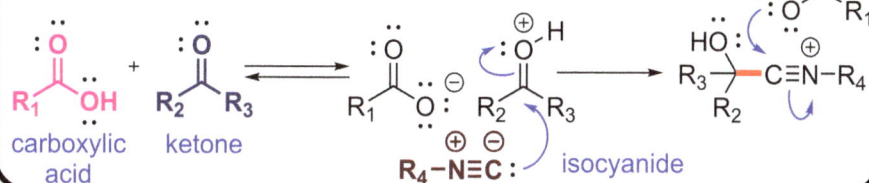

carboxylic acid   ketone

isocyanide

Fig. 91.2: The **Passerini** reaction mechanism.[329]

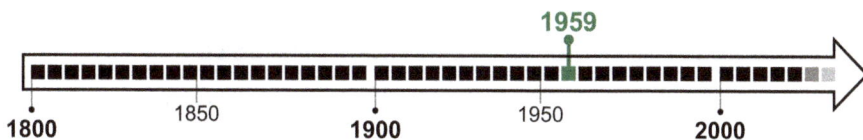

1959

1800   1850   1900   1950   2000

Fig. 91.3: The discovery of the **Ugi reaction**.[330]

---

329 The **Passerini reaction** is mechanistically related to the **Ugi reaction** [91a, 91b]. The product formation can be rationalized either via (1) the *concerted* mechanism or (2) the *ionic* mechanism. Other 3-CR's were also mentioned in this book: the **Gewald reaction** (Chapter 41), the **Mannich reaction** (Chapter 56), the **Petasis reaction** (Chapter 62), and the **Pauson–Khand reaction** (Chapter 73).
330 The reaction was likely first described around 1959 [91c].

92 Ullmann Aryl–Aryl Coupling

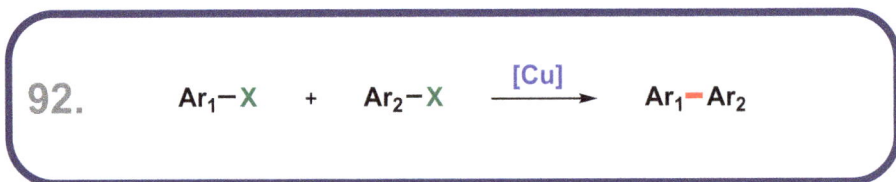

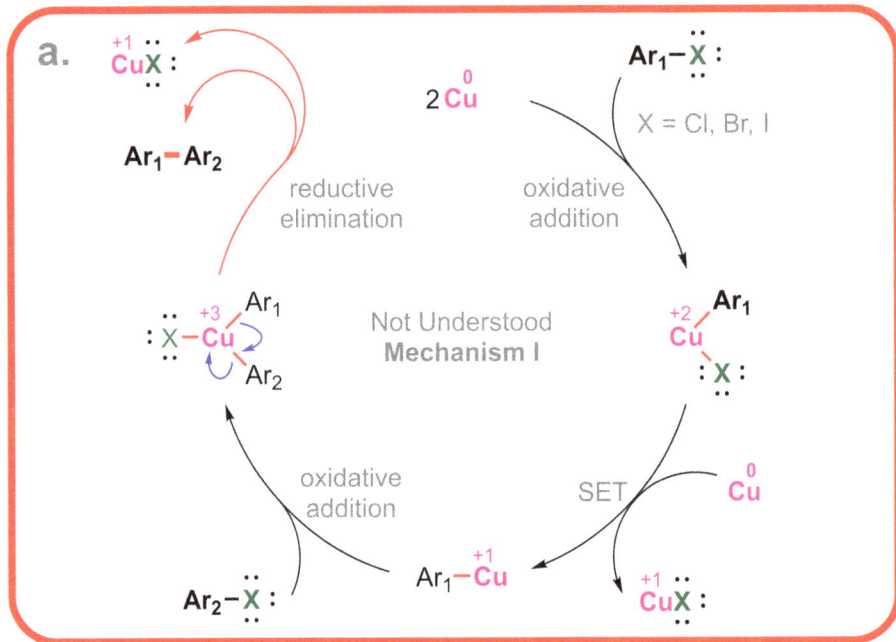

$$92. \quad Ar_1-X \quad + \quad Ar_2-X \quad \xrightarrow{[Cu]} \quad Ar_1-Ar_2$$

a.

Fig. 92.1: The **Ullmann** *aryl–aryl coupling* mechanism I.[331]

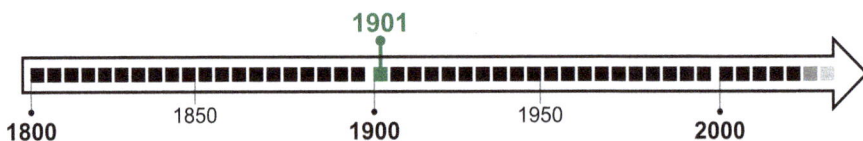

Fig. 92.2: The discovery of the **Ullmann** *aryl–aryl coupling*.[332]

---

331 The **Ullmann** *aryl–aryl coupling* or the **Ullmann** *reaction* is a **Cu-mediated coupling** (C–C bond formation using *aryl halides*). The mechanism is not fully understood. A possible formation of *organo-copper* intermediates (Cu(I) or Cu(II)) is postulated: mechanism I (**a**).
332 The reaction was likely first described around 1901 [92a, 92b].

https://doi.org/10.1515/9783110786835-092

b.

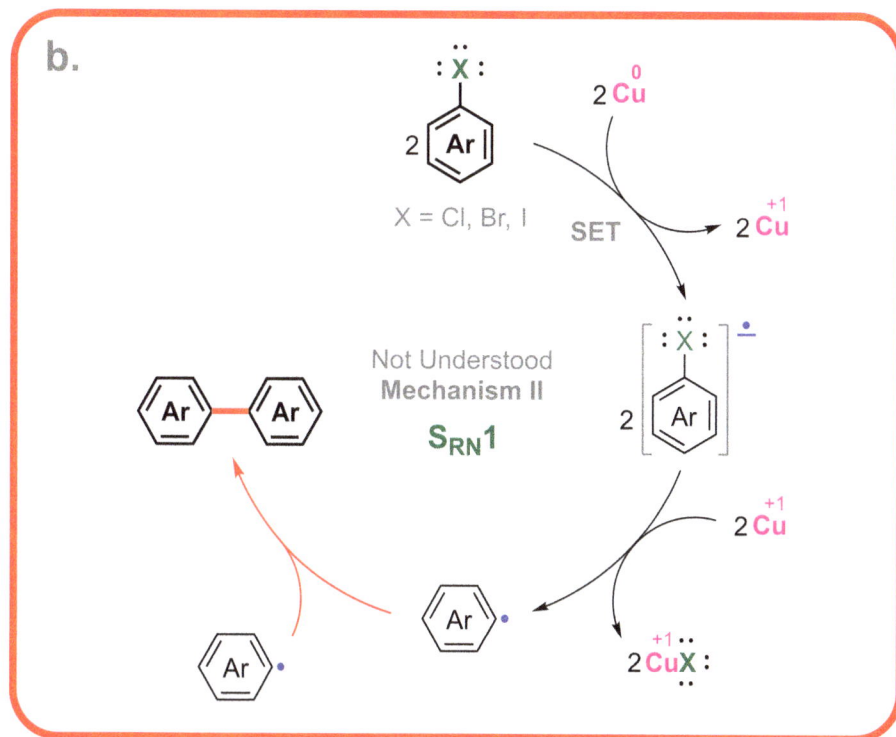

Fig. 92.3: The ***Ullmann*** *aryl–aryl coupling* mechanism II.[333]

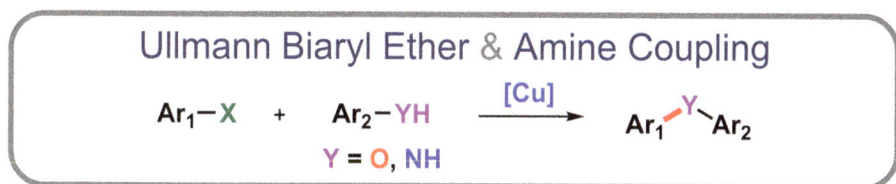

Ullmann Biaryl Ether & Amine Coupling

$$Ar_1{-}X \quad + \quad Ar_2{-}YH \quad \xrightarrow{[Cu]} \quad Ar_1{-}Y{-}Ar_2$$

$$Y = O, NH$$

Fig. 92.4: The ***Ullmann*** *biaryl ether and amine coupling.*[334]

---

333 The **aromatic radical nucleophilic substitution** ($S_{RN}1$) mechanism (Chapter 5) is another explanation for the formation of the *symmetrical* or *asymmetrical biaryl* products: mechanism II (**b**).

334 The ***Ullmann*** *biaryl ether* and *biaryl amine coupling* reaction is more synthetically useful [92c, 92d]. It is also a ***Cu****-mediated coupling* (C–O and C–N bond formation using *aryl halides* with *phenols* or *anilines*) [92e]. An alternative way to synthesize *aryl ethers* and *amines* is via the ***Chan–Evans–Lam*** *crosscoupling* (Chapter 23).

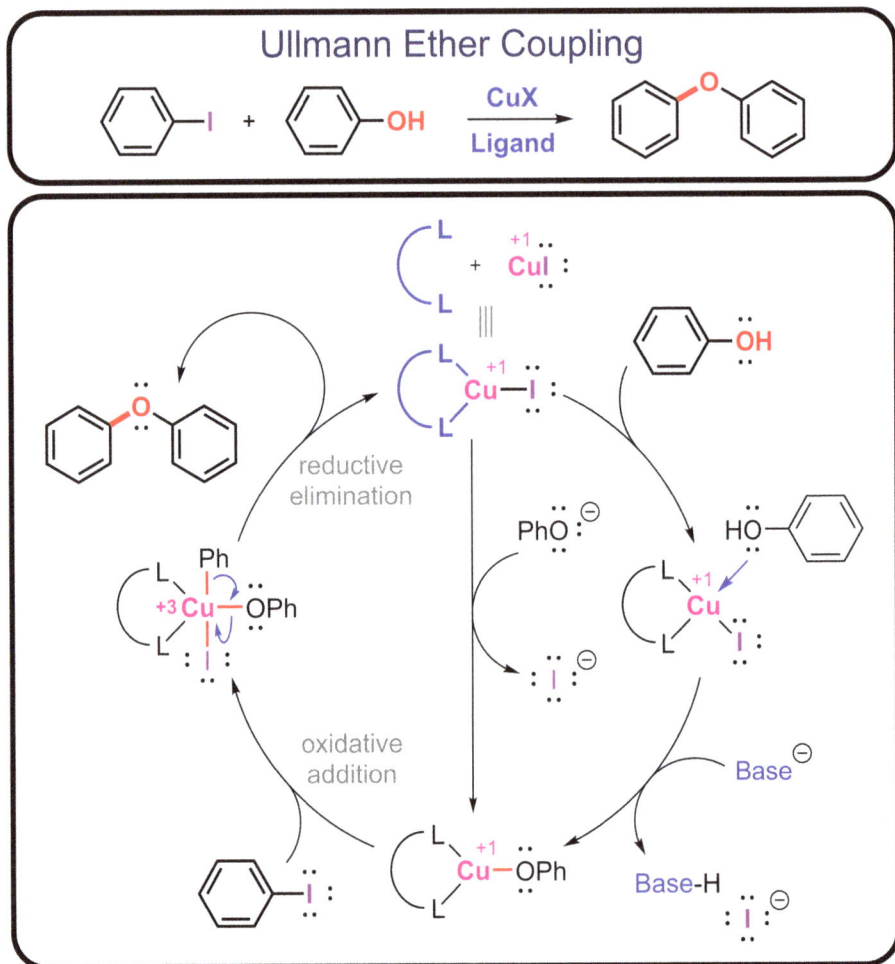

Fig. 92.5: The **Ullmann** *biaryl ether coupling* mechanism (neutral ligand).[335]

---

335 The mechanism varies and depends on the type of substrates, ligands, and other factors. Here is an example of the **Ullmann** *biaryl ether* coupling catalyzed by Cu(I) with a neutral bidentate ligand (often *N,N*-bidentate ligand) [92f, 92g].

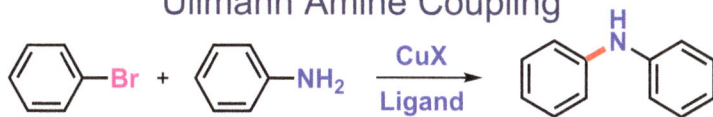

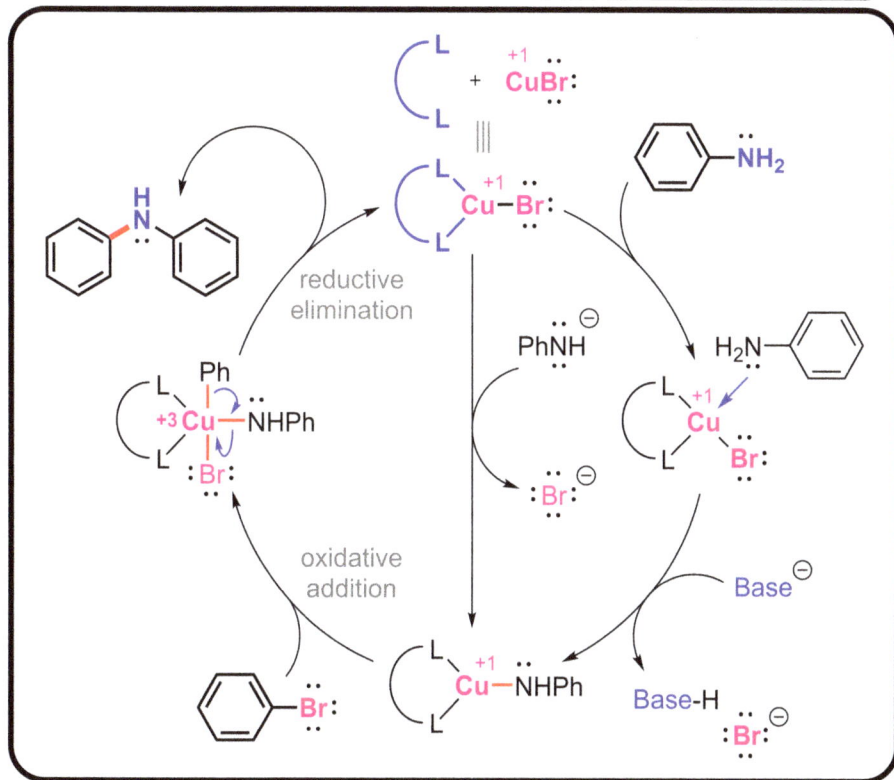

Fig. 92.6: The **Ullmann** *biaryl amine coupling* mechanism (neutral ligand).[336]

---

336 The mechanism varies and depends on the type of substrates, ligands, and other factors. Here is an example of the **Ullmann** *biaryl amine* coupling catalyzed by Cu(I) with a neutral bidentate ligand (often *N,N*-bidentate ligand) [92f, 92g].

93 Upjohn Dihydroxylation

93.

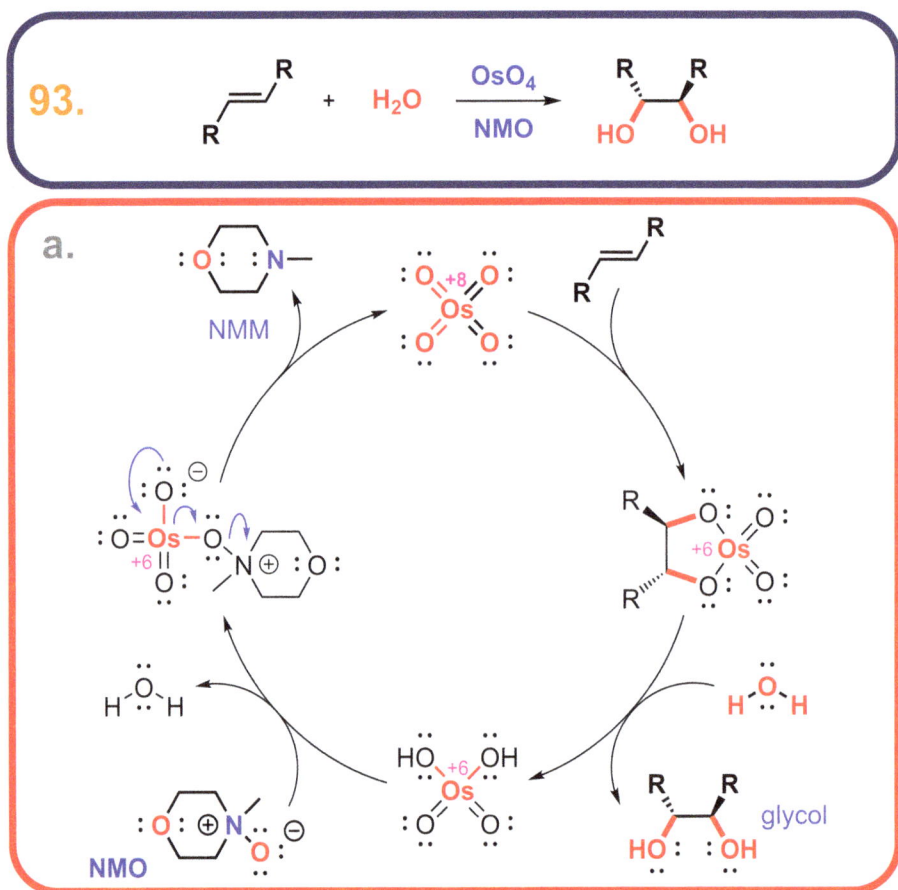

Fig. 93.1: The **Upjohn** *dihydroxylation* mechanism (a).[337]

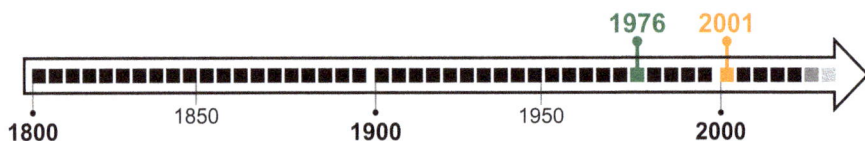

Fig. 93.2: The discovery of the **Upjohn** *dihydroxylation*.[338]

---

337 The **Upjohn** *dihydroxylation* (**a**) yields <u>racemic</u> products (*cis-1,2-glycols* = *cis*-1,2-diols) [93a].
338 The reaction was likely first described around 1976 [93f]. In **2001**, K. Barry Sharpless (together with William S. Knowles and Ryoji Noyori) received the Nobel Prize in Chemistry for the development of chirally catalyzed oxidation and hydrogenation reactions [93g].

https://doi.org/10.1515/9783110786835-093

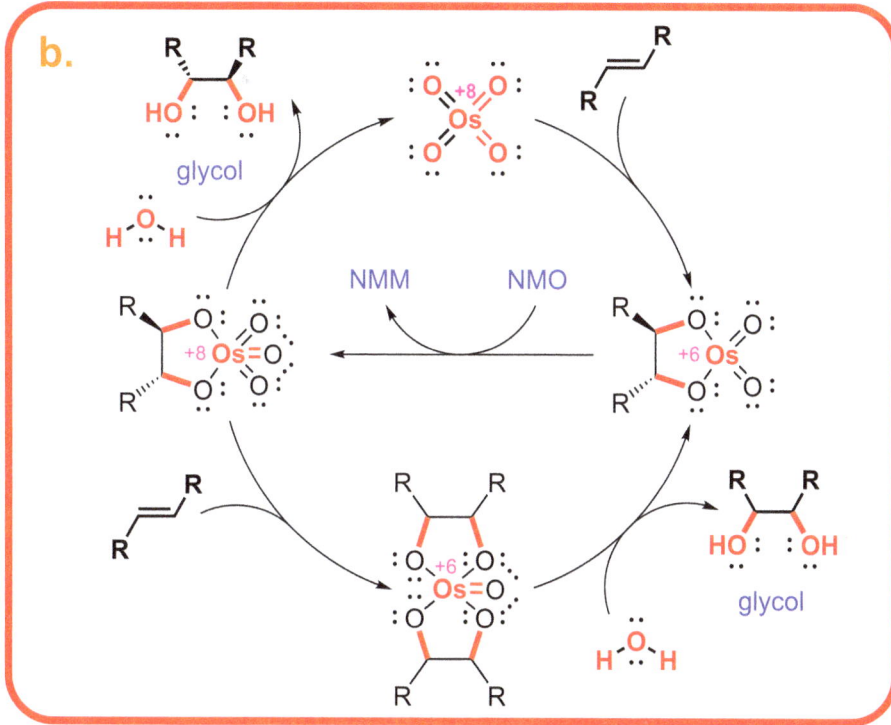

Fig. 93.3: The **Upjohn** *dihydroxylation* mechanism (b).[339]

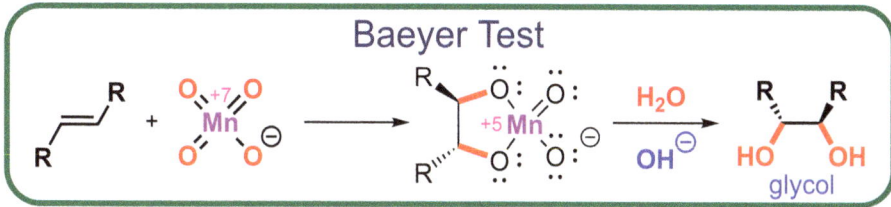

Fig. 93.4: The **Baeyer** *test*.[340]

---

339 The **Sharpless** *asymmetric dihydroxylation* is exemplified in a simplified mechanism (**b**). It is an asymmetric variation of the **Upjohn** *dihydroxylation*, and it yields enironmentally pure products [93b, 93c, 93d].

340 The **Baeyer** *test (Baeyer's test)* (potassium permanganate-based TLC stain) is a reaction related to the **Upjohn** *dihydroxylation*. It is used to detect the presence of *double bonds (unsaturation)* [93e].

94 Vilsmeier–Haack Reaction

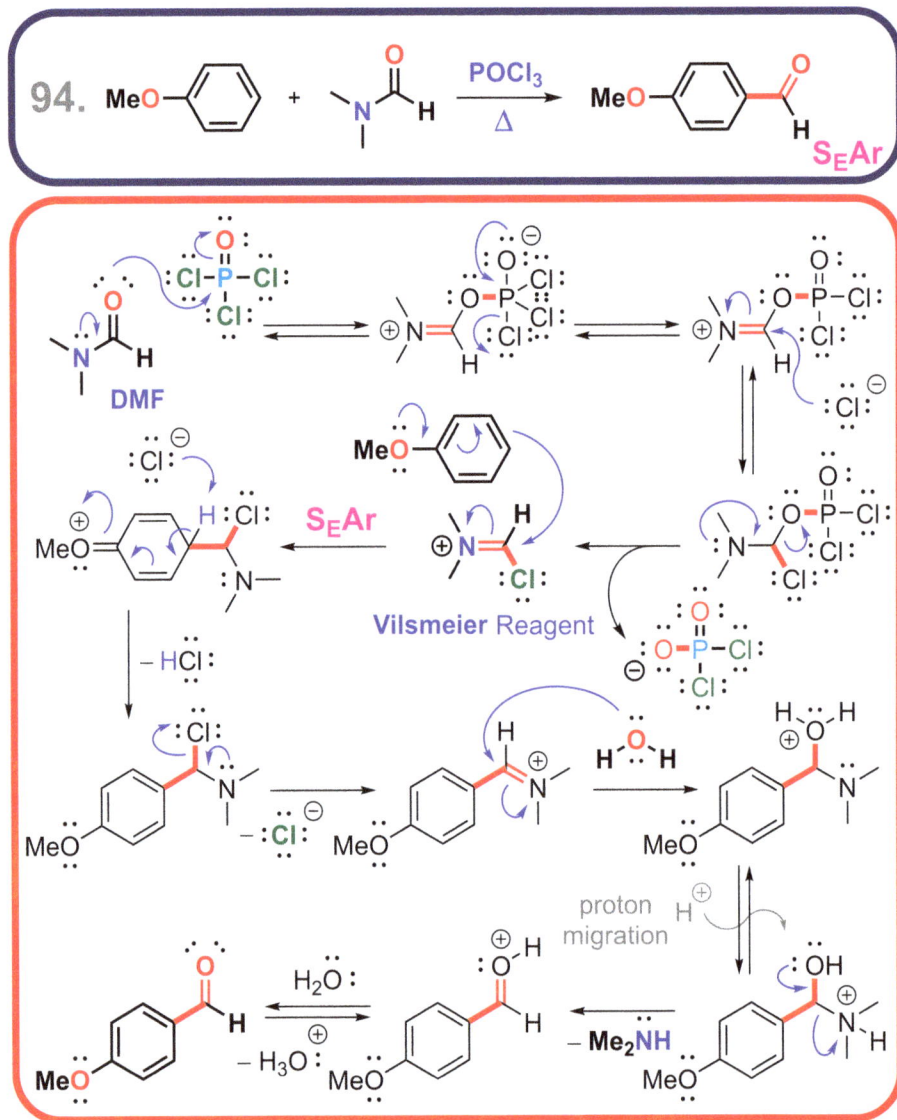

Fig. 94.1: The *Vilsmeier–Haack* reaction mechanism.[341]

---

341 The *Vilsmeier–Haack* reaction or the *Vilsmeier–Haack* formylation is a classic example of **aromatic electrophilic substitution** (the *arenium ion* mechanism = S$_E$Ar, covered in Chapter 3).

https://doi.org/10.1515/9783110786835-094

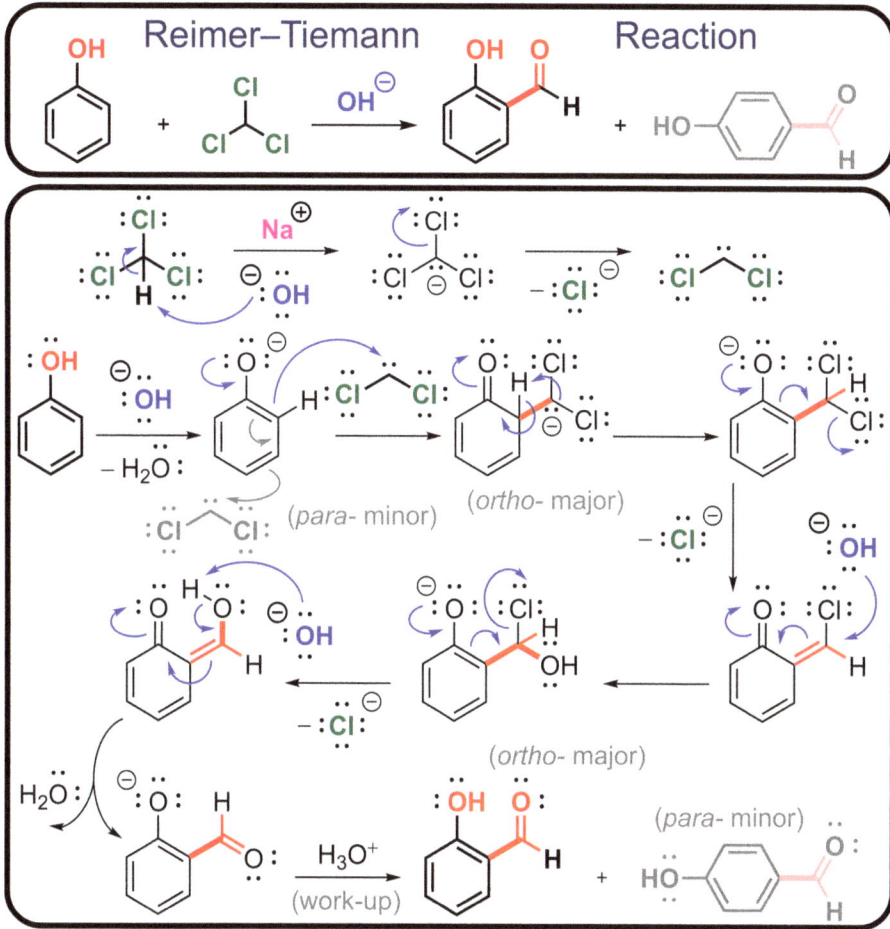

Fig. 94.2: The **Reimer–Tiemann** reaction mechanism.[342]

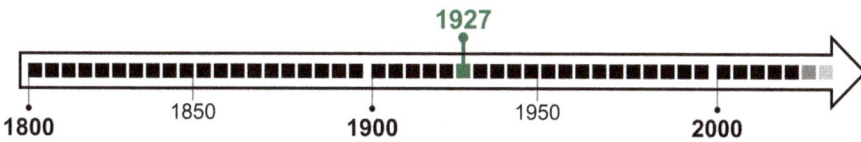

Fig. 94.3: The discovery of the **Vilsmeier–Haack reaction**.[343]

---

342 A few named reactions are related to the **Vilsmeier–Haack** reaction: the **Friedel–Crafts** formylation using *dichloro(methoxy)methane* (covered in Chapter 39) and the **Reimer–Tiemann** reaction using *chloroform* (limited to the *ortho*-formylation of *phenols*) [94a].
343 The reaction was likely first described around 1927 [94b].

95 Wacker Oxidation

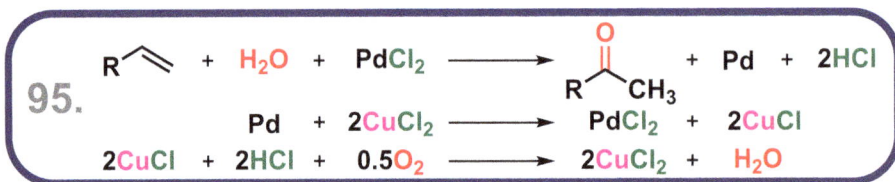

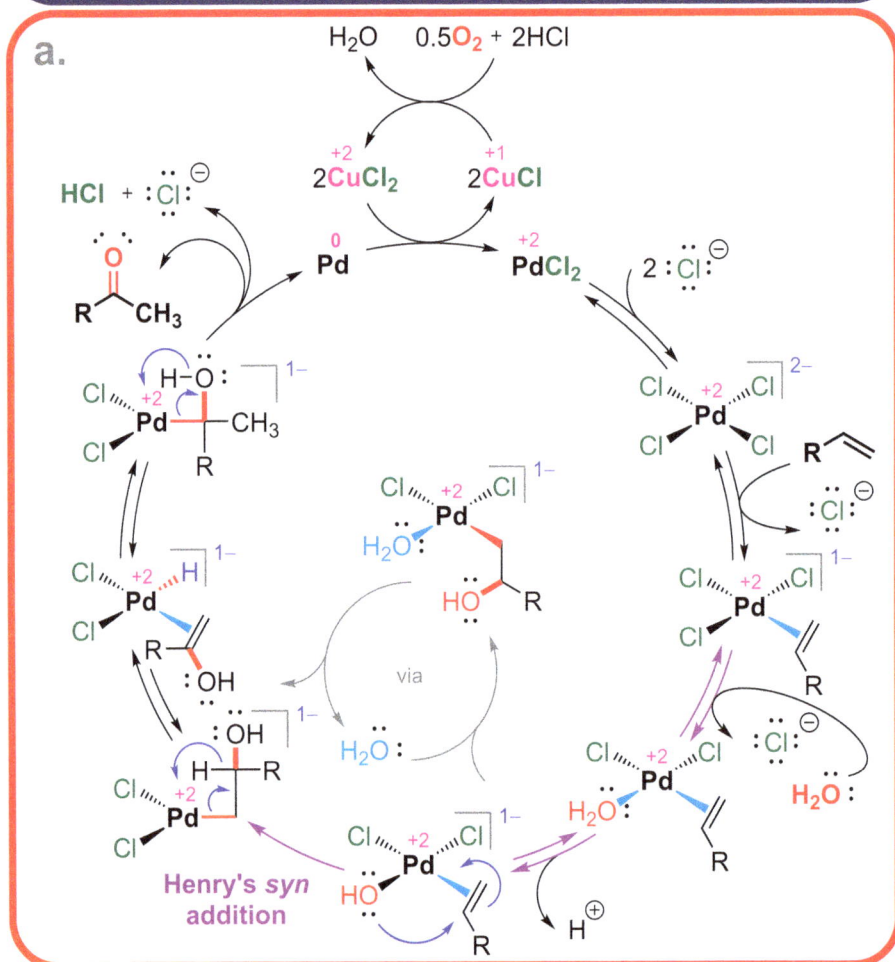

95.

$$R\text{—CH=CH}_2 + H_2O + PdCl_2 \longrightarrow R\text{—C(=O)—}CH_3 + Pd + 2HCl$$

$$Pd + 2CuCl_2 \longrightarrow PdCl_2 + 2CuCl$$

$$2CuCl + 2HCl + 0.5O_2 \longrightarrow 2CuCl_2 + H_2O$$

a.

Fig. 95.1: The **Wacker** oxidation mechanism (a).[344]

---

344 The **Wacker** oxidation or the **Wacker** process is a **Pd**-catalyzed and **Cu**-co-catalyzed alkene (olefin) oxidation. The mechanism can vary: mechanism (a) is proposed by Henry: **Henry's** syn-addition (inner-sphere) [95a, 95b].

https://doi.org/10.1515/9783110786835-095

b.

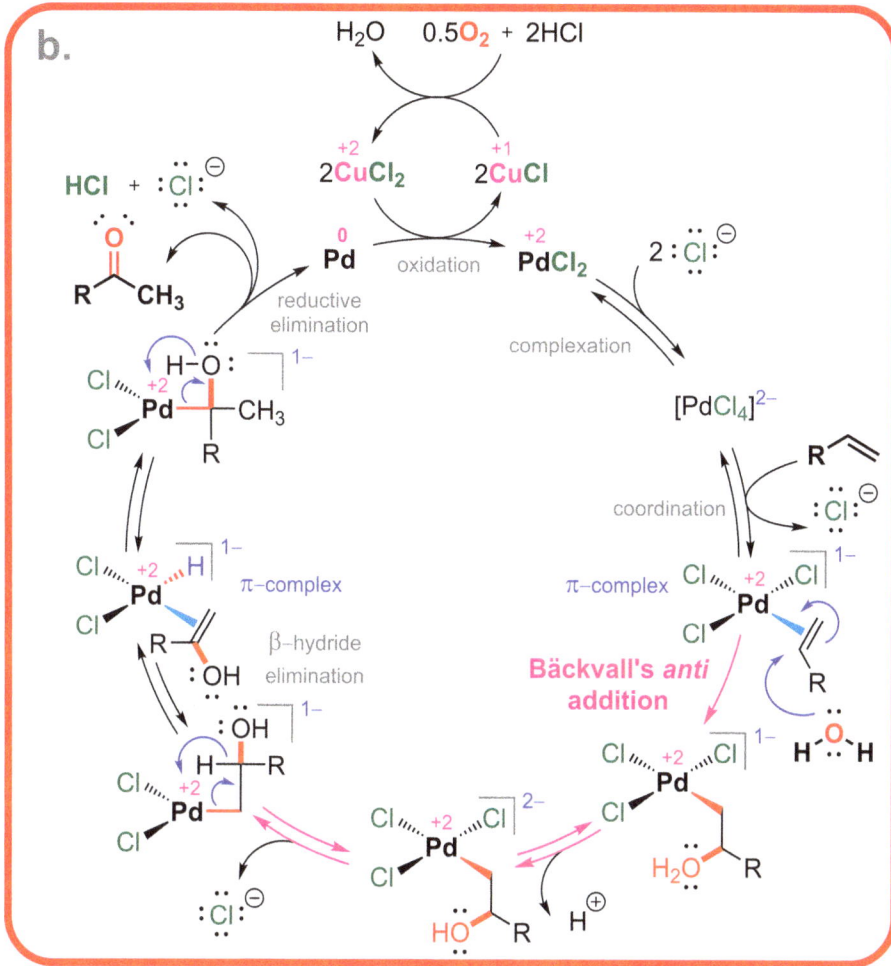

Fig. 95.2: The **Wacker** oxidation mechanism (b).[345]

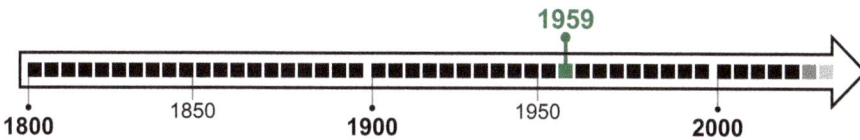

Fig. 95.3: The discovery of the **Wacker** oxidation.[346]

---

345 Mechanism (**b**) is proposed by Bäckvall: **Bäckvall's anti-addition** (outer-sphere) [95a, 95b].
346 The reaction was likely first described around 1959 [95c].

96 Wagner–Meerwein Rearrangement

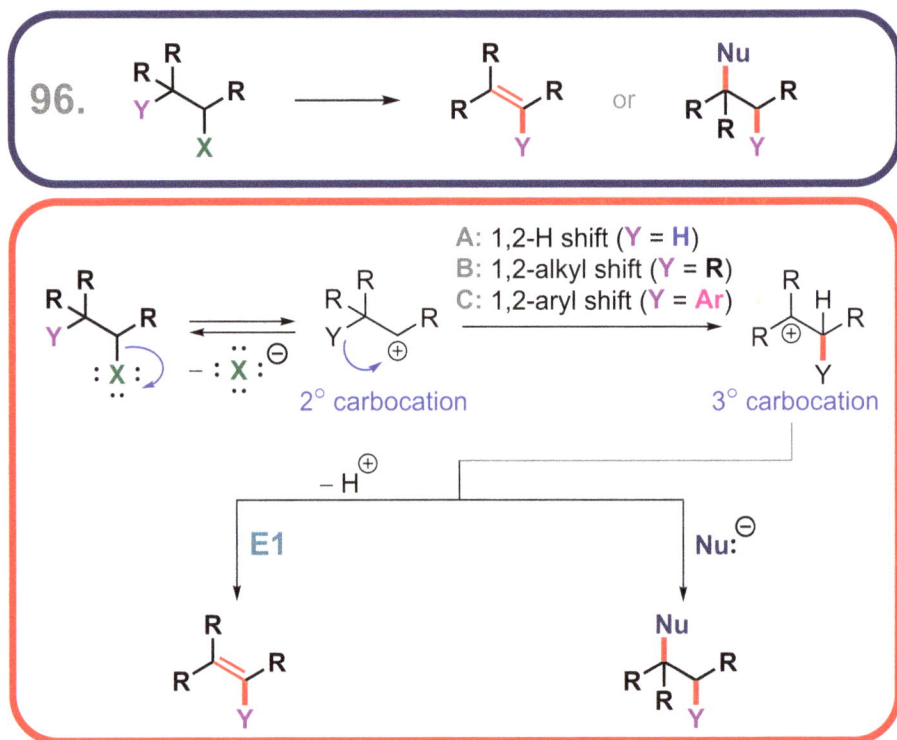

Fig. 96.1: The general **Wagner–Meerwein** rearrangement mechanism.[347]

A: 1,2-H shift (Y = H)
B: 1,2-alkyl shift (Y = R)
C: 1,2-aryl shift (Y = Ar)

2° carbocation

3° carbocation

E1

Nu:

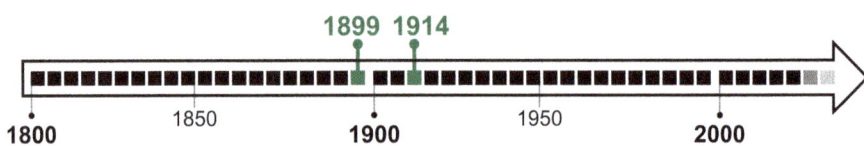

Fig. 96.2: The discovery of the **Wagner–Meerwein** rearrangement.[348]

---

347 The **Wagner–Meerwein** rearrangement is a rearrangement of newly formed carbocations into more stable carbocations (1° → 2° → 3°). This reaction is related to the pinacol–pinacolone rearrangement and the **Tiffeneau–Demjanov** rearrangement (Chapter 76).

348 The reaction was likely first described around 1899 by Wagner [96a, 96b] and around 1914 by Meerwein [96c].

https://doi.org/10.1515/9783110786835-096

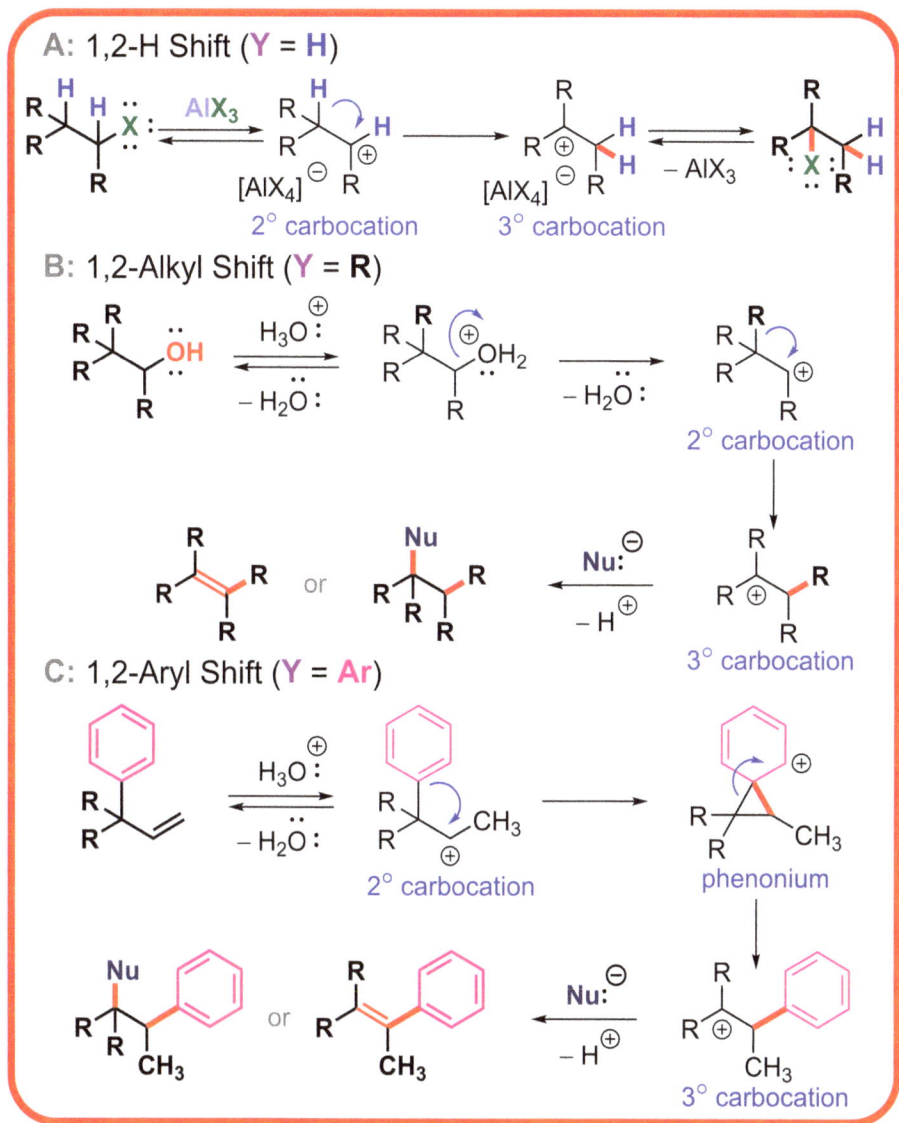

Fig. 96.3: The **Wagner–Meerwein** *rearrangement* mechanism (A, B, and C).[349]

---

[349] The generated *carbocations* rearrange into more stable species via either (a) 1,2-H shift (Y = H); (b) 1,2-alkyl shift (Y = R); or (c) 1,2-aryl shift (Y = Ar). **β–Elimination** reactions (**E1**) often accompany the **Wagner–Meerwein** *rearrangement* [1].

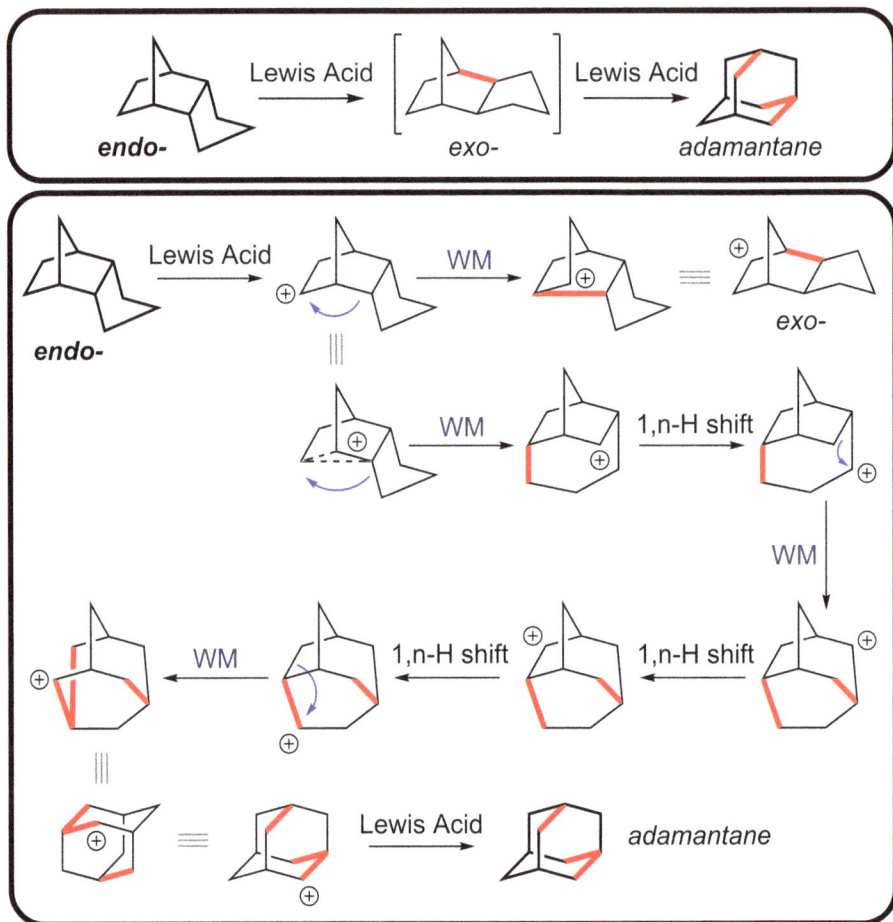

Fig. 96.4: A possible mechanism of *adamantane* rearrangement (pathway A).[350]

---

350 An **adamantane** (tricyclo[3.3.1.1^{3,7}]decane) can be prepared from various saturated polycyclic compounds via a sequence of isomerizations catalyzed by Lewis acids [96d, 96e]. There are many possible pathways [96f]. The key step in each pathway is the *Wagner–Meerwein rearrangement* (here, WM = 1,2-alkyl shift) and 1,2-H shift or 1,n-H shift (here, multiple sequential 1,2-H shifts) [96g].

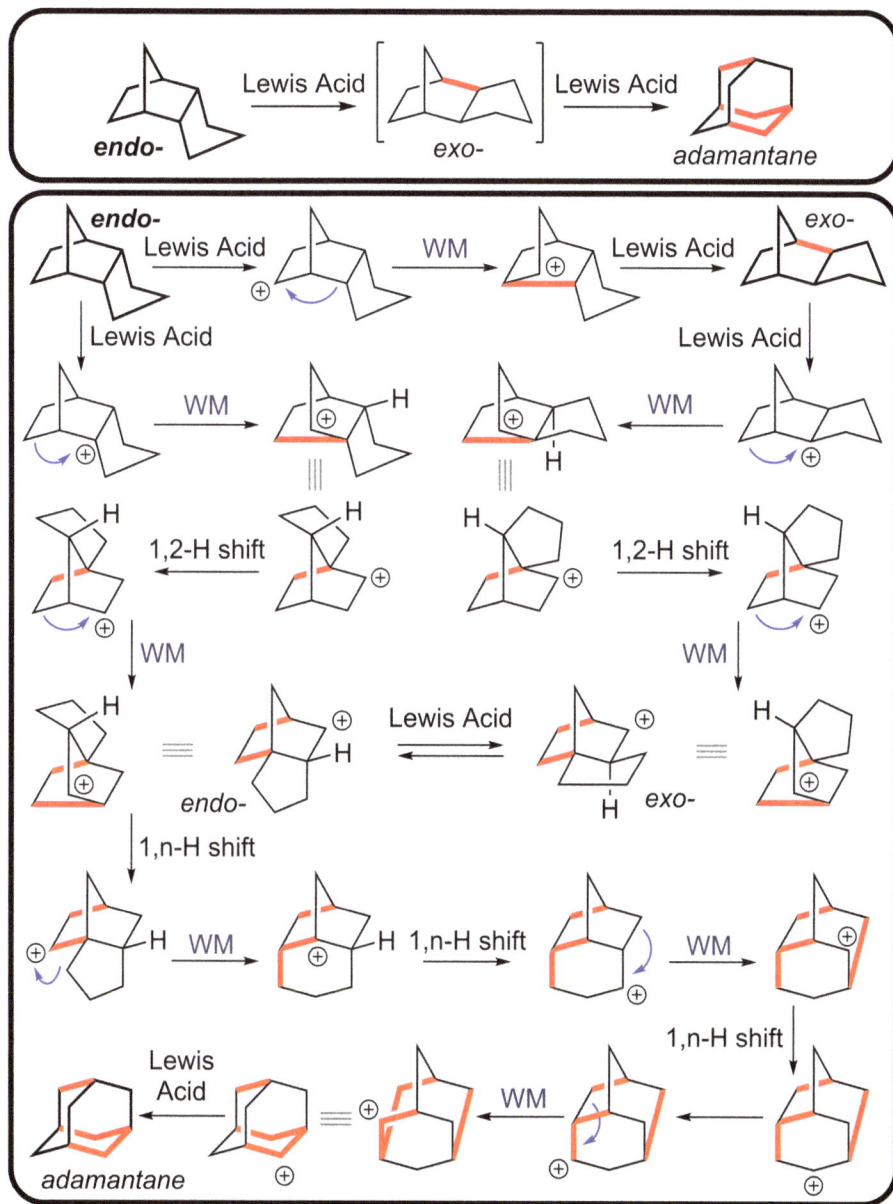

Fig. 96.5: A possible mechanism of *adamantane* rearrangement (pathway B).[351]

---

351 An **adamantane** (tricyclo[3.3.1.1^{3,7}]decane) can be prepared via a sequence of the *Wagner–Meerwein rearrangements* (WM) and 1,2-H shifts or 1,n-H shifts [96f, 96g].

97 Weinreb Ketone Synthesis

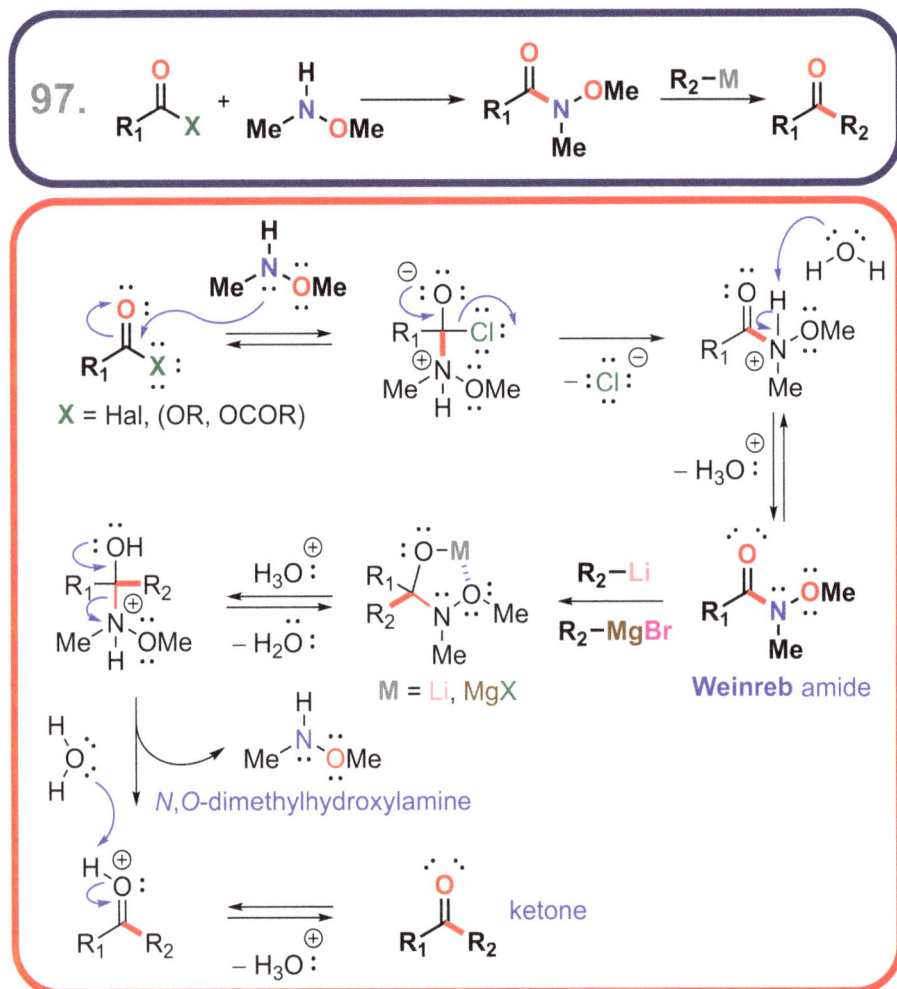

Fig. 97.1: The **Weinreb** *ketone synthesis* mechanism.[352]

---

352 The **Weinreb** *ketone synthesis* is a synthetic procedure (preparation of *ketones*) based on the use of a named reagent: the **Weinreb** *amide (**Weinreb–Nahm** amide)* [97a].

https://doi.org/10.1515/9783110786835-097

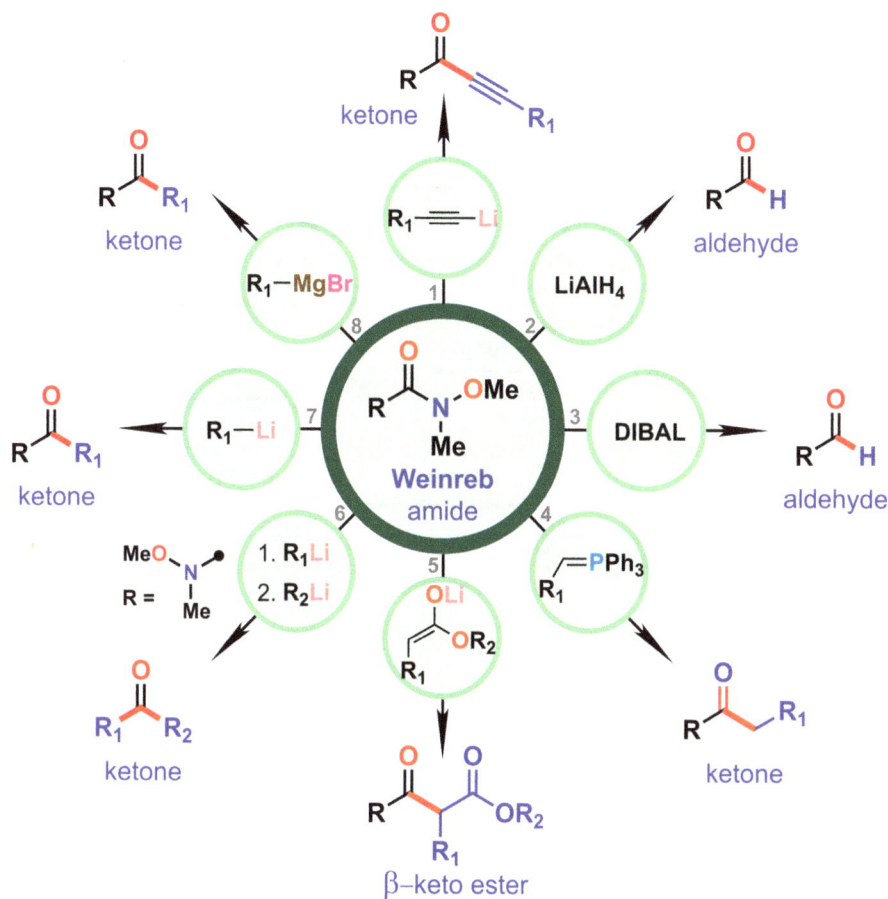

Fig. 97.2: Synthetic versatility of the **Weinreb** amide.[353]

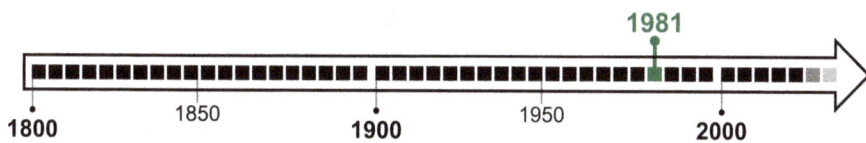

Fig. 97.3: The discovery of the **Weinreb** ketone synthesis.[354]

---

353 The **Weinreb** amide has wide synthetic application, and it can react with a variety of nucleophilic reagents: (a) *organolithium* and *organomagnesium* = **Grignard** *reagents*; (b) reducing reagents like DIBAL; (c) *phosphorus ylides* or *phosphoranes* [97b]; and others [1].
354 The reaction was likely first described around 1981 [97c].

98 Wittig Reaction

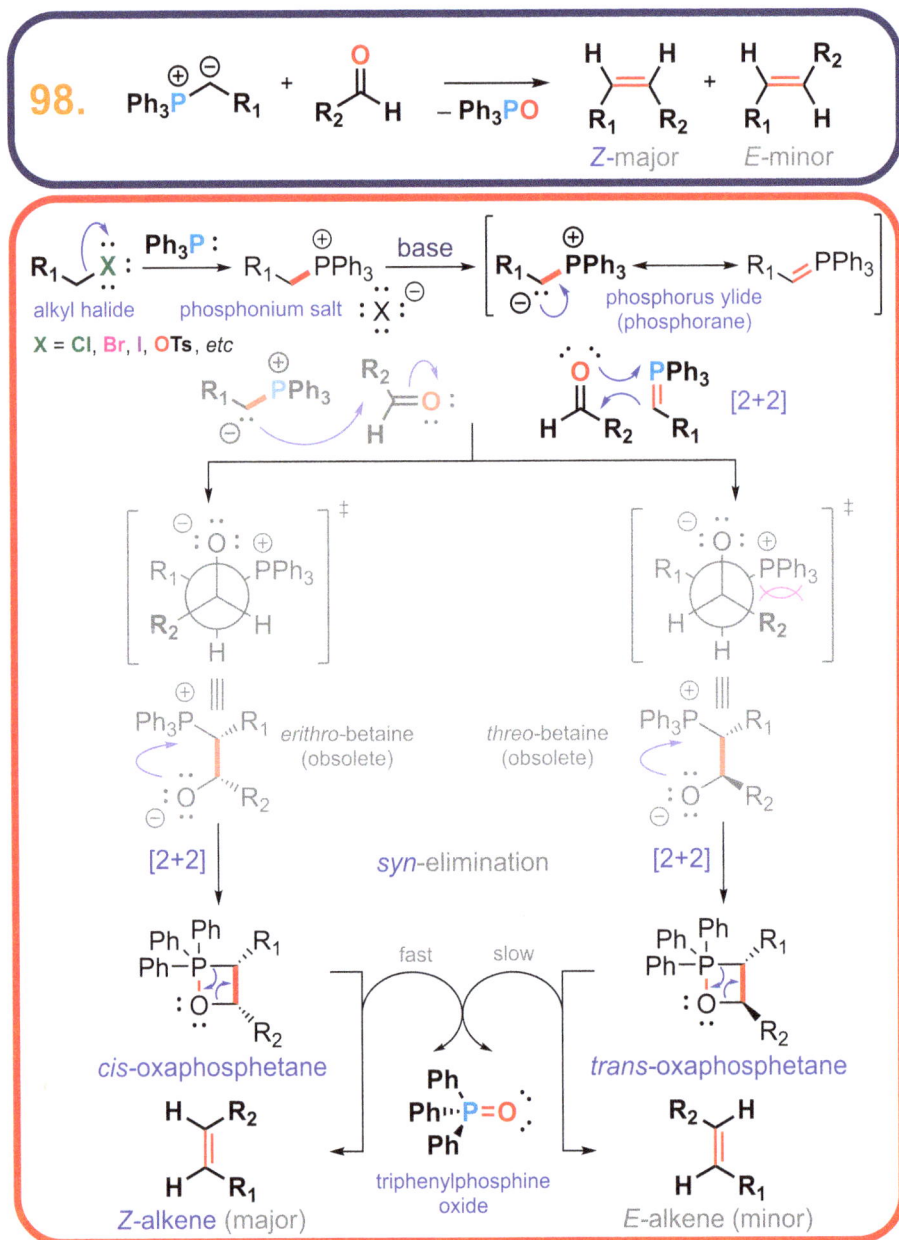

Fig. 98.1: The *Wittig reaction* mechanism.[355]

https://doi.org/10.1515/9783110786835-098

Wittig–Schlosser Modification

Horner–Wadsworth–Emmons Olefination

R_1 = COR, CO_2R, CN, SO_2R, etc

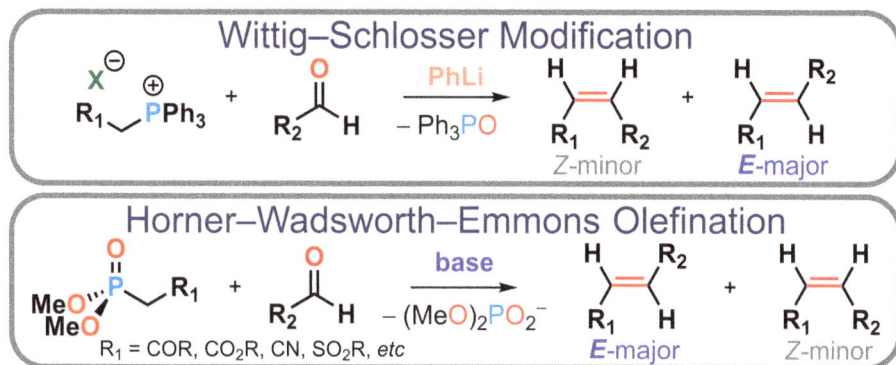

Fig. 98.2: Reactions related to the **Wittig reaction**.[356]

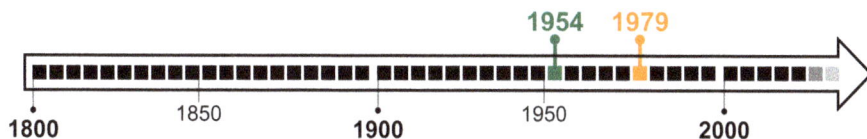

Fig. 98.3: The discovery of the **Wittig reaction**.[357]

---

355 The **Wittig reaction** or the **Wittig olefination** relies on the use of *phosphorus ylides* or *phosphoranes* formed from the *phosphonium salts* [98a].

356 Several reactions are closely related to the **Wittig reaction**: the **Wittig–Schlosser** modification (favoring *E*-alkenes with an excess of **PhLi** as a base) [98b]. The **Horner–Wadsworth–Emmons** olefination (Chapter 50) relies on the use of *phosphonates* [PO(OR)$_2$R], often made via the **Arbuzov reaction** (Chapter 9).

357 The reaction was likely first described around 1954 [98c, 98d]. In **1979**, Georg Wittig (jointly with Herbert C. Brown) received the Nobel Prize in Chemistry for the development of phosphorus (and boron) chemistry [20c].

99 Wohl–Ziegler Reaction

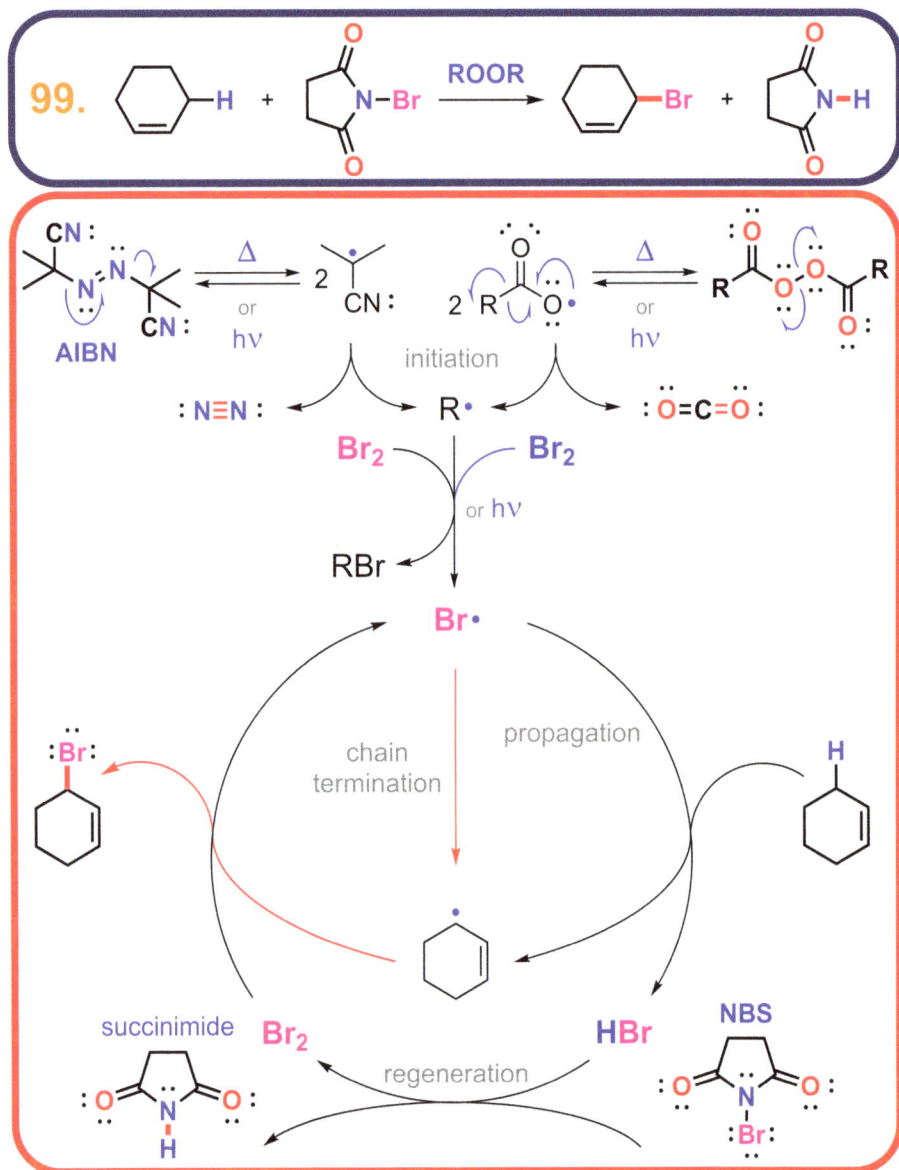

Fig. 99.1: The *Wohl–Ziegler* reaction mechanism.[358]

---

[358] The *Wohl–Ziegler* reaction, or the *Wohl–Ziegler* bromination, is a type of the **free radical substitution** (see the *Minisci* reaction in Chapter 60).

https://doi.org/10.1515/9783110786835-099

Free Radical Substitution

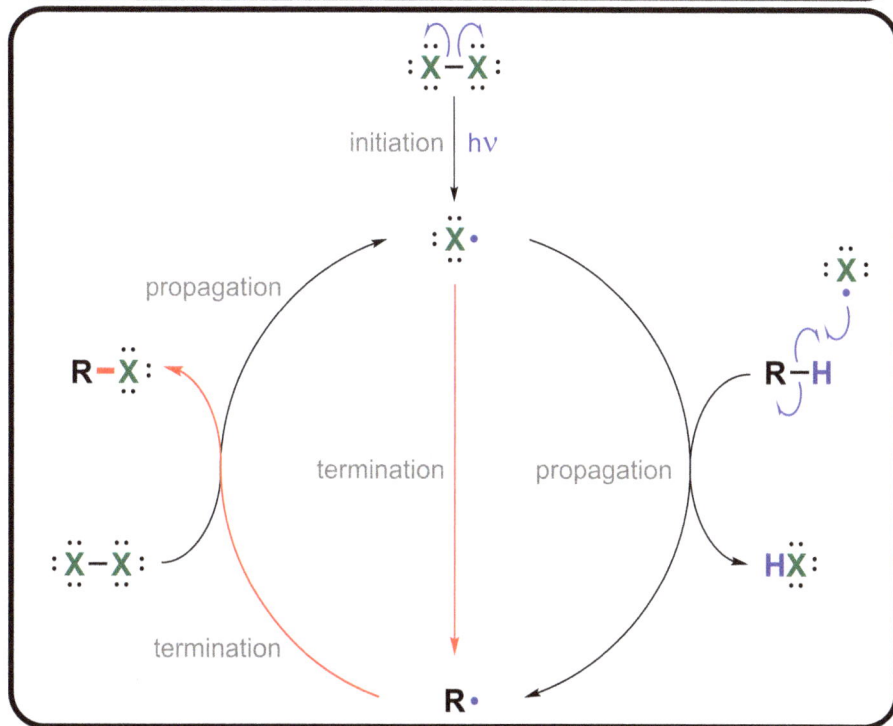

$$R-H \;+\; X_2 \quad \xrightarrow{\;h\nu\;} \quad R-X \;+\; HX$$

Fig. 99.2: The *free radical substitution* mechanism.[359]

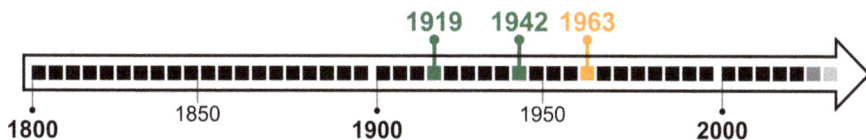

Fig. 99.3: The discovery of the **Wohl–Ziegler** *reaction*.[360]

---

359 The *free radical substitution* mechanisms usually feature three major steps: (a) *initiation*, (b) chain *propagation*, and (c) chain *termination*. A *free radical chlorination* of *alkanes* is a typical example [1].
360 The reaction was likely first described around 1919 by Wohl [99a] and around 1942 by Ziegler [99b]. In **1963**, Karl Ziegler (jointly with Giulio Natta) received the Nobel Prize in Chemistry [99c].

100 Wolff–Kishner Reduction

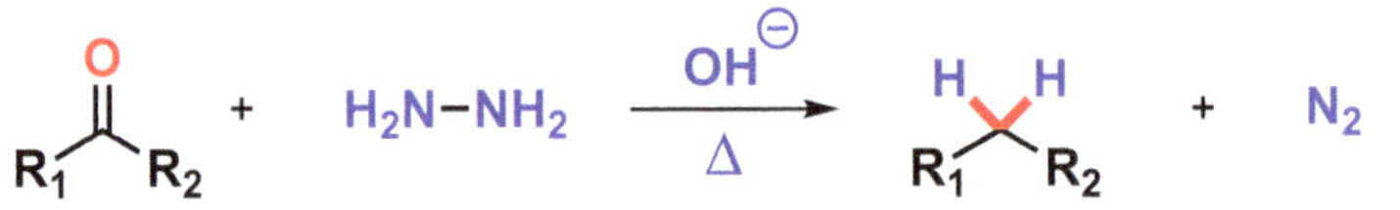

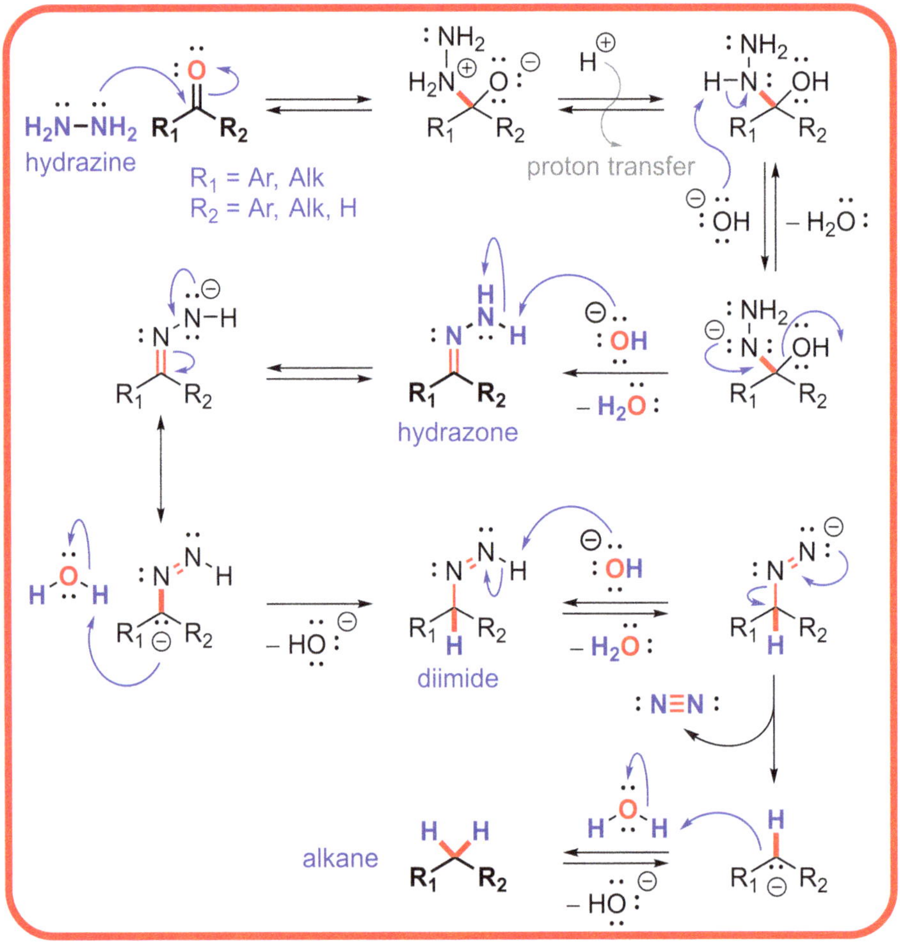

Fig. 100.1: The *Wolff–Kishner* reduction mechanism.[361]

---

361 There are many modifications of the *Wolff–Kishner reduction*: for example, the *Huang–Minlon modification* and many others (not shown) [100a].

https://doi.org/10.1515/9783110786835-100

Clemmensen Reduction

1. Carbenoid Mechanism:

zinc carbenoid

2. Carbanion Mechanism:

H_2O + $ZnCl_2$ +

carbanion

carbanion

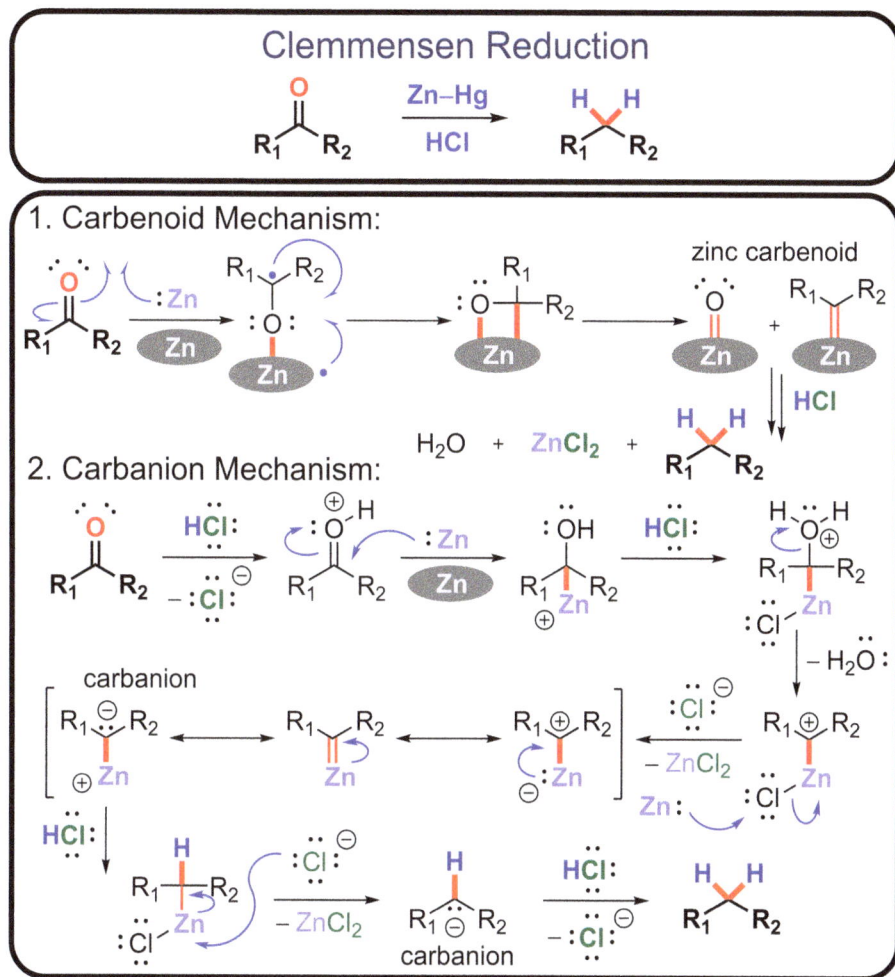

Fig. 100.2: The *Clemmensen reduction* mechanism.[362]

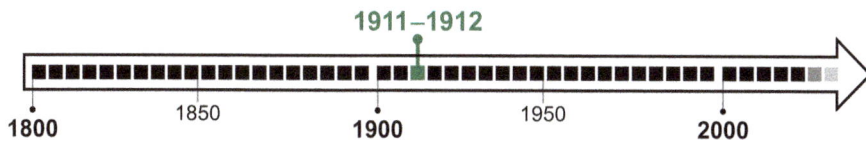

1911–1912

1800   1850   1900   1950   2000

Fig. 100.3: The discovery of the *Wolff–Kishner reduction*.[363]

---

362 The *Clemmensen reduction* is closely related to the *Wolff–Kishner reduction* in terms of the product-type formation but not the mechanism [100b].
363 The reaction was likely first described around 1911 by Kishner [100c] and around 1912 by Wolff [100d].

Acknowledgments

I envision this reference book to be one part of the intellectual and physical library that the developing chemist builds as they gain experience and expertise. This immersion, in conjunction with further learning, can provide an invaluable scientific intuition. Mechanisms have become an integral part of my continued study, research, and learning in organic chemistry, and I hope this book imparts some of that to the field.

The images and infographic summaries ("*MechanoGraphics*") in this book were prepared by the author. Any reference to trademarks or service marks is based on common usage or reference in the discipline or practice, and each such mark is the property of its respective owner and use here does not mean and is not intended to suggest an endorsement. All the "*MechanoGraphic*" schemes are based on common knowledge in the discipline and are supported either by original references, where they were first mentioned, or by other articles and reviews in the literature (where necessary). The reader should be sure to conduct primary research and utilize their own professional judgment, including consultation with peers and more experienced scientists, before planning any studies or research as the illustrations alone are not sufficient as a sole reference.

Special thanks are owned to my family, friends, and peers in the scientific community, and my students. The confluence of an idea and diligence makes possible great progress. But this combination must also be fueled by the compassion, support, and patience of those around us. I am very grateful that this constellation aligned for me and made possible the environment in which I could prepare this book. Moreover, as always, I am most grateful to my students and followers on social media. As I share and teach, I also continue to learn about the discipline as well as myself, always striving to understand, grow, and get closer to truth.

https://doi.org/10.1515/9783110786835-101

Bibliography and References

[1]   (a) March J. Advanced Organic Chemistry: Reactions, Mechanisms, and Structure (Fourth Edition). New York, NY, USA, J. Wiley & Sons, 1992. (b) Carey FA, and Sundberg RJ. Advanced Organic Chemistry: Part A: Structure and Mechanisms (Third Edition). New York, NY, USA, Plenum Press, 1990. (c) Carey FA, and Sundberg RJ. Advanced Organic Chemistry: Part B: Reactions and Synthesis (Third Edition). New York, NY, USA, Plenum Press, 1990.

[2]   (a) Hartwig JF. Organotransition Metal Chemistry: From Bonding to Catalysis. Mill Valley, CA, USA, University Science Books, 2010. (b) Turro NJ, Ramamurthy V, and Scaiano JC. Modern Molecular Photochemistry of Organic Molecules. Sausalito, CA, USA, University Science Books, 2010.

[3]   (a) Fleming I. Molecular Orbitals and Organic Chemical Reactions (Reference Edition). Chichester, West Sussex, UK, J. Wiley & Sons, 2010. (b) Fleming I. Molecular Orbitals and Organic Chemical Reactions (Student Edition). Chichester, West Sussex, UK, J. Wiley & Sons, 2009.

[4]   (a) Kürti L, and Czakó B. Strategic Applications of Named Reactions in Organic Synthesis: Background and Detailed Mechanisms. Burlington, MA, USA, Elsevier Academic Press, 2005. (b) Li JJ. Name Reactions: A Collection of Detailed Mechanisms and Synthetic Applications (Fifth Edition). Heidelberg, Germany, Springer, 2014.

[5]   (a) https://www.organic-chemistry.org/ (accessed December 5, 2019). (b) http://www.name-reaction.com/ (accessed December 5, 2019). (c) https://www.synarchive.com/ (accessed December 5, 2019). (d) https://en.wikipedia.org/wiki/Main_Page (accessed December 5, 2019).

[6]   e-EROS Encyclopedia of Reagents for Organic Synthesis. Chichester, NY, USA John Wiley & Sons, 2001. The on-line e-EROS database of reagents for organic synthesis can be found at https://onlinelibrary.wiley.com/doi/book/10.1002/047084289X (accessed December 5, 2019).

[7]   (a) Bouveault L, and Blanc G. Préparation des alcools primaires au moyen des acides correspondants. Compt. Rend. 1903, 136, 1676–1678. The reference is in French and can be found at https://gallica.bnf.fr/ark:/12148/bpt6k3091c/f1676.image.langFR (accessed December 5, 2019). (b) Bouveault L, and Locquin R. Action du sodium sur les éthers des acides monobasiques à fonction simple de la série grasse. Compt. Rend. 1905, 140, 1593–1595. The reference is in French and can be found at https://gallica.bnf.fr/ark:/12148/bpt6k30949/f1689.image (accessed December 5, 2019).

[8]   (a) Faworsky A. Isomerisationserscheinungen der Kohlen-wasserstoffe C_nH_{2n-2}. Erste Abhandlung. J. Prakt. Chem. 1888, 37 (1), 382–395. (b) Faworsky A. Isomerisationserscheinungen der Kohlen-wasserstoffe C_nH_{2n-2}. Zweite Abhandlung. J. Prakt. Chem. 1888, 37 (1), 417–431. (c) Faworsky A. Isomerisationserscheinungen der Kohlen-wasserstoffe C_nH_{2n-2}. Dritte Abhandlung. J. Prakt. Chem. 1888, 37 (1), 531–536. Please note that these references were originally published in Russian and they are difficult to locate. They can be found in German at https://doi.org/10.1002/prac.18880370133; https://doi.org/10.1002/prac.18880370138; and https://doi.org/10.1002/prac.18880370147, respectively (all accessed December 5, 2019). (d) Brown CA, and Yamashita A. Saline hydrides and superbases in organic reactions. IX. Acetylene zipper. Exceptionally facile contrathermodynamic multipositional isomerization of alkynes with potassium 3-aminopropylamide. J. Am. Chem. Soc. 1975, 97 (4), 891–892.

[9]   (a) See more at https://en.wikipedia.org/wiki/Organophosphorus_compound (accessed December 5, 2019). (b) Michaelis A, and Kaehne R. Ueber das Verhalten der Jodalkyle gegen die sogen. Phosphorigsäureester oder O-Phosphine. Ber. Dtsch. Chem. Ges. 1898, 31 (1), 1048–1055. (c) Arbuzov AE. On the structure of phosphonic acid and its derivatives: Isomerization and transition of bonds from trivalent to pentavalent phosphorus. J. Russ. Phys. Chem. Soc. 1906, 38, 687. Note: the original reference is in Russian and it is difficult to locate. Additional references can be found in [1a].

https://doi.org/10.1515/9783110786835-102

[10]   (a) Wolff L. Ueber Diazoanhydride. *Justus Liebigs Ann. Chem*. **1902**, 325 (2), 129–195. (b) Wolff L, and Krüche R. Über Diazoanhydride (1,2,3-Oxydiazole oder Diazoxyde) und Diazoketone. *Justus Liebigs Ann. Chem*. **1912**, 394 (1), 23–59. (c) Arndt F, and Eistert B. Ein Verfahren zur Überführung von Carbonsäuren in ihre höheren Homologen bzw. deren Derivate. *Ber. Dtsch. Chem. Ges*. A/B **1935**, 68 (1), 200–208.

[11]   (a) Dakin HD. The oxidation of hydroxy derivatives of benzaldehyde, acetophenone and related substances. *Am. Chem. J*. **1909**, 42 (6), 477–498. Note: the original reference is difficult to locate. (b) Baeyer A, and Villiger V. Einwirkung des Caro'schen Reagens auf Ketone. *Ber. Dtsch. Chem. Ges*. **1899**, 32 (3), 3625–3633. (c) The Nobel Prize in Chemistry 1905. NobelPrize.org. Nobel Media AB (2019 accessed December 5, 2019, at https://www.nobelprize.org/prizes/chemistry/1905/summary/).

[12]   (a) Barton DHR, Dowlatshahi HA, Motherwell WB, and Villemin D. A new radical decarboxylation reaction for the conversion of carboxylic acids into hydrocarbons. *J. Chem. Soc., Chem. Commun*. **1980**, (15), 732–733. (b) Barton DHR, Crich D, and Motherwell WB. New and improved methods for the radical decarboxylation of acids. *J. Chem. Soc., Chem. Commun*. **1983**, (17), 939–941. (c) Barton DHR, and McCombie SW. A new method for the deoxygenation of secondary alcohols. *J. Chem. Soc., Perkin Trans. 1* **1975**, (16), 1574–1585. (d) Barrett AGM, Prokopiou PA, and Barton DHR. Novel method for the deoxygenation of alcohols. *J. Chem. Soc., Chem. Commun*. **1979**, (24), 1175. (e) The Nobel Prize in Chemistry 1969. NobelPrize.org. Nobel Media AB (2019 accessed December 5, 2019, at https://www.nobelprize.org/prizes/chemistry/1969/summary/).

[13]   (a) Baylis AB, and Hillman MED. German Patent 2155113, **1972**. (b) Morita K, Suzuki Z, and Hirose H. *Bull. Chem. Soc. Jpn*. **1968**, 41 (11), 2815. Note: the original reference can be found at https://www.journal.csj.jp/doi/abs/10.1246/bcsj.41.2815 (accessed December 5, 2019).

[14]   (a) Beckmann E. Zur Kenntniss der Isonitrosoverbindungen. *Ber. Dtsch. Chem. Ges*. **1886**, 19 (1), 988–993. (b) Additional information on caprolactam can be found at https://en.wikipedia.org/wiki/Caprolactam (accessed January 14, 2023).

[15]   (a) Matsumoto T, Ohishi M, and Inoue S. Selective cross-acyloin condensation catalyzed by thiazolium salt. Formation of 1-hydroxy 2-one from formaldehyde and other aldehydes. *J. Org. Chem*. **1985**, 50 (5), 603–606. (b) Breslow R, and Kool E. A γ-cyclodextrin thiazolium salt holoenzyme mimic for the benzoin condensation. *Tetrahedron Lett*. **1988**, 29 (14), 1635–1638. For the original discovery and additional references please see [1a]. (c) Wöhler, and Liebig. Untersuchungen über das Radikal der Benzoesäure. *Ann. Pharm*. **1832**, 3 (3), 249–282. (d) Lapworth A. XCVI. – Reactions involving the addition of hydrogen cyanide to carbon compounds. *J. Chem. Soc., Trans*. **1903**, 83, 995–1005.

[16]   (a) Tadross PM, and Stoltz BM. A Comprehensive History of Arynes in Natural Product Total Synthesis. *Chem. Rev*. **2012**, 112 (6), 3550–3577. (b) Additional references can be found at https://en.wikipedia.org/wiki/Aryne (accessed December 5, 2019) and in [1, 2b]. (c) Roberts JD, Simmons HE, Carlsmith LA, and Vaughan CW. Rearrangement in the reaction of chlorobenzene-1-C^{14} with potassium amid. *J. Am. Chem. Soc*. **1953**, 75 (13), 3290–3291. Note: the reaction itself was known for a very long time. It is difficult to locate the original references.

[17]   (a) Jones RR, and Bergman RG. *p*-Benzyne. Generation as an intermediate in a thermal isomerization reaction and trapping evidence for the 1,4-benzenediyl structure. *J. Am. Chem. Soc*. **1972**, 94 (2), 660–661. (b) Bergman RG. Reactive 1,4-dehydroaromatics. *Acc. Chem. Res*. **1973**, 6 (1), 25–31. (c) Myers AG, Proteau PJ, Handel TM. Stereochemical assignment of neocarzinostatin chromophore. Structures of neocarzinostatin chromophore-methyl thioglycolate adducts. *J. Am. Chem. Soc*. **1988**, 110 (21), 7212–7214.

[18]   (a) Kraus CA. Solutions of Metals in Non-Metallic Solvents; I. General Properties of Solutions of Metals in Liquid Ammonia. *J. Am. Chem. Soc*. **1907**, 29 (11), 1557–1571. (b) Additional references can be found at https://en.wikipedia.org/wiki/Solvated_electron (accessed December 5, 2019). (c) Birch AJ. 117. Reduction by dissolving metals. Part I. *J. Chem. Soc*. **1944**, (0), 430–436.

[19] (a) Pomeranz C. Über eine neue Isochinolinsynthese. *Monatshefte für Chemie* **1893**, 14 (1), 116–119. https://doi.org/10.1007/BF01517862 (accessed December 5, 2019). (b) Fritsch P. Synthesen in der Isocumarin- und Isochinolinreihe. *Ber. Dtsch. Chem. Ges.* **1893**, 26 (1), 419–422. (c) Bischler A, and Napieralski B. Zur Kenntniss einer neuen Isochinolinsynthese. *Ber. Dtsch. Chem. Ges.* **1893**, 26 (2), 1903–1908.

[20] (a) Soderquist JA, Roush WR, and Heo J. (2004). 9-Borabicyclo[3.3.1]nonane Dimer. In e-EROS Encyclopedia of Reagents for Organic Synthesis, (Ed.). doi:10.1002/047084289X.rb235.pub2. (b) Brown HC, and Rao BCS. A new technique for the conversion of olefins into organoboranes and related alcohols. *J. Am. Chem. Soc.* **1956**, 78 (21), 5694–5695. (c) The Nobel Prize in Chemistry 1979. NobelPrize.org. Nobel Media AB (2019 accessed December 5, 2019, at https://www.nobelprize.org/prizes/chemistry/1979/summary/).

[21] (a) Guram AS, and Buchwald SL. Palladium-Catalyzed Aromatic Aminations with in situ Generated Aminostannanes. *J. Am. Chem. Soc.* **1994**, 116 (17), 7901–7902. (b) Paul F, Patt J, and Hartwig JF. Palladium-catalyzed formation of carbon-nitrogen bonds. Reaction intermediates and catalyst improvements in the hetero cross-coupling of aryl halides and tin amides. *J. Am. Chem. Soc.* **1994**, 116 (13), 5969–5970.

[22] (a) Cannizzaro S. Ueber den der Benzoësäure entsprechenden Alkohol. *Justus Liebigs Ann. Chem.* **1853**, 88 (1), 129–130. (b) For additional references please see: Geissman TA. (2011). The Cannizzaro Reaction. In Organic Reactions, (Ed.). doi:10.1002/0471264180.or002.03.

[23] (a) King AE, Brunold TC, and Stahl SS. Mechanistic Study of Copper-Catalyzed Aerobic Oxidative Coupling of Arylboronic Esters and Methanol: Insights into an Organometallic Oxidase Reaction. *J. Am. Chem. Soc.* **2009**, 131 (14), 5044–5045. (b) King AE, Ryland BL, Brunold TC, and Stahl SS. Kinetic and Spectroscopic Studies of Aerobic Copper(II)-Catalyzed Methoxylation of Arylboronic Esters and Insights into Aryl Transmetalation to Copper(II). *Organometallics* **2012**, 31 (22), 7948–7957. (c) Vantourout JC, Miras HN, Isidro-Llobet A, Sproules S, and Watson AJB. Spectroscopic Studies of the Chan–Lam Amination: A Mechanism-Inspired Solution to Boronic Ester Reactivity. *J. Am. Chem. Soc.* **2017**, 139 (13), 4769–4779. (d) Chan DMT, Monaco KL, Wang RP, and Winter MP. New N- and O-arylations with phenylboronic acids and cupric acetate. *Tetrahedron Lett.* **1998**, 39 (19), 2933–2936. (e) Evans DA, Katz JL, and West TR. Synthesis of diaryl ethers through the copper-promoted arylation of phenols with arylboronic acids. An expedient synthesis of thyroxine. *Tetrahedron Lett.* **1998**, 39 (19), 2937–2942. (f) Lam PYS, Clark CG, Saubern S, Adams J, Winters MP, Chan DMT, and Combs A. New aryl/heteroaryl C-N bond cross-coupling reactions via arylboronic acid/cupric acetate arylation. *Tetrahedron Lett.* **1998**, 39 (19), 2941–2944.

[24] (a) Chichibabin AE, and Zeide OA. New reaction for compounds containing the pyridine nucleus. *J. Russ. Phys. Chem. Soc.* **1914**, 46, 1216–1236. Note: the original reference is published in Russian and it is difficult to locate: Чичибабин А. Е., Зейде О. А. ЖРФХО, 46, 1216 (1914). This reference can be found at https://ru.wikipedia.org/wiki/Реакция_Чичибабина (accessed December 5, 2019). (b) Tschitschibabin AE. Eine neue Darstellungsmethode von Oxyderivaten des Pyridins, Chinolins und ihrer Homologen. *Ber. Dtsch. Chem. Ges.* A/B **1923**, 56 (8), 1879–1885. (c) Tondys H, van der Plas HC, Woźniak M. On the chichibabin amination of quinoline and some nitroquinolines. *J. Heterocyclic Chem.* **1985**, 22 (2), 353–355.

[25] (a) Dieckmann W. Ueber ein ringförmiges Analogon des Ketipinsäureesters. *Ber. Dtsch. Chem. Ges.* **1894**, 27 (1), 965–966. (b) Claisen L, and Lowman O. Ueber eine neue Bildungsweise des Benzoylessigäthers. *Ber. Dtsch. Chem. Ges.* **1887**, 20 (1), 651–654.

[26] (a) Additional references and good summary can be found at https://en.wikipedia.org/wiki/Claisen_rearrangement (accessed December 5, 2019). (b) Claisen L. Über Umlagerung von Phenol-allyläthern in C-Allyl-phenole. *Ber. Dtsch. Chem. Ges.* **1912**, 45 (3), 3157–3166.

[27] (a) Sharpless KB, Lauer RF, and Teranishi AY. Electrophilic and nucleophilic organoselenium reagents. New routes to α,β-unsaturated carbonyl compounds. *J. Am. Chem. Soc.* **1973**, 95 (18),

6137–6139. (b) Sharpless KB, Young MW, and Lauer RF. Reactions of selenoxides: Thermal *syn*-elimination and $H_2^{18}O$ exchange. *Tetrahedron Lett.* **1973**, 14 (22), 1979–1982. (c) Cope AC, Foster TT, and Towle PH. Thermal Decomposition of Amine Oxides to Olefins and Dialkylhydroxylamines. *J. Am. Chem. Soc.* **1949**, 71 (12), 3929–3934.

[28]   (a) Additional references can be found at https://en.wikipedia.org/wiki/Cope_rearrangement (accessed December 5, 2019). (b) Cope AC, and Hardy EM. The Introduction of Substituted Vinyl Groups. V. A Rearrangement Involving the Migration of an Allyl Group in a Three-Carbon System. *J. Am. Chem. Soc.* **1940**, 62 (2), 441–444.

[29]   (a) Criegee R. Eine oxydative Spaltung von Glykolen (II. Mitteil. über Oxydationen mit Blei(IV)-salzen). *Ber. Dtsch. Chem. Ges.* A/B **1931**, 64 (2), 260–266. (b) Malaprade L. Action of polyalcohols on periodic acid. Analytical application. *Bull. Soc. Chim. Fr.* **1928**, 43, 683–696. (c) Malaprade L. A study of the action of polyalcohols on periodic acid and alkaline periodates. *Bull. Soc. Chim. Fr.* **1934**, 5, 833–852. Note: the original references are published in French and they are difficult to locate.

[30]   (a) IUPAC. Compendium of Chemical Terminology, 2nd ed. (the "Gold Book"). Compiled by A. D. McNaught and A. Wilkinson. Blackwell Scientific Publications, Oxford (1997). Online version (2019) created by S. J. Chalk. ISBN 0-9678550-9-8. https://doi.org/10.1351/goldbook. The link to the page can be found at https://goldbook.iupac.org/terms/view/C01496 (accessed December 5, 2019). (b) Huisgen R, and Eckell A. 1.3-Dipolare additionen der azomethin-imine. *Tetrahedron Lett.* **1960**, 1 (33), 5–8. (c) Grashey R, Huisgen R, and Leitermann H. 1.3-Dipolare additionen der nitrone. *Tetrahedron Lett.* **1960**, 1 (33), 9–13. (d) Tornøe CW, Christensen C, and Meldal M. Peptidotriazoles on Solid Phase: [1,2,3]-Triazoles by Regiospecific Copper(I)-Catalyzed 1,3-Dipolar Cycloadditions of Terminal Alkynes to Azides. *J. Org. Chem.* **2002**, 67 (9), 3057–3064. (e) Rostovtsev VV, Green LG, Fokin VV, and Sharpless KB. A Stepwise Huisgen Cycloaddition Process: Copper(I)-Catalyzed Regioselective "Ligation" of Azides And Terminal Alkynes. *Angew. Chem. Int. Ed.* **2002**, 41 (14), 2596–2599. (f) Worrell BT, Malik JA, and Fokin VV. Direct Evidence of a Dinuclear Copper Intermediate in Cu(I)-Catalyzed Azide-Alkyne Cycloadditions. *Science* **2013**, 340 (6131), 457–460. (g) The Nobel Prize in Chemistry 2022. NobelPrize.org. Nobel Prize Outreach AB (2023 accessed January 15, 2023, at https://www.no belprize.org/prizes/chemistry/2022/summary/). (h) Bertozzi C. A Special Virtual Issue Celebrating the 2022 Nobel Prize in Chemistry for the Development of Click Chemistry and Bioorthogonal Chemistry. *ACS Central Science* **2022**. The reference can be found at https://doi.org/10.1021/acs centsci.2c01430.

[31]   (a) Curtius T. Ueber Stickstoffwasserstoffsäure (Azoimid) N_3H. *Ber. Dtsch. Chem. Ges.* **1890**, 23 (2), 3023–3033. (b) Curtius T. 20. Hydrazide und Azide organischer Säuren I. Abhandlung. *J. Prakt. Chem.* **1894**, 50 (1), 275–294. (c) Schmidt KF. Aus den Sitzungen der Abteilungen. *Angew. Chem.* **1923**, 36 (57), 506–523. The reference can be found at https://doi.org/10.1002/ange.19230366703 (accessed December 5, 2019). (d) Schmidt KF. Über den Imin-Rest. *Ber. Dtsch. Chem. Ges.* A/B **1924**, 57, 704–706. (e) Hofmann AW. Ueber die Einwirkung des Broms in alkalischer Lösung auf Amide. *Ber. Dtsch. Chem. Ges.* **1881**, 14 (2), 2725–2736. (f) Lossen W. Ueber Benzoylderivate des Hydroxylamins. *Justus Liebigs Ann. Chem.* **1872**, 161 (2-3), 347–362.

[32]   (a) In 1990 Elias James Corey received the Nobel Prize in Chemistry: The Nobel Prize in Chemistry 1990. NobelPrize.org. Nobel Media AB (2019 accessed December 5, 2019, at https://www.nobelprize.org/prizes/chemistry/1990/summary/). (b) Corey EJ, and Chaykovsky M. Dimethyloxosulfonium Methylide ($(CH_3)_2SOCH_2$) and Dimethylsulfonium Methylide ($(CH_3)_2SCH_2$). Formation and Application to Organic Synthesis. *J. Am. Chem. Soc.* **1965**, 87 (6), 1353–1364. (c) Darzens G. Method generale de synthese des aldehyde a l'aide des acides glycidique substitues. *Compt. Rend.* **1904**, 139, 1214–1217. The reference is in French and can be accessed at https://gallica.bnf.fr/ark:/12148/bpt6k30930/f1214.image.langFR (accessed December 5, 2019).

[33]   (a) Boeckman RJ, and George KM. (2009). 1,1,1-Triacetoxy-1,1-dihydro-1,2-benziodoxol-3(1*H*)-one. In e-EROS Encyclopedia of Reagents for Organic Synthesis, (Ed.). doi:10.1002/047084289X.rt157m.pub2.

(b) Zhdankin VV. Organoiodine(V) Reagents in Organic Synthesis. *J. Org. Chem*. **2011**, 76 (5), 1185–1197. (c) Dess DB, and Martin JC. Readily accessible 12-I-5 oxidant for the conversion of primary and secondary alcohols to aldehydes and ketones. *J. Org. Chem*. **1983**, 48 (22), 4155–4156.

[34]   (a) This term appears in some publications and can be found at https://en.wikipedia.org/wiki/Diazo nium_compound (accessed December 5, 2019). (b) Griess P. Vorläufige Notiz über die Einwirkung von salpetriger Säure auf Amidinitro- und Aminitrophenylsäure. *Annalen der Chemie und Pharmacie* **1858**, 106 (1), 123–125. The original reference is in German and can be found at https://babel.hathi trust.org/cgi/pt?id=njp.32101044011037&view=1up&seq=541 (accessed December 5, 2019).

[35]   (a) Diels O, and Alder K. Synthesen in der hydroaromatischen Reihe. *Justus Liebigs Ann. Chem*. **1928**, 460 (1), 98–122. (b) Additional summary of 27 references can be found at https://en.wikipedia.org/ wiki/Diels%E2%80%93Alder_reaction (accessed December 5, 2019). (c) The Nobel Prize in Chemistry 1950. NobelPrize.org. Nobel Media AB (2019 accessed December 5, 2019, at https://www.nobelprize. org/prizes/chemistry/1950/summary/).

[36]   (a) Zimmerman HE, and Grunewald GL. The chemistry of Barrelene. III. A Unique Photoisomerization to Semibullvalene. *J. Am. Chem. Soc*. **1966**, 88 (1), 183–184. (b) Zimmerman HE, Binkley RW, Givens RS, and Sherwin MA. Mechanistic organic photochemistry. XXIV. The mechanism of the conversion of barrelene to semibullvalene. A general photochemical process. *J. Am. Chem. Soc*. **1967**, 89 (15), 3932–3933.

[37]   (a) Smissman EE, and Hite G. The Quasi-Favorskii rearrangement. I. The Preparation of Demerol and β-Pethidine. *J. Am. Chem. Soc*. **1959**, 81 (5), 1201–1203. (b) Smissman EE, and Hite G. The Quasi-Favorskii Rearrangement. II. Stereochemistry and Mechanism. *J. Am. Chem. Soc*. **1960**, 82 (13), 3375–3381. (c) Favorskii AE. *J. Russ. Phys. Chem. Soc*. **1894**, 26, 590. Note: the original reference is published in Russian and it is difficult to locate: Favorskii AE. *Zh. Russ. Khim. Obshch*. **1894**, 26, 590. (d) Faworsky A. Über die Einwirkung von Phosphorhalogenverbindungen auf Ketone, Bromketone und Ketonalkohole. *J. Prakt. Chem*. **1913**, 88 (1), 641–698. (e) Bliese M, and Tsanaktsidis J. Dimethyl Cubane-1,4-dicarboxylate: a Practical Laboratory Scale Synthesis. *Australian Journal of Chemistry* **1997**, 50 (3), 189–192. (f) Eaton PE, and Cole TW. The Cubane System. *J. Am. Chem. Soc*. **1964**, 86 (5), 962–964.

[38]   (a) Shine HJ, Zmuda H, Park KH, Kwart H, Horgan AG, Collins C, and Maxwell BE. Mechanism of the benzidine rearrangement. Kinetic isotope effects and transition states. Evidence for concerted rearrangement. *J. Am. Chem. Soc*. **1981**, 103 (4), 955–956. (b) Fischer E, and Jourdan F. Ueber die Hydrazine der Brenztraubensäure. *Ber. Dtsch. Chem. Ges*. **1883**, 16 (2), 2241–2245. (c) Fischer E, and Hess O. Synthese von Indolderivaten. *Ber. Dtsch. Chem. Ges*. **1884**, 17 (1), 559–568. (d) The Nobel Prize in Chemistry 1902. NobelPrize.org. Nobel Media AB (2019 accessed December 5, 2019, at https://www.nobelprize.org/prizes/chemistry/1902/summary/).

[39]   (a) Ador E, and Crafts J. Ueber die Einwirkung des Chlorkohlenoxyds auf Toluol in Gegenwart von Chloraluminium. *Ber. Dtsch. Chem. Ges*. **1877**, 10 (2), 2173–2176. (b) Friedel C, and Crafts JM. A New General Synthetical Method of Producing Hydrocarbons. *J. Chem. Soc*. **1877**, 32 (0), 725–791. The reference can be found at DOI: 10.1039/JS8773200725 (accessed December 5, 2019).

[40]   (a) Ing HR, and Manske RHF. CCCXII – a modification of the Gabriel synthesis of amines. *J. Chem. Soc*. **1926**, 129 (0), 2348–2351. (b) Delépine M. Sur l'hexaméthylene-amine (suite). Solubilités, hydrate, bromure, sulfate, phosphate. *Bull. Soc. Chim. Fr*. **1895**, 13 (3), 352–361. Note: the original reference is published in French and it is difficult to locate. (c) Gabriel S. Ueber eine Darstellungsweise primärer Amine aus den entsprechenden Halogenverbindungen. *Ber. Dtsch. Chem. Ges*. **1887**, 20 (2), 2224–2236.

[41]   (a) Knoevenagel E. Condensation von Malonsäure mit aromatischen Aldehyden durch Ammoniak und Amine. *Ber. Dtsch. Chem. Ges*. **1898**, 31 (3), 2596–2619. (b) Gewald K, Schinke E, and Böttcher H. Heterocyclen aus CH-aciden Nitrilen, VIII. 2-Amino-thiophene aus methylenaktiven Nitrilen, Carbonylverbindungen und Schwefel. *Chem. Ber*. **1966**, 99 (1), 94–100.

[42]   (a) Eglinton G, and Galbraith AR. 182. Macrocyclic acetylenic compounds. Part I. *Cyclo*tetradeca-1:3-diyne and related compounds. *J. Chem. Soc.* **1959**, (0), 889–896. (b) Behr OM, G. Eglinton G, Galbraith AR, and Raphael RA. 722. Macrocyclic acetylenic compounds. Part II. 1, 2:7,8-Dibenzocyclododeca-1,7-diene-3,5,9,11-tetrayne. *J. Chem. Soc.* **1960**, (0), 3614–3625. (c) Glaser C. Beiträge zur Kenntniss des Acetenylbenzols. *Ber. Dtsch. Chem. Ges.* **1869**, 2 (1), 422–424. (d) Hay A. Communications – Oxidative Coupling of Acetylenes. *J. Org. Chem.* **1960**, 25 (7), 1275–1276. (e) Hay AS. Oxidative Coupling of Acetylenes. II *J. Org. Chem.* **1962**, 27 (9), 3320–3321. (f) Shi W, and Lei A. 1,3-Diyne chemistry: synthesis and derivations. *Tetrahedron Lett.* **2014**, 55 (17), 2763–2772.
See also references therein.

[43]   (a) Grignard V. Sur quelques nouvelles combinaisons organométaliques du magnésium et leur application à des synthèses d'alcools et d'hydrocabures. *Compt. Rend.* **1900**, 130, 1322–1324. The original reference is in French and can be accessed at https://gallica.bnf.fr/ark:/12148/bpt6k3086n/f1322.table (accessed December 5, 2019). (b) The Nobel Prize in Chemistry 1912. NobelPrize.org. Nobel Media AB (2019 accessed December 5, 2019, at https://www.nobelprize.org/prizes/chemistry/1912/summary/).

[44]   (a) Prantz K, and Mulzer J. Synthetic Applications of the Carbonyl Generating Grob Fragmentation. *Chem. Rev.* **2010**, 110 (6), 3741–3766. (b) Grob CA, and Baumann W. Die 1,4-Eliminierung unter Fragmentierung. *Helv. Chim. Acta* **1955**, 38 (3), 594–610. (c) Grob CA, and Schiess PW. Heterolytic Fragmentation. A Class of Organic Reactions. *Angew. Chem. Int. Ed. Engl.* **1967**, 6 (1), 1–15.

[45]   (a) Surellas GS. Notes sur l'Hydriodate de potasse et l'Acide hydriodique. – Hydriodure de carbone; moyen d'obtenir, à l'instant, ce composé triple. **1822**. The original reference is in French and can be accessed at https://gallica.bnf.fr/ark:/12148/bpt6k6137757n/f2.image (accessed December 5, 2019). (b) Liebig J. Ueber die Verbindungen, welche durch die Einwirkung des Chlors auf Alkohol, Aether, ölbildendes Gas und Essiggeist entstehen. *Ann. Phys.* **1832**, 100 (2), 243–295. (c) Lieben A. Ueber Entstehung von Jodoform und Anwendung dieser Reaction in der chemischen Analyse. *Annalen der Chemie. Supplementband.* **1870**, 7, 218–236. The original reference is in German and can be accessed at https://babel.hathitrust.org/cgi/pt?id=uiug.30112018225695&view=1up&seq=230 (accessed December 5, 2019). Note: additional references can be found at https://en.wikipedia.org/wiki/Haloform_reaction (accessed December 5, 2019).

[46]   (a) Heck RF. Acylation, methylation, and carboxyalkylation of olefins by Group VIII metal derivatives. *J. Am. Chem. Soc.* **1968**, 90 (20), 5518–5526. (b) Mizoroki T, Mori K, and Ozaki A. Arylation of Olefin with Aryl Iodide Catalyzed by Palladium. *Bull. Chem. Soc. Jpn.* **1971**, 44 (2), 581. (c) The Nobel Prize in Chemistry 2010. NobelPrize.org. Nobel Media AB (2019 accessed December 5, 2019, at https://www.nobelprize.org/prizes/chemistry/2010/summary/).

[47]   (a) Hell C. Ueber eine neue Bromirungsmethode organischer Säuren. *Ber. Dtsch. Chem. Ges.* **1881**, 14 (1), 891–893. (b) Volhard J. 4) Ueber Darstellung α-bromirter Säuren. *Justus Liebigs Ann. Chem.* **1887**, 242 (1-2), 141–163. (c) Zelinsky N. Ueber eine bequeme Darstellungsweise von α-Brompropionsäureester. *Ber. Dtsch. Chem. Ges.* **1887**, 20 (1), 2026. (d) Chow AW, Jakas DR, Hoover JRE. Preparation of 1-substituted bicyclo[3.2.1]octanes by a rearrangement reaction. *Tetrahedron Lett.* **1966**, 7 (44), 5427–5431.

[48]   (a) Denmark SE, and Regens CS. Palladium-Catalyzed Cross-Coupling Reactions of Organosilanols and Their Salts: Practical Alternatives to Boron- and Tin-Based Methods. *Acc. Chem. Res.* **2008**, 41 (11), 1486–1499. (b) Hatanaka Y, Hiyama T. Cross-coupling of organosilanes with organic halides mediated by a palladium catalyst and tris(diethylamino)sulfonium difluorotrimethylsilicate. *J. Org. Chem.* **1988**, 53 (4), 918–920.

[49]   (a) Saytzeff A. Zur Kenntniss der Reihenfolge der Analgerung und Ausscheidung der Jodwasserstoffelemente in organischen Verbindungen. *Justus Liebigs Ann. Chem.* **1875**, 179 (3), 296–301. (b) Cope AC, and Trumbull ER. (2011). Olefins from Amines: The Hofmann Elimination

Reaction and Amine Oxide Pyrolysis. In Organic Reactions, (Ed.). doi:10.1002/0471264180.or011.05.
(c) Hofmann AW. Beiträge zur Kenntniss der flüchtigen organischen Basen. *Justus Liebigs Ann. Chem.* **1851**, 78 (3), 253–286. (d) Hofmann AW. Beiträge zur Kenntniss der flüchtigen organischen Basen. *Justus Liebigs Ann. Chem.* **1851**, 79 (1), 11–39.

[50] (a) Maryanoff BE, and Reitz AB. The Wittig olefination reaction and modifications involving phosphoryl-stabilized carbanions. Stereochemistry, mechanism, and selected synthetic aspects. *Chem. Rev.* **1989**, 89 (4), 863–927. (b) Peterson DJ. Carbonyl olefination reaction using silyl-substituted organometallic compounds. *J. Org. Chem.* **1968**, 33 (2), 780–784. (c) Horner L, Hoffmann H, and Wippel HG. Phosphororganische Verbindungen, XII. Phosphinoxyde als Olefinierungsreagenzien. *Chem. Ber.* **1958**, 91 (1), 61–63. (d) Horner L, Hoffmann H, Wippel HG, and Klahre G. Phosphororganische Verbindungen, XX. Phosphinoxyde als Olefinierungsreagenzien. *Chem. Ber.* **1959**, 92 (10), 2499–2505. (e) Wadsworth WS, and Emmons WD. The Utility of Phosphonate Carbanions in Olefin Synthesis. *J. Am. Chem. Soc.* **1961**, 83 (7), 1733–1738.

[51] (a) Freeman F. (2001). Chromic Acid. In e-EROS Encyclopedia of Reagents for Organic Synthesis, (Ed.). doi:10.1002/047084289X.rc164. (b) Piancatelli G, and Luzzio FA. (2007). Pyridinium Chlorochromate. In e-EROS Encyclopedia of Reagents for Organic Synthesis, (Ed.). doi:10.1002/9780470842898.rp288.pub2. (c) Piancatelli G. (2001). Pyridinium Dichromate. In e-EROS Encyclopedia of Reagents for Organic Synthesis, (Ed.). doi:10.1002/047084289X.rp290. (d) Bowden K, Heilbron IM, Jones ERH, Weedon BCL. 13. Researches on acetylenic compounds. Part I. The preparation of acetylenic ketones by oxidation of acetylenic carbinols and glycols. *J. Chem. Soc.* **1946**, (0), 39–45.

[52] (a) Markownikoff W. I. Ueber die Abhängigkeit der verschiedenen Vertretbarkeit des Radicalwasserstoffs in den isomeren Buttersäuren. *Justus Liebigs Ann. Chem.* **1870**, 153 (2), 228–259. (b) Kutscheroff M. Ueber eine neue Methode direkter Addition von Wasser (Hydratation) an die Kohlenwasserstoffe der Acetylenreihe. *Ber. Dtsch. Chem. Ges.* **1881**, 14 (1), 1540–1542.

[53] (a) Corriu RJP, and Masse JP. Activation of Grignard reagents by transition-metal complexes. A new and simple synthesis of *trans*-stilbenes and polyphenyls. *J. Chem. Soc., Chem. Commun.* **1972**, (3), 144a. (b) Tamao K, Sumitani K, and Kumada M. Selective carbon-carbon bond formation by cross-coupling of Grignard reagents with organic halides. Catalysis by nickel-phosphine complexes. *J. Am. Chem. Soc.* **1972**, 94 (12), 4374–4376. (c) Tamao K, Kiso Y, Sumitani K, and Kumada M. Alkyl group isomerization in the cross-coupling reaction of secondary alkyl Grignard reagents with organic halides in the presence of nickel-phosphine complexes as catalysts. *J. Am. Chem. Soc.* **1972**, 94 (26), 9268–9269.

[54] (a) Ley SV, Norman J, and Wilson AJ. (2011). Tetra-*n*-propylammonium Perruthenate. In e-EROS Encyclopedia of Reagents for Organic Synthesis, (Ed.). doi:10.1002/047084289X.rt074.pub2. (b) Griffith WP, Ley SV, Whitcombe GP, and White AD. Preparation and use of tetra-n-butylammonium per-ruthenate (TBAP reagent) and tetra-n-propylammonium per-ruthenate (TPAP reagent) as new catalytic oxidants for alcohols. *J. Chem. Soc., Chem. Commun.* **1987**, (21), 1625–1627.

[55] (a) Liebeskind LS, and Srogl J. Thiol Ester–Boronic Acid Coupling. A Mechanistically Unprecedented and General Ketone Synthesis. *J. Am. Chem. Soc.* **2000**, 122 (45), 11260–11261. (b) Cheng HG, Chen H, Liu Y, and Zhou Q. The Liebeskind–Srogl Cross-Coupling Reaction and its Synthetic Applications. *Asian J. Org. Chem.* **2018**, 7 (3), 490–508. The open access paper can be found at https://onlinelibrary.wiley.com/doi/epdf/10.1002/ajoc.201700651 (accessed December 5, 2019).

[56] (a) Kleinman EF. (2001). Dimethyl(methylene)ammonium Iodide. In e-EROS Encyclopedia of Reagents for Organic Synthesis, (Ed.). doi:10.1002/047084289X.rd346. (b) Mannich C, and Krösche W. Ueber ein Kondensationsprodukt aus Formaldehyd, Ammoniak und Antipyrin. *Arch. Pharm. Pharm. Med. Chem.* **1912**, 250 (1), 647–667. (c) Kabachnik MI, and Medved TY. Новый метод синтеза аминофосфиновых кислот (New synthesis of aminophosphonic acids). *Dokl Akad Nauk SSSRKhim*

1952, 83, 689–692. (d) Fields EK. The Synthesis of Esters of Substituted Amino Phosphonic Acids. *J. Am. Chem. Soc*. **1952**, 74 (6), 1528–1531.

[57] (a) Fittig R. Ueber einige Producte der trockenen Destillation essigsaurer Salze. *Justus Liebigs Ann. Chem*. **1859**, 110 (1), 17–23. (b) Demselben. Ueber einige Metamorphosen des Acetons der Essigsäure. *Justus Liebigs Ann. Chem*. **1859**, 110 (1), 23–45. (c) McMurry JE, and Fleming MP. New method for the reductive coupling of carbonyls to olefins. Synthesis of β–carotene. *J. Am. Chem. Soc*. **1974**, 96 (14), 4708–4709.

[58] (a) Meerwein H, and Schmidt R. Ein neues Verfahren zur Reduktion von Aldehyden und Ketonen. *Justus Liebigs Ann. Chem*. **1925**, 444 (1), 221–238. (b) Verley A. Sur l'échange de groupements fonctionnels entre deux molécules. Passage de la fonction alcool à la fonction aldéhyde et inversement. *Bull. Soc. Chim. Fr*. **1925**, 37, 537–542. Note: the original reference is published in French and it is difficult to locate. (c) Ponndorf W. Der reversible Austausch der Oxydationsstufen zwischen Aldehyden oder Ketonen einerseits und primären oder sekundären Alkoholen anderseits. *Angew. Chem*. **1926**, 39 (5), 138–143.

[59] (a) Stetter H, and Schreckenberg M. A New Method for Addition of Aldehydes to Activated Double Bonds. *Angew. Chem. Int. Ed. Engl*. **1973**, 12 (1), 81. (b) Michael A. Ueber die Addition von Natriumacetessig- und Natriummalonsäureäthern zu den Aethern ungesättigter Säuren. *J. Prakt. Chem*. **1887**, 35 (1), 349–356.

[60] (a) Hong Y. (2001). Hydrogen Peroxide–Iron(II) Sulfate. In e-EROS Encyclopedia of Reagents for Organic Synthesis, (Ed.). doi:10.1002/047084289X.rh043. (b) Mihailović ML, Čeković Ž, and Mathes BM. (2005). Lead(IV) Acetate. In e-EROS Encyclopedia of Reagents for Organic Synthesis, (Ed.). doi:10.1002/047084289X.rl006.pub2. (c) Kolbe H. Untersuchungen über die Elektrolyse organischer Verbindungen. *Justus Liebigs Ann. Chem*. **1849**, 69 (3), 257–294. (d) Minisci F, Galli R, Cecere M, Malatesta V, and Caronna T. Nucleophilic character of alkyl radicals: new syntheses by alkyl radicals generated in redox processes. *Tetrahedron Lett*. **1968**, 9 (54), 5609–5612. (e) Minisci F, Bernardi R, Bertini F, Galli R, and Perchinummo M. Nucleophilic character of alkyl radicals – VI: A new convenient selective alkylation of heteroaromatic bases. *Tetrahedron* **1971**, 27 (15), 3575–3579.

[61] (a) Jenkins ID, and Mitsunobu O. (2001). Triphenylphosphine–Diethyl Azodicarboxylate. In e-EROS Encyclopedia of Reagents for Organic Synthesis, (Ed.). doi:10.1002/047084289X.rt372. (b) Hughes DL. (2004). The Mitsunobu Reaction. In Organic Reactions, (Ed.). doi:10.1002/0471264180.or042.02. (c) Swamy KCK, Kumar NNB, Balaraman E, and Kumar KVPP. Mitsunobu and Related Reactions: Advances and Applications. *Chem. Rev*. **2009**, 109 (6), 2551–2651. (d) Mitsunobu O, Yamada M, and Mukaiyama T. Preparation of Esters of Phosphoric Acid by the Reaction of Trivalent Phosphorus Compounds with Diethyl Azodicarboxylate in the Presence of Alcohols. *Bull. Chem. Soc. Jpn*. **1967**, 40 (4), 935–939. The open access paper can be found at https://doi.org/10.1246/bcsj.40.935 (accessed December 5, 2019). (e) Mitsunobu O, and Yamada M. Preparation of Esters of Carboxylic and Phosphoric Acid via Quaternary Phosphonium Salts. *Bull. Chem. Soc. Jpn*. **1967**, 40 (10), 2380–2382. The open access paper can be found at https://doi.org/10.1246/bcsj.40.2380 (accessed December 5, 2019). (f) Fischer E, and Speier A. Darstellung der Ester. *Ber. Dtsch. Chem. Ges*. **1895**, 28 (3), 3252–3258.

[62] (a) Ishiyama T, Chen H, Morken JP, Mlynarski SN, Ferris GE, Xu S, and Wang J. (2018). 4,4,4′,4′,5,5,5′,5′-Octamethyl-2,2′-bi-1,3,2-dioxaborolane. In e-EROS Encyclopedia of Reagents for Organic Synthesis. doi:10.1002/047084289X.rn00188.pub4. (b) Petasis NA, and Akritopoulou I. The boronic acid Mannich reaction: A new method for the synthesis of geometrically pure allylamines. *Tetrahedron Lett*. **1993**, 34 (4), 583–586. (c) Ishiyama T, Murata M, and Miyaura N. Palladium(0)-Catalyzed Cross-Coupling Reaction of Alkoxydiboron with Haloarenes: A Direct Procedure for Arylboronic Esters. *J. Org. Chem*. **1995**, 60 (23), 7508–7510.

[63] (a) Tokuyasu T, Kunikawa S, Masuyama A, and Nojima M. Co(III)−Alkyl Complex- and Co(III)−Alkylperoxo Complex-Catalyzed Triethylsilylperoxidation of Alkenes with Molecular Oxygen and

Triethylsilane. *Org. Lett.* **2002**, 4 (21), 3595–3598. (b) Mukaiyama T, Isayama S, Inoki S, Kato K, Yamada T, and Takai T. Oxidation-Reduction Hydration of Olefins with Molecular Oxygen and 2-Propanol Catalyzed by Bis(acetylacetonato)cobalt(II). *Chem. Lett.* **1989**, 18 (3), 449–452. (c) Inoki S, Kato K, Takai T, Isayama S, Yamada T, and Mukaiyama T. Bis(trifluoroacetylacetonato)cobalt(II) Catalyzed Oxidation-Reduction Hydration of Olefins Selective Formation of Alcohols from Olefins. *Chem. Lett.* **1989**, 18 (3), 515–518. (d) Isayama S, and Mukaiyama T. A New Method for Preparation of Alcohols from Olefins with Molecular Oxygen and Phenylsilane by the Use of Bis(acetylacetonato) cobalt(II). *Chem. Lett.* **1989**, 18 (6), 1071–1074.

[64]   (a) Woodward RB, and Hoffmann R. Stereochemistry of Electrocyclic Reactions. *J. Am. Chem. Soc.* **1965**, 87 (2), 395–397. (b) Woodward RB, and Hoffmann R. The Conservation of Orbital Symmetry. *Angew. Chem. Int. Ed. Engl.* **1969**, 8 (11), 781–853. (c) In 1965 Robert Burns Woodward received the Nobel Prize in Chemistry (accessed December 5, 2019, at https://www.nobelprize.org/prizes/chemistry/1965/sum mary/). In 1981 Roald Hoffmann (jointly with Kenichi Fukui) received the Nobel Prize in Chemistry (accessed December 5, 2019, at https://www.nobelprize.org/prizes/chemistry/1981/summary/). (d) Nazarov IN, and Zaretskaya II. Acetylene derivatives. XVII. Hydration of Hydrocarbons of the Divinylacetylene Series. *Izv. Akad. Nauk. SSSR, Ser. Khim.* **1941**, 211–224. (e) Nazarov IN, and Zaretskaya II. Derivatives of acetylene. XXVII. Hydration of divinylacetylene. *Bull. acad. sci. U.R.S.S., Classe Sci. Chim.* **1942**, 200–209. Note: the original references are published in Russian and they are difficult to locate. See also (f) Santelli-Rouvier C, and Santelli M. The Nazarov Cyclisation. *Synthesis* **1983** (6), 429–442. (g) Frontier AJ, and Collison C. The Nazarov cyclization in organic synthesis. Recent advances. *Tetrahedron* **2005**, 61 (32), 7577–7606. Other review references can be found in [5c].

[65]   (a) Kornblum N, and Brown RA. The Action of Acids on Nitronic Esters and Nitroparaffin Salts. Concerning the Mechanisms of the Nef and the Hydroxamic Acid Forming Reactions of Nitroparaffins. *J. Am. Chem. Soc.* **1965**, 87 (8), 1742–1747. (b) Konovalov MI. *J. Russ. Phys. Chem. Soc.* **1893**, 25, 509. Note: the original reference is published in Russian and it is difficult to locate: Konovalov MI. *Zhur. Russ. Khim. Obshch.* **1893**, 25, 389, 472, 509. (c) Nef JU. Ueber die Constitution der Salze der Nitroparaffine. *Justus Liebigs Ann. Chem.* **1894**, 280 (2-3), 263–291. (d) Nef JU. Ueber das zweiwerthige Kohlenstoffatom. *Justus Liebigs Ann. Chem.* **1894**, 280 (2-3), 291–342.

[66]   (a) King AO, Okukado N, and Negishi E. Highly general stereo-, regio-, and chemo-selective synthesis of terminal and internal conjugated enynes by the Pd-catalysed reaction of alkynylzinc reagents with alkenyl halides. *J. Chem. Soc., Chem. Commun.* **1977**, (19), 683–684. (b) Negishi E, King AO, Okukado N. Selective carbon-carbon bond formation via transition metal catalysis. 3. A highly selective synthesis of unsymmetrical biaryls and diarylmethanes by the nickel- or palladium-catalyzed reaction of aryl- and benzylzinc derivatives with aryl halides. *J. Org. Chem.* **1977**, 42 (10), 1821–1823.

[67]   (a) Norrish RGW, and Kirkbride FW. 204. Primary photochemical processes. Part I. The decomposition of formaldehyde. *J. Chem. Soc.* **1932**, (0), 1518–1530. (b) Norrish RGW, and Appleyard MES. 191. Primary photochemical reactions. Part IV. Decomposition of methyl ethyl ketone and methyl butyl ketone. *J. Chem. Soc.* **1934**, (0), 874–880. (c) Norrish RGW, Crone HG, and Saltmarsh OD. 318. Primary photochemical reactions. Part V. The spectroscopy and photochemical decomposition of acetone. *J. Chem. Soc.* **1934**, (0), 1456–1464. (d) Bamford CH, and Norrish RGW. 359. Primary photochemical reactions. Part VII. Photochemical decomposition of *iso*valeraldehyde and di-*n*-propyl ketone. *J. Chem. Soc.* **1935**, (0), 1504–1511. (e) Norrish RGW, and Bamford CH. Photodecomposition of Aldehydes and Ketones. *Nature* **1936**, 138, 1016. (f) Norrish RGW, and Bamford CH. Photo-decomposition of Aldehydes and Ketones. *Nature* **1937**, 140, 195–196. (g) The Nobel Prize in Chemistry 1967. NobelPrize.org. Nobel Media AB (2019 accessed December 5, 2019, at https://www.nobelprize.org/prizes/chemistry/1967/summary/).

[68]   (a) Nelson DJ, Manzini S, Urbina-Blanco CA, and Nolan SP. Key processes in ruthenium-catalysed olefin metathesis. *Chem. Commun.* **2014**, 50 (72), 10355–10375. (b) Ziegler K, Holzkamp E, Breil H,

and Martin H. Polymerisation von Äthylen und anderen Olefinen. *Angew. Chem.* **1955**, 67 (16), 426. (c) Anderson AW, and Merckling NG. Polymeric bicyclo-(2,2,1)-2-heptene. US Patent Office 2721189, **1955** (E. I. du Pont de Nemours and Company). (d) The Nobel Prize in Chemistry 2005. NobelPrize.org. Nobel Media AB (2019 accessed December 5, 2019, at https://www.nobelprize.org/prizes/ chemistry/2005/summary/). (e) Takacs JM, and Atkins JM. (2002). Dichloro-bis(tricyclohexylphosphine) methyleneruthenium. In e-EROS Encyclopedia of Reagents for Organic Synthesis, (Ed.). doi:10.1002/ 047084289X.rn00110. (f) Diver ST, and Middleton MD. (2010). Ruthenium, [1,3-Bis(2,4,6- trimethylphenyl)-2-imidazolidinylidene]dichloro(phenylmethylene)(tricyclohexylphosphine) (Grubbs' Second-Generation Catalyst). In e-EROS Encyclopedia of Reagents for Organic Synthesis, (Ed.). doi:10.1002/047084289X.rn01100. (g) Garber SB, Khan RKM, Mann TJ, and Hoveyda AH. (2013). Dichloro[[2-(1-methylethoxy-O)phenyl]-methylene] (tricyclohexylphosphine) Ruthenium. In e-EROS Encyclopedia of Reagents for Organic Synthesis, (Ed.). doi:10.1002/047084289X.rn00129.pub2.

[69]   Oppenauer RV. Eine Methode der Dehydrierung von Sekundären Alkoholen zu Ketonen. I. Zur Herstellung von Sterinketonen und Sexualhormonen. *Recl. Trav. Chim. Pays-Bas* **1937**, 56 (2), 137–144.

[70]   (a) Criegee R, and Wenner G. Die Ozonisierung des 9,10-Oktalins. *Justus Liebigs Ann. Chem.* **1949**, 564 (1), 9–15. (b) Criegee R. Mechanism of Ozonolysis. *Angew. Chem. Int. Ed. Engl.* **1975**, 14 (11), 745–752. (c) Criegee R. Mechanismus der Ozonolyse. *Angew. Chem.* **1975**, 87 (21), 765–771. (d) Wee AG, Liu B, Jin Z, and Shah AK. (2012). Sodium Periodate–Osmium Tetroxide. In e-EROS Encyclopedia of Reagents for Organic Synthesis, (Ed.). https://onlinelibrary.wiley.com/doi/abs/10.1002/047084289X. rs095m.pub2 (accessed December 5, 2019). (e) Berglund RA, and Kreilein MM. (2006). Ozone. In e-EROS Encyclopedia of Reagents for Organic Synthesis, (Ed.). doi:10.1002/047084289X.ro030.pub2. (f) Wee AG, and Liu B. (2001). Sodium Periodate–Potassium Permanganate. In e-EROS Encyclopedia of Reagents for Organic Synthesis, (Ed.). doi:10.1002/047084289X.rs096. (g) Harries C. Ueber die Einwirkung des Ozons auf organische Verbindungen. *Justus Liebigs Ann. Chem.* **1905**, 343 (2-3), 311–344. Note: the original reference is very old (1840) and it is difficult to locate. See also references in [70g]. (h) Martínez AG, Vilar ET, Fraile AG, de la Moya Cerero S, Maroto BL. A new type of anomalous ozonolysis in strained allylic bicycloalkan-1-ols. *Tetrahedron Lett.* **2005**, 46 (31), 5157–5159.

[71]   (a) Voss J. (2006). 2,4-Bis(4-methoxyphenyl)-1,3,2,4-dithiadiphosphetane 2,4-Disulfide. In e-EROS Encyclopedia of Reagents for Organic Synthesis, (Ed.). doi:10.1002/047084289X.rb170.pub2. (b) Paal C. Ueber die Derivate des Acetophenonacetessigesters und des Acetonylacetessigesters. *Ber. Dtsch. Chem. Ges.* **1884**, 17 (2), 2756–2767. (c) Knorr L. Synthese von Furfuranderivaten aus dem Diacetbernsteinsäureester. *Ber. Dtsch. Chem. Ges.* **1884**, 17 (2), 2863–2870.

[72]   (a) Paternò E, and Chieffi G. Sintesi in chimica organica per mezzo della luce. Nota II. Composti degli idrocarburi non saturi con aldeidi e chetoni. *Gazz. Chim. Ital.* **1909**, 39, 341. Note: the original reference is published in Italian and it is difficult to locate. (b) Büchi G, Inman CG, and Lipinsky ES. Light-catalyzed Organic Reactions. I. The Reaction of Carbonyl Compounds with 2-Methyl-2-butene in the Presence of Ultraviolet Light. *J. Am. Chem. Soc.* **1954**, 76 (17), 4327–4331.

[73]   (a) Brummonda KM, and Kent JL. Recent Advances in the Pauson–Khand Reaction and Related [2+2 +1] Cycloadditions. *Tetrahedron* **2000**, 56 (21), 3263–3283. Please check additional review articles in [5]. (b) Khand IU, Knox GR, Pauson PL, and Watts WE. Organocobalt complexes. Part I. Arene complexes derived from dodecacarbonyltetracobalt. *J. Chem. Soc., Perkin Trans. 1* **1973**, (0), 975–977. (c) Khand IU, Knox GR, Pauson PL, Watts WE, and Foreman MI. Organocobalt complexes. Part II. Reaction of acetylenehexacarbonyldicobalt complexes, $(R^1C_2R^2)Co_2(CO)_6$, with norbornene and its derivatives. *J. Chem. Soc., Perkin Trans. 1* **1973**, (0), 977–981. (d) Pauson PL, and Khand IU. Uses of Cobalt-Carbonyl Acetylene Complexes in Organic Synthesis. *Ann. N. Y. Acad. Sci.* **1977**, 295 (1), 2–14.

[74]   (a) Valeur E, and Bradley M. Amide bond formation: beyond the myth of coupling reagents. *Chem. Soc. Rev.* **2009**, 38 (2), 606–631. (b) El-Faham A, and Albericio F. Peptide Coupling Reagents, More

than a Letter Soup. *Chem. Rev.* **2011**, 111 (11), 6557–6602. (c) Carpino LA, Imazumi H, El-Faham A, Ferrer FJ, Zhang C, Lee Y, Foxman BM, Henklein P, Hanay C, Mügge C, Wenschuh H, Klose J, Beyermann M, and Bienert M. The Uronium/Guanidinium Peptide Coupling Reagents: Finally the True Uronium Salts. *Angew. Chem. Int. Ed.* **2002**, 41 (3), 441–445. (d) Albert JS, Hamilton AD, Hart AC, Feng X, Lin L, and Wang Z. (2017). 1,3-Dicyclohexylcarbodiimide. In e-EROS Encyclopedia of Reagents for Organic Synthesis. doi:10.1002/047084289X.rd146.pub3. (e) Pottorf RS, Szeto P, and Srinivasarao M. (2017). 1-Ethyl-3-(3′-dimethylaminopropyl)carbodiimide Hydrochloride. In e-EROS Encyclopedia of Reagents for Organic Synthesis. doi:10.1002/047084289X.re062.pub2. (f) Albericio F, and Kates SA. (2001). *O*-Benzotriazol-1-yl-N,N,N′,N′-tetramethyluronium Hexafluorophosphate. In e-EROS Encyclopedia of Reagents for Organic Synthesis, (Ed.). doi:10.1002/047084289X.rb038. (g) Albericio F, Kates SA, and Carpino LA. (2001). *N*-[(Dimethylamino)-1*H*-1,2,3-triazolo[4,5-*b*]pyridin-1-ylmethylene]-*N*-methylmethanaminium Hexafluorophosphate *N*-Oxide. In e-EROS Encyclopedia of Reagents for Organic Synthesis, (Ed.). https://onlinelibrary.wiley.com/doi/10.1002/047084289X.rd312s (accessed December 5, 2019). (h) Coste J, and Jouin P. (2003). (1*H*-Benzotriazol-1-yloxy) tripyrrolidino-phosphonium Hexafluorophosphate. In e-EROS Encyclopedia of Reagents for Organic Synthesis, (Ed.). doi:10.1002/047084289X.rn00198. (i) Lygo B, and Pelletier G. (2013). 1-Hydroxybenzotriazole. In e-EROS Encyclopedia of Reagents for Organic Synthesis, (Ed.). doi:10.1002/047084289X.rh052.pub2. (j) Fischer E, and Fourneau E. Ueber einige Derivate des Glykocolls. *Ber. Dtsch. Chem. Ges.* **1901**, 34 (2), 2868–2877. (k) Sheehan JC, and Hess GP. A New Method of Forming Peptide Bonds. *J. Am. Chem. Soc.* **1955**, 77 (4), 1067–1068. (l) Dourtoglou V, Ziegler JC, and Gross B. L'hexafluorophosphate de O-benzotriazolyl-N,N-tetramethyluronium: Un reactif de couplage peptidique nouveau et efficace. *Tetrahedron Lett.* **1978**, 19 (15), 1269–1272.

[75] (a) Baldwin JE. Rules for ring closure. *J. Chem. Soc., Chem. Commun.* **1976**, (18), 734–736. (b) Pictet A, and Spengler T. Über die Bildung von Isochinolin-derivaten durch Einwirkung von Methylal auf Phenyl-äthylamin, Phenyl-alanin und Tyrosin. *Ber. Dtsch. Chem. Ges.* **1911**, 44 (3), 2030–2036.

[76] (a) Demjanov NJ, and Lushnikov M. *J. Russ. Phys. Chem. Soc.* **1903**, 35, 26–42. Note: the original reference is published in Russian and it is difficult to locate: Demjanov NJ, and Lushnikov M. *Zhur. Russ. Khim. Obshch.* **1903**, 35, 26–42. (b) Tiffeneau M, Weill P, and Tchoubar B. Isomérisation de l'oxyde de méthylène cyclohexane en hexahydrobenzaldéhyde et désamination de l'aminoalcool correspondant en cycloheptanone. *Compt. Rend.* **1937**, 205, 54–56. The reference is in French and can be viewed at https://gallica.bnf.fr/ark:/12148/bpt6k3157c.image.f54 (accessed December 5, 2019). (c) Fittig R. 41. Ueber einige Derivate des Acetons. *Justus Liebigs Ann. Chem.* **1860**, 114 (1), 54–63.

[77] (a) Cave A, Kan-Fan C, Potier P, and Men JL. Modification de la reaction de polonovsky: Action de l'anhydride trifluoroacetique sur un aminoxide. *Tetrahedron* **1967**, 23 (12), 4681–4689. (b) Ahond A, Cave A, Kan-Fan C, Husson HP, Rostolan J, and Potier P. Facile N-O bond cleavages of amine oxides. *J. Am. Chem. Soc.* **1968**, 90 (20), 5622–5623. (c) Polonovski M, and Polonovski M. Sur les aminoxydes des alcaloides. III. Action des anhydrides et chlorules d'acides organiques. Preparations des bases nor. *Bull. Soc. Chim. Fr.* **1927**, 41, 1190–1208. Note: the original reference is published in French and it is difficult to locate.

[78] (a) Katsuki T, and Sharpless KB. The first practical method for asymmetric epoxidation. *J. Am. Chem. Soc.* **1980**, 102 (18), 5974–5976. (b) Tu Y, Wang ZX, and Shi Y. An Efficient Asymmetric Epoxidation Method for *trans*-Olefins Mediated by a Fructose-Derived Ketone. *J. Am. Chem. Soc.* **1996**, 118 (40), 9806–9807. (c) Prileschajew N. Oxydation ungesättigter Verbindungen mittels organischer Superoxyde. *Ber. Dtsch. Chem. Ges.* **1909**, 42 (4), 4811–4815.

[79] (a) Dobbs AP, Guesné SJJ, Parker RJ, Skidmore J, Stephensond RA, and Mike B. Hursthouse MB. A detailed investigation of the aza-Prins reaction. *Org. Biomol. Chem.* **2010**, 8 (5), 1064–1080. (b) Reddy BVS, Nair PN, Antony A, Lalli C, and Grée R. The Aza-Prins Reaction in the Synthesis of Natural Products and Analogues. *Eur. J. Org. Chem.* **2017**, (14), 1805–1819. (c) Rajasekaran P, Singh GP, Hassam M, and Vankar YD. A Cascade "Prins-Pinacol-Type Rearrangement and C4-OBn

Participation" on Carbohydrate Substrates: Synthesis of Bridged Tricyclic Ketals, Annulated Sugars and C2-Branched Heptoses. *Chem. Eur. J.* **2016**, 22 (51), 18383–18387. (d) Prins HJ. Over de condensatie van formaldehyd met onverzadigde verbindingen. *Chem. Weekblad* **1919**, 16, 1072–1073. (e) Prins HJ.
The reciprocal condensation of unsaturated organic compounds. *Chem. Weekblad* **1919**, 16, 1510–1526. Note: the original references are difficult to locate.

[80] (a) Pummerer R. Über Brom-Additionsprodukte von Aryl-thioglykolsäuren. *Ber. Dtsch. Chem. Ges.* **1909**, 42 (2), 2275–2282. (b) McCamley K, Ripper JA, Singer RD, Scammells PJ. Efficient N-Demethylation of Opiate Alkaloids Using a Modified Nonclassical Polonovski Reaction. *J. Org. Chem.* **2003**, 68 (25), 9847–9850. (c) Ruda AM, Papadouli S, Thangavadivale V, Moseley JD. Application of the Polonovski Reaction: Scale-up of an Efficient and Environmentally Benign Opioid Demethylation. *Org. Process Res. Dev.* **2022**, 26 (5), 1398–1404.

[81] (a) See more at https://en.wikipedia.org/wiki/Cheletropic_reaction (accessed December 5, 2019). (b) Philips JC, and Morales O. Sulphur dioxide extrusion from substituted thiiren 1,1-dioxides. *J. Chem. Soc., Chem. Commun.* **1977**, (20), 713–714. (c) Ramberg L, and Bäcklund B. *Ark. Kemi. Mineral. Geol.* **1940**, 27, Band 13A, 1–50. Note: the original reference is difficult to locate (see also *Chem. Abstr.* **1940**, 34, 4725).

[82] (a) Blaise EE. *Comp. Rend. Hebd. Seances Acad. Sci.* **1901**, 132, 478–480. Note: the original reference is difficult to locate. (b) Cason J, Rinehart KL, and Thornton SD. The Preparation of β-Keto Esters from Nitriles and α-Bromoesters. *J. Org. Chem.* **1953**, 18 (11), 1594–1600. (c) Reformatsky S. Neue Synthese zweiatomiger einbasischer Säuren aus den Ketonen. *Ber. Dtsch. Chem. Ges.* **1887**, 20 (1), 1210–1211.

[83] (a) Rapson WS, and Robinson R. 307. Experiments on the synthesis of substances related to the sterols. Part II. A new general method for the synthesis of substituted cyclohexenones. *J. Chem. Soc.* **1935**, (0), 1285–1288. (b) The Nobel Prize in Chemistry 1947. NobelPrize.org. Nobel Media AB (2019 accessed December 5, 2019, at https://www.nobelprize.org/prizes/chemistry/1947/summary/).

[84] (a) Bamford WR, and Stevens TS. 924. The decomposition of toluene-*p*-sulphonylhydrazones by alkali. *J. Chem. Soc.* **1952**, (0), 4735–4740. (b) Shapiro RH, and Heath MJ. Tosylhydrazones. V. Reaction of Tosylhydrazones with Alkyllithium Reagents. A New Olefin Synthesis. *J. Am. Chem. Soc.* **1967**, 89 (22), 5734–5735. (c) Shapiro RH, Lipton MF, Kolonko KJ, Buswell RL, Capuano LA. Tosylhydrazones and alkyllithium reagents: More on the regiospecificity of the reaction and the trapping of three intermediates. **1975**, 16 (22–23), 1811–1814. (d) Shapiro RH. (2011). Alkenes from Tosylhydrazones. In Organic Reactions, (Ed.). doi:10.1002/0471264180.or023.03.

[85] (a) Stephens RD, and Castro CE. The Substitution of Aryl Iodides with Cuprous Acetylides. A Synthesis of Tolanes and Heterocyclics. *J. Org. Chem.* **1963**, 28 (12), 3313–3315. (b) Sonogashira K, Tohda Y, Hagihara N. A convenient synthesis of acetylenes: catalytic substitutions of acetylenic hydrogen with bromoalkenes, iodoarenes and bromopyridines. *Tetrahedron Lett.* **1975**, 16 (50), 4467–4470.

[86] (a) Staudinger H. Zur Kenntniss der Ketene. Diphenylketen. *Justus Liebigs Ann. Chem.* **1907**, 356 (1-2), 51–123. (b) Staudinger H. Über Ketene. 4. Mitteilung: Reaktionen des Diphenylketens. *Ber. Dtsch. Chem. Ges.* **1907**, 40 (1), 1145–1148. (c) Saxon E, and Bertozzi CR. Cell Surface Engineering by a Modified Staudinger Reaction. *Science* **2000**, 287 (5460), 2007–2010. (d) Saxon E, Armstrong JI, and Bertozzi CR. A "Traceless" Staudinger Ligation for the Chemoselective Synthesis of Amide Bonds. *Org. Lett.* **2000**, 2 (14), 2141–2143. (e) Wang ZPA, Tiana CL, and Zheng JS. The recent developments and applications of the traceless-Staudinger reaction in chemical biology study. *RSC Adv.* **2015**, 5 (130), 107192–107199. (f) Staudinger H, and Meyer J. Über neue organische Phosphorverbindungen III. Phosphinmethylenderivate und Phosphinimine. *Helv. Chim. Acta* **1919**, 2 (1), 635–646. (g) The Nobel Prize in Chemistry 1953. NobelPrize.org. Nobel Media AB (2019 accessed December 5, 2019, at https://www.nobelprize.org/prizes/chemistry/1953/summary/). (h) Bednarek C, Wehl I, Jung N, Schepers U, Bräse S. The Staudinger Ligation. *Chem. Rev.* **2020**, 120 (10), 4301–4354.

[87] (a) Albert JS, and Hamilton AD. (2001). 1,3-Dicyclohexylcarbodiimide–4-Dimethylaminopyridine. In e-EROS Encyclopedia of Reagents for Organic Synthesis, (Ed.). doi:10.1002/047084289X.rd147. (b) Neises B, and Steglich W. Simple Method for the Esterification of Carboxylic Acids. *Angew. Chem. Int. Ed. Engl.* **1978**, 17 (7), 522–524.

[88] (a) Merrifield JH, Godschalx JP, and Stille JK. Synthesis of unsymmetrical diallyl ketones: the palladium-catalyzed coupling of allyl halides with allyltin reagents in the presence of carbon monoxide. *Organometallics* **1984**, 3 (7), 1108–1112. (b) Tokuyama H, Yokoshima S, Yamashita T, and Fukuyama T. A novel ketone synthesis by a palladium-catalyzed reaction of thiol esters and organozinc reagents. *Tetrahedron Lett.* **1998**, 39 (20), 3189–3192. (c) Milstein D, and Stille JK. A general, selective, and facile method for ketone synthesis from acid chlorides and organotin compounds catalyzed by palladium. *J. Am. Chem. Soc.* **1978**, 100 (11), 3636–3638. (d) Milstein D, and Stille JK. Palladium-catalyzed coupling of tetraorganotin compounds with aryl and benzyl halides. Synthetic utility and mechanism. *J. Am. Chem. Soc.* **1979**, 101 (17), 4992–4998.

[89] (a) Carrow BP, and Hartwig JF. Distinguishing Between Pathways for Transmetalation in Suzuki–Miyaura Reactions. *J. Am. Chem. Soc.* **2011**, 133 (7), 2116–2119. (b) Thomas AA, and Denmark SE. Pre-transmetalation intermediates in the Suzuki-Miyaura reaction revealed: The missing link. *Science* **2016**, 352 (6283), 329–332. (c) Miyaura N, Yamada K, and Suzuki A. A new stereospecific cross-coupling by the palladium-catalyzed reaction of 1-alkenylboranes with 1-alkenyl or 1-alkynyl halides. *Tetrahedron Lett.* **1979**, 20 (36), 3437–3440. (d) Miyaura N, and Suzuki A. Stereoselective synthesis of arylated (*E*)-alkenes by the reaction of alk-1-enylboranes with aryl halides in the presence of palladium catalyst. *J. Chem. Soc., Chem. Commun.* **1979**, (19), 866–867. (e) Please check https://www.sigmaaldrich.com/US/en/product/aldrich/697230 (accessed January 20, 2023). (f) Please check https://www.sigmaaldrich.com/US/en/product/aldrich/216666 (accessed January 20, 2023). (g) Gildner PG, and Colacot TJ. Reactions of the 21st Century: Two Decades of Innovative Catalyst Design for Palladium-Catalyzed Cross-Couplings. *Organometallics* **2015**, 34 (23), 5497–5508.

[90] (a) Omura K, Sharma AK, and Swern D. Dimethyl sulfoxide-trifluoroacetic anhydride. New reagent for oxidation of alcohols to carbonyls. *J. Org. Chem.* **1976**, 41 (6), 957–962. (b) Tidwell TT. (2001). Dimethyl Sulfoxide–Dicyclohexylcarbodiimide. In e-EROS Encyclopedia of Reagents for Organic Synthesis, (Ed.). doi:10.1002/047084289X.rd375. Please also see (c) Omura K, and Swern D. Oxidation of alcohols by "activated" dimethyl sulfoxide. A preparative, steric and mechanistic study. *Tetrahedron* **1978**, 34 (11), 1651–1660. (d) Mancuso AJ, Brownfain DS, and Swern D. Structure of the dimethyl sulfoxide-oxalyl chloride reaction product. Oxidation of heteroaromatic and diverse alcohols to carbonyl compounds. *J. Org. Chem.* **1979**, 44 (23), 4148–4150.

[91] (a) Passerini M, and Simone L. Sopra gli isonitrili (I). Composto del *p*-isonitril-azobenzolo con acetone ed acido acetico. *Gazz. Chim. Ital.* **1921**, 51, 126–129. (b) Passerini M. Sopra gli isonitrili (II). Composti con aldeidi o con chetoni ed acidi organici monobasici. *Gazz. Chim. Ital.* **1921**, 51, 181–189. Note: the original references are published in Italian and they are difficult to locate. (c) Ugi I, Meyr R, and Fetzer U. Versammlungsberichte. *Angew. Chem.* **1959**, 71 (11), 373–388. The citation can be found at https://onlinelibrary.wiley.com/doi/abs/10.1002/ange.19590711110 (accessed December 5, 2019).

[92] (a) Ullmann F, and Bielecki J. Ueber Synthesen in der Biphenylreihe. *Ber. Dtsch. Chem. Ges.* **1901**, 34 (2), 2174–2185. (b) Ullmann F. Ueber symmetrische Biphenylderivate. *Justus Liebigs Ann. Chem.* **1904**, 332 (1-2), 38–81. (c) Ullmann F. Ueber eine neue Bildungsweise von Diphenylaminderivaten. *Ber. Dtsch. Chem. Ges.* **1903**, 36 (2), 2382–2384. (d) Ullmann F, and Sponagel P. Ueber die Phenylirung von Phenolen. *Ber. Dtsch. Chem. Ges.* **1905**, 38 (2), 2211–2212. (e) Sambiagio C, Marsden SP, Blackera AJ, and McGowan PC. Copper catalysed Ullmann type chemistry: from mechanistic aspects to modern development. *Chem. Soc. Rev.* **2014**, 43 (10), 3525–3550. (f) Yang Q, Zhao Y, Ma D. Cu-Mediated Ullmann-Type Cross-Coupling and Industrial Applications in Route Design, Process Development, and Scale-up of Pharmaceutical and Agrochemical Processes. *Org. Process Res. Dev.* **2022**, 26 (6), 1690–1750. (g) Giri R, Brusoe A, Troshin K, Wang JY, Font M, Hartwig JF. Mechanism of the Ullmann

Biaryl Ether Synthesis Catalyzed by Complexes of Anionic Ligands: Evidence for the Reaction of Iodoarenes with Ligated Anionic Cu[I] Intermediates. *J. Am. Chem. Soc.* **2018**, 140 (2), 793–806.

[93] (a) Gao Y, and Cheun Y. (2013). Osmium Tetroxide–*N*-Methylmorpholine *N*-Oxide. In e-EROS Encyclopedia of Reagents for Organic Synthesis, (Ed.). doi:10.1002/047084289X.ro009.pub2. (b) Hentges SG, and Sharpless KB. Asymmetric induction in the reaction of osmium tetroxide with olefins. *J. Am. Chem. Soc.* **1980**, 102 (12), 4263–4265. (c) Wai JSM, Marko I, Svendsen JS, Finn MG, Jacobsen EN, and Sharpless KB. A mechanistic insight leads to a greatly improved osmium-catalyzed asymmetric dihydroxylation process. *J. Am. Chem. Soc.* **1989**, 111 (3), 1123–1125. (d) Ogino Y, Chen H, Kwong HL, and Sharpless KB. On the timing of hydrolysis/reoxidation in the osmium-catalyzed asymmetric dihydroxylation of olefins using potassium ferricyanide as the reoxidant. *Tetrahedron Lett.* **1991**, 32 (32), 3965–3968. (e) Lee DG, Ribagorda M, and Adrio J. (2007). Potassium Permanganate. In e-EROS Encyclopedia of Reagents for Organic Synthesis, (Ed.). doi:10.1002/9780470842898.rp244.pub2. (f) VanRheenen V, Kelly RC, and Cha DY. An improved catalytic OsO$_4$ oxidation of olefins to cis-1,2-glycols using tertiary amine oxides as the oxidant. *Tetrahedron Lett.* **1976**, 17 (23), 1973–1976. (g) The Nobel Prize in Chemistry 2001. NobelPrize.org. Nobel Media AB (2019 accessed December 5, 2019, at https://www.nobelprize.org/prizes/chemistry/2001/summary/).

[94] (a) Reimer K, and Tiemann F. Ueber die Einwirkung von Chloroform auf alkalische Phenolate. *Ber. Dtsch. Chem. Ges.* **1876**, 9 (1), 824–828. (b) Vilsmeier A, and Haack A. Über die Einwirkung von Halogenphosphor auf Alkyl-formanilide. Eine neue Methode zur Darstellung sekundärer und tertiärer *p*-Alkylamino-benzaldehyde. *Ber. Dtsch. Chem. Ges.* A/B **1927**, 60 (1), 119–122.

[95] (a) Keith JA, Nielsen RJ, Oxgaard J, and Goddard WA. Unraveling the Wacker Oxidation Mechanisms. *J. Am. Chem. Soc.* **2007**, 129 (41), 12342–12343. (b) Keith JA, and Henry PM. The Mechanism of the Wacker Reaction: A Tale of Two Hydroxypalladations. *Angew. Chem. Int. Ed.* **2009**, 48 (48), 9038–9049. (c) Smidt J, Hafner W, Jira R, Sedlmeier J, Sieber R, Rüttinger R, and Kojer H. Katalytische Umsetzungen von Olefinen an Platinmetall-Verbindungen Das Consortium-Verfahren zur Herstellung von Acetaldehyd. *Angew. Chem.* **1959**, 71 (5), 176–182.

[96] (a) Wagner G. *J. Russ. Phys. Chem. Soc.* **1899**, 31, 690–693. Note: the original reference is published in Russian and it is difficult to locate. (b) Wagner G, and Brickner W. Ueber die Beziehung der Pinenhaloïdhydrate zu den Haloïdanhydriden des Borneols. *Ber. Dtsch. Chem. Ges.* **1899**, 32 (2), 2302–2325. (c) Meerwein H. Über den Reaktionsmechanismus der Umwandlung von Borneol in Camphen; [Dritte Mitteilung über Pinakolinumlagerungen.]. *Justus Liebigs Ann. Chem.* **1914**, 405 (2), 129–175. (d) von R. Schleyer P. A simple preparation of adamantane. *J. Am. Chem. Soc.* **1957**, 79 (12), 3292–3292. (e) Olah GA, Wu AH, Farooq O, Prakash GS. Single-Step Reductive Isomerization of Unsaturated Polycyclics to C$_{4n+6}$H$_{4n+12}$ Diamondoid Cage Hydrocarbons with Sodium Borohydride/Triflic Acid. *J. Org. Chem.* **1989**, 54 (6), 1450–1451. (f) Engler EM, Farcasiu M, Sevin A, Cense JM, Schleyer PVR. Mechanism of Adamantane Rearrangements. *J. Am. Chem. Soc.* **1973**, 95 (17), 5769–5771. (g) "Organic Chemistry Masterclasses by S. Ranganathan" (accessed January 20, 2023, at https://www.ias.ac.in/Publications/e-Books/Organic_Chemistry_Masterclasses).

[97] (a) Weinreb SM, Folmer JJ, Ward TR, and Georg GI. (2006). *N*, *O*-Dimethylhydroxylamine. In e-EROS Encyclopedia of Reagents for Organic Synthesis, (Ed.). https://onlinelibrary.wiley.com/doi/10.1002/047084289X.rd341.pub2 (accessed December 5, 2019). (b) Hisler K, Tripoli R, and Murphy JA. Reactions of Weinreb amides: formation of aldehydes by Wittig reactions. *Tetrahedron Lett.* **2006**, 47 (35), 6293–6295. (c) Nahm S, and Weinreb SM. N-methoxy-N-methylamides as effective acylating agents. *Tetrahedron Lett.* **1981**, 22 (39), 3815–3818.

[98] (a) Byrnea PA, and Gilheany DG. The modern interpretation of the Wittig reaction mechanism. *Chem. Soc. Rev.* **2013**, 42 (16), 6670–6696. (b) Schlosser M, and Christmann KF. Trans-Selective Olefin Syntheses. *Angew. Chem. Int. Ed. Engl.* **1966**, 5 (1), 126–126. (c) Wittig G, and Schöllkopf U. Über Triphenyl-phosphin-methylene als olefinbildende Reagenzien (I. Mitteil.). *Chem. Ber.* **1954**, 87 (9),

1318–1330. (d) Wittig G, and Haag W. Über Triphenyl-phosphinmethylene als olefinbildende Reagenzien (II. Mitteil.). *Chem. Ber.* **1955**, 88 (11), 1654–1666.

[99] (a) Wohl A. Bromierung ungesättigter Verbindungen mit *N*-Brom-acetamid, ein Beitrag zur Lehre vom Verlauf chemischer Vorgänge. *Ber. Dtsch. Chem. Ges.* A/B **1919**, 52 (1), 51–63. (b) Ziegler K, Späth A, Schaaf E, Schumann W, and Winkelmann E. Die Halogenierung ungesättigter Substanzen in der Allylstellung. *Justus Liebigs Ann. Chem.* **1942**, 551 (1), 80–119. (c) The Nobel Prize in Chemistry 1963. NobelPrize.org. Nobel Media AB (2019 accessed December 5, 2019, at https://www.nobelprize.org/prizes/chemistry/1963/summary/).

[100] (a) Huang-Minlon. A Simple Modification of the Wolff-Kishner Reduction. *J. Am. Chem. Soc.* **1946**, 68 (12), 2487–2488. (b) Clemmensen E. Reduktion von Ketonen und Aldehyden zu den entsprechenden Kohlenwasserstoffen unter Anwendung von amalgamiertem Zink und Salzsäure. *Ber. Dtsch. Chem. Ges.* **1913**, 46 (2), 1837–1843. (c) Kishner N. *J. Russ. Phys. Chem. Soc.* **1911**, 43, 582–592. Note: the original reference is published in Russian and it is difficult to locate. (d) Wolff L. Chemischen Institut der Universität Jena: Methode zum Ersatz des Sauerstoffatoms der Ketone und Aldehyde durch Wasserstoff. [Erste Abhandlung.]. *Justus Liebigs Ann. Chem.* **1912**, 394 (1), 86–108.

www.ingramcontent.com/pod-product-compliance
Lightning Source LLC
Chambersburg PA
CBHW061347210326
41598CB00035B/5904